Studies in Organic Chemistry 29

BIOCATALYSIS IN ORGANIC MEDIA

PROCEEDINGS OF AN INTERNATIONAL SYMPOSIUM ORGANIZED UNDER AUSPICES OF THE WORKING PARTY ON APPLIED BIOCATALYSIS OF THE EUROPEAN FEDERATION OF BIOTECHNOLOGY AND THE WORKING PARTY ON BIOCATALYSTS OF THE AGRICULTURAL UNIVERSITY, WAGENINGEN

Wageningen, The Netherlands, 7—10 December 1986

Studies in Organic Chemistry

Other titles in this series:

1 **Complex Hydrides** by A. Hajós
2 **Proteoglycans — Biological and Chemical Aspects in Human Life** by J.F. Kennedy
3 **New Trends in Heterocyclic Chemistry** edited by R.B. Mitra, N.R. Ayyangar, V.N. Gogte, R.M. Acheson and N. Cromwell
4 **Inositol Phosphates: Their Chemistry, Biochemistry and Physiology** by D.J. Cosgrove
5 **Comprehensive Carbanion Chemistry. Part A. Structure and Reactivity** edited by E. Buncel and T. Durst
Comprehensive Carbanion Chemistry. Part B. Selectivity in Carbon-Carbon Bond Forming Reactions edited by E. Buncel and T. Durst
6 **New Synthetic Methodology and Biologically Active Substances** edited by Z.-I Yoshida
7 **Quinonediazides** by V.V. Ershov, G.A. Nikiforov and C.R.H.I. de Jonge
8 **Synthesis of Acetylenes, Allenes and Cumulenes: A Laboratory Manual** by L. Brandsma and H.D. Verkruijsse
9 **Electrophilic Additions to Unsaturated Systems** by P.B.D. de la Mare and R. Bolton
10 **Chemical Approaches to Understanding Enzyme Catalysis: Biomimetic Chemistry and Transition-State Analogs** edited by B.S. Green, Y. Ashani and D. Chipman
11 **Flavonoids and Bioflavonoids 1981** edited by L. Farkas, M. Gábor, F. Kállay and H. Wagner
12 **Crown Compounds: Their Characteristics and Applications** by M. Hiraoka
13 **Biomimetic Chemistry** edited by Z.-I. Yoshida and N. Ise
14 **Electron Deficient Aromatic- and Heteroaromatic-Base Interactions. The Chemistry of Anions Sigma Complexes** by E. Buncel, M.R. Crampton, M.J. Strauss and F. Terrier
15 **Ozone and its Reactions with Organic Compounds** by S.D. Razumovskii and G.E. Zaikov
16 **Non-benzenoid Conjugated Carbocyclic Compounds** by D. Lloyd
17 **Chemistry and Biotechnology of Biologically Active Natural Products** edited by Cs. Szántay, Á. Gottsegen and G. Kovács
18 **Bio-Organic Heterocycles: Synthetic, Physical Organic and Pharmacological Aspects** edited by H.C. van der Plas, L. Ötvös and M. Simonyi
19 **Organic Sulfur Chemistry: Theoretical and Experimental Advances** edited by F. Bernardi, I.G. Czismadia and A. Mangini
20 **Natural Products Chemistry 1984** edited by R.I. Zalewski and J.J. Skolik
21 **Carbocation Chemistry** by P. Vogel
22 **Biocatalysts in Organic Syntheses** edited by J. Tramper, H.C. van der Plas and P. Linko
23 **Flavonoids and Bioflavonoids 1985** edited by L. Farkas, M. Gábor and F. Kállay
24 **The Organic Chemistry of Nucleic Acids** by Y. Mizuno
25 **New Synthetic Methodology and Functionally Interesting Compounds** edited by Z.-I. Yoshida
26 **New Trends in Natural Products Chemistry 1986** edited by A.-ur-Rahman and P.W. Le Quesne
27 **Bio-Organic Heterocycles 1986. Synthesis, Mechanisms and Bioactivity** edited by H.C. van der Plas, M. Simonyi, F.C. Alderweireldt and J.A. Lepoivre
28 **Perspectives in the Organic Chemistry of Sulfur** edited by B. Zwanenburg and A.J.H. Klunder

Studies in Organic Chemistry 29

BIOCATALYSIS IN ORGANIC MEDIA

PROCEEDINGS OF AN INTERNATIONAL SYMPOSIUM ORGANIZED UNDER AUSPICES OF THE WORKING PARTY ON APPLIED BIOCATALYSIS OF THE EUROPEAN FEDERATION OF BIOTECHNOLOGY AND THE WORKING PARTY ON BIOCATALYSTS OF THE AGRICULTURAL UNIVERSITY, WAGENINGEN

Wageningen, The Netherlands, 7—10 December 1986

Edited by

C. Laane
Unilever Research Laboratory, Vlaardingen, The Netherlands

J. Tramper
Department of Food Science, Agricultural University, Wageningen, The Netherlands

and

M.D. Lilly
Department of Chemical and Biochemical Engineering, University College, London, England

ELSEVIER

Amsterdam — Oxford — New York — Tokyo 1987

ELSEVIER SCIENCE PUBLISHERS B.V.
Sara Burgerhartstraat 25,
P.O. Box 211, 1000 AE Amsterdam, The Netherlands

Distributors for the United States and Canada:

ELSEVIER SCIENCE PUBLISHING COMPANY, INC.
52 Vanderbilt Avenue
New York, NY 10017, U.S.A.

ISBN 0-444-42785-6 (Vol. 29)
ISBN 0-444-41737-0 (Series)

Printed in The Netherlands

FOREWORD

One of the main objectives of the European Federation of Biotechnology Working Party on Applied Biocatalysis is to stimulate interest and activity in new areas of biocatalysis of relevance to industry. In April 1985 the Working Party organised a successful symposium on "Biocatalysts in Organic Syntheses" (ref. 1) which brought together organic chemists and biotechnologists to exchange views on the potential of biocatalysts in organic syntheses. As a result several of us proposed a further symposium on "Biocatalysis in Organic Media" to discuss in more detail one particularly exciting aspect of biocatalysis. This met with the unanimous approval of the Working Party. When the symposium was proposed we expected to attract about 80–100 participants and were pleasantly surprised by the interest it generated.

For the first time it was possible to bring together over two hundred people to listen to and see almost sixty oral and poster presentations on biocatalysts in organic media. This allowed detailed consideration of the behaviour of biocatalysts in the presence of water–miscible and water–immiscible organic solvents, in reversed micelles and in reactions with low water activity.

During the symposium it became clear that much progress has been made in the last few years but that there is still much more to learn before we understand enough about the effects of organic media on biocatalysts to predict satisfactorily their behaviour in such reaction systems. It is important to recognise the broad range of biocatalyst forms which may be used for the diverse reactions of interest to industry. It seems unlikely that the same criteria will apply, for instance, for an enzyme being used for hydrolysis as for a microbial cell catalysing a complex series of reactions. Nevertheless, parameters such as the logarithm of the partition coefficient of a solvent in a standard octanol–water system have been used with some success to predict the effect of that solvent on the biocatalyst. Above all, it was most encouraging to learn of the many situations where enzymes or cells are sufficiently resilient to operate in conditions which traditionally would have been considered rather harmful.

It is inevitable, with our limited knowledge of the field, that the papers and discussions centred on fundamental scientific problems, some of which may require other scientific skills such as those of the genetic engineer and protein engineer. However, on many occasions during the meeting reactions of interest to the food, pharmaceutical and chemical industries were described and references made to pilot– or large–scale processes involving biocatalysts in organic media. Some of these are operating commercially and others will reach that stage in the next few years. We are, therefore, at a very exciting point in the development of biocatalysis in organic media. We hope that the symposium and this book which contains most of the papers presented at the symposium will make a useful contribution to this key area of applied biocatalysis.

We wish to thank the Working Party for its enthusiastic support of the symposium and the many members of the Agricultural University and the International Agricultural Centre whose efforts made the symposium a memorable event.

The Editors,
January 1987

REFERENCE

1 J. Tramper, H. C. van der Plas and P. Linko (Eds.), Biocatalysts in Organic Syntheses, Elsevier, Amsterdam, 1985

ACKNOWLEDGEMENTS

The organizing committee of the international symposium "Biocatalysis in Organic Media" acknowledges with gratitude the following organizations, who generously contributed to this symposium.

Agricultural University, Wageningen, The Netherlands
Amylum NV, Aalst, Belgium
Applikon BV, Schiedam, The Netherlands
CCA Biochem BV, Gorinchem, The Netherlands
Chemie Linz AG, Austria
Diosynth BV, Oss, The Netherlands
DSM NV, Heerlen, The Netherlands
Duphar, Weesp, The Netherlands
Gist-Brocades NV, Delft, The Netherlands
De Kuyper, Schiedam, The Netherlands
Naarden International, Naarden, The Netherlands
Rhone-Poulenc, Courbevoie, France
Royal Netherlands Academy of Sciences, Amsterdam, The Netherlands
Royal Netherlands Chemical Society, The Hague, The Netherlands
Royal Shell Laboratory, Amsterdam, The Netherlands
Town of Wageningen, The Netherlands
Unilever Research Laboratory, Vlaardingen, The Netherlands

CONTENTS

FOREWORD v

ACKNOWLEDGEMENTS vii

SESSION I : OPENING OF THE SYMPOSIUM

Biological conversions involving water-insoluble organic compounds
M.D. LILLY, A.J. BRAZIER, M.D. HOCKNULL, A.C. WILLIAMS AND J.M. WOODLEY 3

SESSION II : BIOCATALYST AND MEDIUM ENGINEERING

Immobilization of biocatalysts for bioprocesses in organic solvent media
S. FUKUI, A. TANAKA AND T. IIDA 21

Enantioselective reactions in aqueous and in organic media using carboxyl esterase fractions obtained from crude porcine pancreas lipase preparations
G.M. RAMOS TOMBO, H.-P. SCHAER, X. FERNANDEZ I BUSQUETS AND O. GHISALBA 43

Steroid side chain cleavage with immobilized living cells in organic solvents
H.-J. STEINERT, K.D. VORLOP AND J. KLEIN 51

Optimization of biocatalysis in organic media
C. LAANE, S. BOEREN, R. HILHORST AND C. VEEGER 65

SESSION III : CONVERSIONS I

Log P as a hydrophobicity index for biocatalysis; cofactor regeneration during enzymatic steroid oxidation in organic solvents
A.M. SNIJDER-LAMBERS, H.J. DODDEMA, H.J. GRANDE AND P.H. VAN LELYVELD 87

Multiphase reactors a new opportunity
M.D. LEGOY, M. BELLO, S. PULVIN AND D. THOMAS 97

Co-immobilization - An alternative to biocatalysis in organic media
R. KAUL, P. ADLERCREUTZ AND B. MATTIASSON 107

Enzymatic production of chemicals in organic solvents
A.M. KLIBANOV 115

SESSION IV : ENGINEERING ASPECTS

Enzyme action and enzyme reversal in water-in-oil microemulsions
C. OLDFIELD, G.D. REES, B.H. ROBINSON AND R.B. FREEDMAN 119

Water activity in biphasic reaction systems
P.J. HALLING 125

Design of an organic-liquid-phase/immobilized-cell reactor for the microbial epoxidation of propene
L.E.S. BRINK AND J. TRAMPER 133

Phase toxicity in a water-solvent two-liquid phase microbial system
R. BAR 147

SESSION V : CONVERSIONS II

Biocatalysis in water-organic solvent two-phase systems
G. CARREA 157

SESSION VI : SPECIFIC/FUTURE APPLICATIONS

Commercial aspects of biocatalysis in low-water systems
P. CRITCHLEY 173

Integration of enzyme catalysis in an extractive fermentation process
M.R. AIRES BARROS, A.C. OLIVEIRA AND J.M.S. CABRAL 185

Nicotinamide cofactor-requiring enzymatic synthesis in organic solvent-water biphasic systems
C.-H. WONG 197

Enzymatic synthesis of aspartame in organic solvents
K. OYAMA 209

POSTER PAPERS

Natural flavour esters: Production by Candida cylindracae lipase adsorbed to silica gel
B. GILLIES, H. YAMAZAKI AND D.W. ARMSTRONG 227

Preparation of acyl derivatives of 1-hydroxy aldose by lipase-catalyzed hydrolysis or alcoholysis of fully acylated aldose in organic solvent
J.-F. SHAW AND E.-T. LIAW 233

The anaerobic transformation of linoleic acid by Acetobacterium woodii
H. GIESEL-BÜHLER, O. BARTSCH, H. KNEIFEL, H. SAHM AND R. SCHMID 241

Stability of enzymes in low water activity media
V. LARRETA-GARDE, Z.F. XU, J. BITON AND D. THOMAS 247

Synthesis of polysaccharides bearing a lipophilic chain for the chemical modification of enzymes
M. WAKSELMAN AND D. CABARET 253

Dead mycelium stabilized lipolytic activity in organic media: Application to ester linkage hydrolysis and synthesis in a fixed-bed reactor
C. GANCET AND C. GUIGNARD 261

Thermal stability of immobilized horseradish peroxidase (HRP) in water-organic solvent systems
L. D'ANGIURO, S. GALLIANI AND P. CREMONESI 267

Properties and specificity of an alcohol dehydrogenase from thermophilic archaebacterium Sulfolobus solfataricus
R. RELLA, C.A. RAIA, A. TRINCONE, A. GAMBACORTA, M. DE ROSA AND M. ROSSI 273

Enantioselective hydrolysis and transesterification of glycidyl butyrate by lipase preparations from porcine pancreas
M.CHR. PHILIPPI, J.A. JONGEJAN AND J.A. DUINE 279

Reversed micellar extraction of enzymes; Effect of nonionic surfactants on the distribution and extraction efficiency of α-amylase
M. DEKKER, J.W.A. BALTUSSEN, K. VAN 'T RIET, B.H. BIJSTERBOSCH, C. LAANE AND R. HILHORST 285

Haloperoxidases in reversed micelles: Use in organic synthesis and optimisation of the system
M.C.R. FRANSSEN, J.G.J. WEIJNEN, J.P. VINCKEN, C. LAANE AND H.C. VAN DER PLAS 289

Potential of organic solvents in cultivating micro-organisms on toxic water-insoluble compounds
J.M. REZESSY-SZABÓ, G.N.M. HUIJBERTS AND J.A.M. DE BONT 295

Viability and activity of Flavobacterium dehydrogenans in organic solvent/culture two-liquid-phase-systems
S. BOEREN, C. LAANE AND R. HILHORST 303

The liquid-impelled loop reactor: A new type of density-difference-mixed bioreactor
J. TRAMPER, I. WOLTERS AND P. VERLAAN 311

Algal vanadium(V)-bromoperoxidase, A halogenating enzyme retaining full activity in apolar solvent systems
E. DE BOER, H. PLAT AND R. WEVER 317

Production of L-tryptophan in a two-liquid-phase system
M.H. RIBEIRO, J.M.S. CABRAL AND M.M.R. FONSECA 323

Effect of water-miscible organic solvents on the catalytic activity of penicillin acylase from Kluyvera citrophila
J.M. GUISAN, G. ALVARO AND R.M. BLANCO 331

Activity of staphylococcal nuclease in water-organic media
A. ALCANTARA, A. BALLESTEROS, M.A. HERAS, J.M. MARINAS, J.M.S. MONTERO AND J.V. SINISTERRA 337

Steroid bioconversion in aqueous two-phase systems
R. KAUL AND B. MATTIASSON 343

Organic solvents for bioorganic synthesis. 2. Influence of log P and water solubility in solvents on enzymatic activity
M. RESLOW, P. ADLERCREUTZ AND B. MATTIASSON 349

Study of horse liver alcohol dehydrogenase (HLADH) in AOT-cyclohexane reverse micelles
K. LARSSON, P. ADLERCREUTZ AND B. MATTIASSON 355

Pig liver esterase in asymmetric synthesis. Steric requirements and control of reaction conditions
J. BOUTELJE, M. HJALMARSSON, P. SZMULIK, T. NORIN AND K. HULT 361

A comparison of the enzyme catalyzed formation of peptides and oligosaccharides in various hydroorganic solutions using the non-equilibrium approach
K.G.I. NILSSON 369

Optical resolution of phenylalanine by enzymic transesterification
E. FLASCHEL AND A. RENKEN 375

The effects of solvents on the kinetics of free and immobilized lipase
T. UCAR, H.I. EKIZ, S.S. CELEBI AND A. CAGLAR 381

Stereoselective synthesis of S-(-)-B-blockers via microbially produced epoxide intermediates
S.L. JOHNSTONE, G.T. PHILLIPS, B.W. ROBERTSON, P.D. WATTS, M.A. BERTOLA, H.S. KOGER AND A.F. MARX 387

The Δ'-dehydrogenation of hydrocortisone by Arthrobacter simplex in organic-aqueous two-liquid phase environments
M.D. HOCKNULL AND M.D. LILLY 393

Denaturation and inhibition studies in a two-liquid phase biocatalytic reaction: The hydrolysis of menthyl acetate by pig liver esterase
A.C. WILLIAMS, J.M. WOODLEY, P.A. ELLIS AND M.D. LILLY 399

Effect of water miscible organic solvents on activity and thermostability of thermolysin
P.B. RODGERS, I. DURRANT AND R.J. BEYNON 405

Stereospecific reductions of bicycloheptenones catalysed by 3α,20β-hydroxysteroid dehydrogenase in one, two and three phase systems
J. LEAVER, T.C.C. GARTENMANN, S.M. ROBERTS AND M.K. TURNER 411

L-Phenylalanine production process utilizing enzymatic resolution in the presence of an organic solvent
S.K. DAHOD AND M.W. EMPIE 419

Author Index 425

SESSION I

OPENING OF THE SYMPOSIUM

Chairman : J.A.M. de Bont

C. Laane, J. Tramper and M.D. Lilly (Editors), *Biocatalysis in Organic Media*, Proceedings of an International Symposium held at Wageningen, The Netherlands, 7–10 December 1986.

BIOLOGICAL CONVERSIONS INVOLVING WATER-INSOLUBLE ORGANIC COMPOUNDS

M. D. LILLY, A. J. BRAZIER, M. D. HOCKNULL, A. C. WILLIAMS AND J. M. WOODLEY

Department of Chemical and Biochemical Engineering, University College London, Torrington Place, London WC1E 7JE, England.

SUMMARY

Biocatalysis in organic media is of growing importance. If a liquid organic phase is present various reaction systems are possible. The main characteristics of these are summarised. The potential effects of the organic phase, especially if it contains an organic solvent, on the activity and stability of enzymic and microbial catalysts is described and illustrated by data on menthyl ester hydrolysis by pig liver esterase, steroid Δ^1-dehydrogenation by *Arthrobacter simplex*, and toluene oxidation by *Pseudomonas putida*. The importance of an integrated approach to reactor and downstream product recovery design and operation is emphasized and some guidelines given.

INTRODUCTION

Last year at a meeting organised by the European Federation of Biotechnology Working Party on Applied Biocatalysis, we highlighted the rapidly increasing academic and industrial interest in biological conversions involving organic compounds which have low solubilities in water (ref. 1). It is possible to increase their solubilities in the aqueous phase by the use of water-miscible solvents. For instance methanol, ethylene glycol and dimethyl sulphoxide have been used for this reason in various transformations of steroids (refs 2 and 3). Water-miscible organic solvents may also be used, for instance in peptide synthesis (ref. 4), to alter the reaction equilibrium by reducing the water activity.

In most cases, however, the inclusion of water-miscible solvents only raises the solubilities of many organic compounds to a limited extent in the aqueous phase. An alternative approach is to carry out the reaction with an organic phase present in the reactor. This second liquid phase may consist of either a water-immiscible liquid or a solid dissolved in a water-immiscible solvent. In order to create a coherent approach to the diverse range of reactions which had already been reported, we proposed a simple classification system for reactions involving reactant(s) and/or product(s) which have low solubilities in an aqueous environment (ref. 1). Excluding those reactions involving only an aqueous phase, five main types of reaction system were identified based on the natures of the aqueous phase and the catalyst. These are illustrated diagrammatically in Fig. 1.

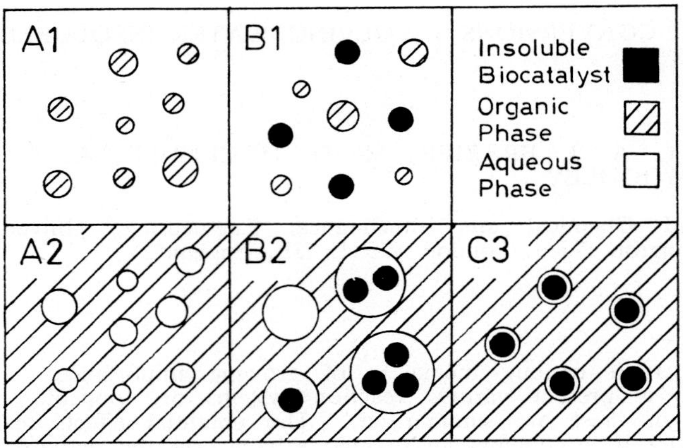

Fig. 1. Diagrammatic representation of reaction systems involving an organic liquid phase.

At the same time we indicated that there are twenty five different distributions of reactants and products between the aqueous and organic phases for reactions involving two to four reaction components. Of these, twenty one have at least one reaction component in the organic phase. The number of possibilities increases greatly if the physical nature of the components at the operating conditions in the reactor are considered. For instance, a reactant or product can be a solid, liquid or gas which may be predominantly soluble in the aqueous or the organic phase. Also in some cases a reactant or product may not partition mainly into a liquid phase, as for example occurs with oxygen. Some examples of the distributions of reactants and products are shown in Table 1. In most of the reactions where an organic solvent is being used the role of water is an essential consideration. In addition to the water essential for catalyst activity, it may be either a reactant or product of the reaction (Table 1). In some cases such as fat interesterification it may participate in the reaction as both reactant and product such that theoretically there is no net gain or loss of water. In practice, however, unwanted net hydrolysis of triglyceride does lead to a reduction in the amount of water present. For certain types of reaction where water is not a reactant or product and all of the reactant(s) and product(s) are predominantly in the organic phase it is possible to operate without a discrete aqueous phase present. Such systems have been reported for free (ref. 5) and immobilised

TABLE 1

EXAMPLES OF PHASE DISTRIBUTIONS OF REACTANTS AND PRODUCTS

REACTION	AQUEOUS		ORGANIC		GAS	WATER
	LIQUID	SOLID	LIQUID	SOLID		
Steroid Δ^1-dehydrogenation				SP		
Cortisone reduction		S_1P_1		S_2P_2		
Menthyl acetate hydrolysis	P_1		S_1	P_2		S_2
Benzyl acetate hydrolysis	P_1		S_1P_2			S_2
Hydroquinone oxidation	S_1	S_2		P_1		P_2
Benzene oxidation		P	S_1		S_2	
Propene oxidation					S_1S_2P	

where S,S_1,S_2 are reactants and P,P_1,P_2 are products

cells (ref. 6). The amount of water present can be reduced further until the water activity is much less than unity as in the case of fat interesterification by immobilised lipase (refs 7 and 8) and transesterification by lipase (ref. 9). The solubility of some enzymes in organic solvents has been increased by forming polyethylene glycol derivatives (ref. 10). This allows the reaction to be done in a single organic phase but addition of another solvent is necessary to precipitate the enzyme to allow its recovery and re-use.

Very little has been published so far on the design of processes for reactions involving sparingly water-soluble reactants and products. Here, some of the factors affecting the biocatalyst in a two-liquid phase reactor are discussed. To ensure good overall process design it is necessary at the same time to consider the recovery of products and a general approach to the downstream processing is presented.

THE REACTOR

In a previous paper (ref. 1) we described a reactor classification based on whether or not a soluble or insoluble biocatalyst was used, and in the latter case whether the catalyst was mobile or stationary in the reactor. The choice of catalyst form and suitable reactor is more complex than for reactions involving only water-soluble reactants and products (refs 11 and 12) because of the additional need to transfer reactant and/or product across the liquid/liquid interface. For those reactions where an insoluble biocatalyst is used without a discrete aqueous phase the interfacial area is determined by the surface area of the catalyst, which may be in the form of spheres or other shaped particles. Similarly in reactors with membranes to separate the two phases (ref. 13), the interfacial area is determined by the membrane surface area.

In most cases where two discrete liquid phases are present agitation has been used to increase the interfacial area and, therefore, the rates of transfer of reactant(s) and/or product(s) between the phases. When *Bacillus subtilis* was used for the stereospecific hydrolysis of menthyl acetate in a 50 ml stirred reactor the reaction rate reached a maximum at 800 rpm and remained constant above that value at each of the organic/aqueous phase ratios tested (ref. 14). However, when the same reaction, albeit non-stereospecifically, was performed with pig liver esterase in a stirred tank the reaction rate declined with time at high agitation speeds (ref. 15). The concentration of menthol produced in one hour at different stirrer speeds is shown in Fig. 2. It would appear, therefore, that the effect of the organic liquid (in this case, substrate) is different for cells and enzymes.

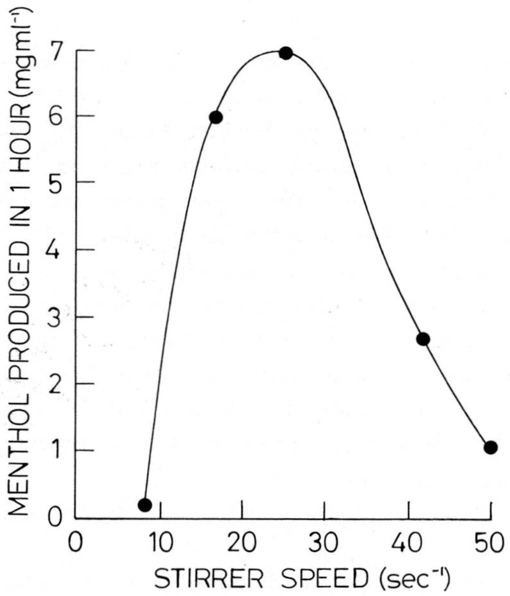

Fig. 2. The effect of stirrer speed on the amount of menthol produced from menthyl acetate in one hour by pig liver esterase. Reaction conditions: Total volume, 75 ml; organic/aqueous phase ratio, 0.5; aqueous phase, 0.1 M phosphate buffer, pH 7.0; enzyme concentration, 0.1 mg ml^{-1}.

Most studies on the influence of organic solvents on microbial biocatalysis have concentrated on the avoidance of cell inactivation by the organic phase. Brink and Tramper (ref. 16) for example felt that high activity retention was favoured by a combination of low solubility parameter, and high molecular weight; Laane *et al*. (ref. 17) considered that activity retention was related to the logarithm of the partition coefficient (Log P) of the organic solvent, and that only solvent of Log P ≥ 4 favoured good activity retention. However, we have shown that organic solvents of Log P ≥ 4 can have an effect on the stability of cell biocatalysts in two-liquid phase environments depending upon the reaction conditions. This is illustrated in Fig. 3 which shows the Δ^1-dehydrogenation of hydrocortisone by *Arthrobacter simplex* in the presence of various solvents. It seems that precisely those conditions which favour high initial activities (e.g. good mass transfer), also result in poor stability of the cell biocatalyst (ref. 18).

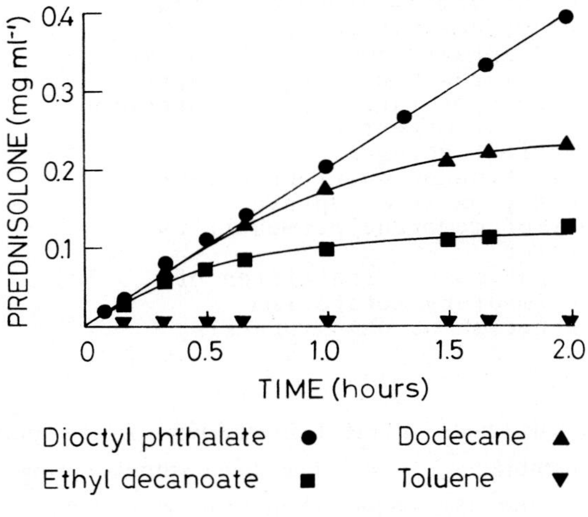

Fig. 3. Δ^1-dehydrogenation of hydrocortisone by *Arthrobacter simplex* in various two-liquid phase environments. Reaction conditions: reaction volume, 70 ml; organic/aqueous phase ratio, 1; stirrer speed, 750 rpm; initial hydrocortisone concentration, 0.5 mg ml^{-1}; aqueous cell concentration, 2 mg wet wt ml^{-1}.

The problem of cell inactivation by organic solvents therefore cannot simply be avoided by selections of appropriate solvents, and ways of enhancing the stability of cells must be sought. In some cases, immobilisation of the cells has proved successful in this context (refs 6,19,20 and 21) but in others no improvement has been observed (e.g. ref. 16).

It is only through an understanding of the mechanisms of solvent-inactivation of cells that rational strategies for enhancing cell stability can be adopted.

Solvents have been shown to cause the effects listed in Table 2. It is possible to rationalise these observations in terms of a non-specific action of solvents at the cell membrane.

TABLE 2

POTENTIAL EFFECTS OF ORGANIC SOLVENTS ON MICROORGANISMS

		References.
1.	Loss of cell viability	22
2.	Changes in cell morphology	
	a) cytoplasmic shrinkage	22
	b) loss of membrane organisation	22, 23, 27, 33
	c) ultrastructural changes	
	i) displacemenmt of the chromosome towards the cell periphery	23
	ii) loss of intracellular electron dense material	22
3.	Physiological changes	
	a) inhibition of nutrient uptake	34, 35
	b) inhibition of O_2 uptake	31
	c) loss of membrane permeability barrier	22, 23, 25, 26, 34
	d) alterations or inhibition of intermediary metabolism	32
	e) inhibition of DNA synthesis	33

Organic solvents cause a rapid breakdown of the permeability barrier function of cell membranes. This has been shown using several solvents: toluene (e.g. refs 22 and 23), phenethyl alcohol (refs 24 and 25), alkanes (ref. 26) and alcohols (ref. 25). As a consequence of this breakdown there is leakage of intracellular material from the cell. Initially only ions such as K^+, Mg^{++} are released, but later low molecular weight organics and finally larger molecules (RNA and protein) are released. The extent and kinetics of this leakage (and hence of membrane damage) by organic solvents appears to be related to the volume of solvent added to the bacterial suspension (refs 22 and 24), and may also be related to solvent properties such as hydrophobicity (ref. 25).

Solvent action at the membrane may inhibit other cellular processes. If the process is structurally associated with the membrane, for example DNA synthesis, then membrane damage would be expected to impair the functioning of that process. DNA synthesis has been shown to be dependent upon the high cellular K^+ concentration in *E. coli* (refs 28 and 29). Hence loss of the permeability barrier may be the mechanism of inhibition of DNA synthesis. Another mechanism for loss of cellular activity is inhibition of nutrient uptake. It has been shown (refs 34 and 35) that the inhibition of sugar uptake in yeast by alcohols is non-competitive. This is consistent with the view that solvent partitioning into the cell membrane causes the inhibition because the protein structure (i.e. substrate binding) is not affected.

Inhibition of oxygen uptake may also cause loss of cellular activity. This may be explained by considering the role of oxygen in the cell. This molecule is used largely as the terminal electron acceptor in oxidative metabolism. The electron transport chain requires the existence of a selectively permeable membrane for normal functioning. Therefore a loss of the permeability barrier will result in the inhibition of the electron transport chain and hence an inhibition of oxygen uptake.

Clearly, attempts to stabilize cell biocatalysts against organic solvents must be directed towards the cell membrane. Some work has shown that the effects of solvents may be reversible by the addition of Mg^{++} or Ca^{++} (ref. 30). In other cases, reversibility may be achieved simply by washing the cells to remove any solvent in the membrane (ref. 24). This may have applications for cell re-use.

The organic phase can be toxic to the biocatalyst in two ways: either the presence of the second phase itself or the (usually small) amount of organic dissolved in the aqueous phase may cause the inactivation. As we have stated previously (Table 1), it is necessary for many reactions to have a second, organic, phase present in the reactor. However, in reactions where the water-immiscible reactant is harmful to the catalyst the concentration of the reactant in the reactor may be minimised while maintaining an acceptable rate of reaction. As reactant in the aqueous phase is consumed it is replaced by solution of the reactant from a discrete organic phase which can be fed into the reactor. If the presence of this phase is toxic to the catalyst then the amount present should only be sufficient to obtain an adequate rate of transfer into the aqueous phase. If, however, the reactant dissolved in the aqueous phase is itself toxic to the catalyst then limitation of the concentration in the aqueous phase by controlling the rate of transfer from the organic phase is advantageous. These points are illustrated in Fig. 4 which shows the influence of toluene feed rate to a stirred tank reactor on the conversion of toluene to toluene *cis*-glycol by a mutant of *Pseudomonas putida*. When the toluene concentration was allowed to rise above saturation level there was a rapid cessation in *cis*-glycol formation. Thus in such conversions continuous monitoring of the level of the reactant in the reactor is essential for good process control in order to achieve extended use of the catalyst and the consequent high productivities (weight of product formed/weight of catalyst used).

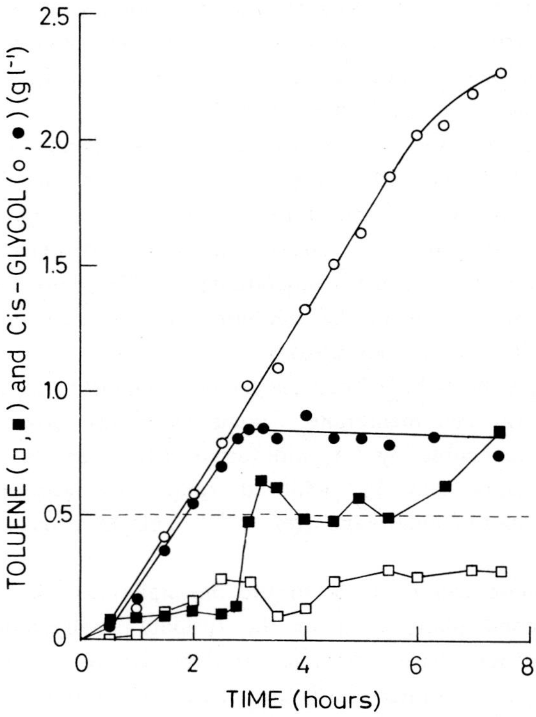

Fig. 4. Oxidation of toluene to the corresponding *cis*–glycol (5,6–dihydroxy–cyclohexa–1,3–diene) by a mutant of *Pseudomonas putida*. Reaction conditions: reaction volume, 1.7 l; stirrer speed, 600 rpm; air flow rate, 50 l h^{-1}; cell concentration, 2.8 g wet wt l^{-1}. Toluene was normally fed to maintain a low concentration in the reactor. In the experiment shown by closed symbols the toluene feed rate was deliberately increased after 2.8 h.

The above considerations are concerned with the effect of the organic phase on the biocatalyst. For an industial process there are many other factors which influence the choice of the organic phase, especially if this contains an organic solvent. These are listed in Table 3. Similar criteria are relevant when considering the subsequent downstream product recovery operations (ref. 35).

TABLE 3

FACTORS INFLUENCING THE CHOICE OF ORGANIC PHASE FOR THE REACTOR AND PRODUCT RECOVERY

REACTANT/PRODUCT CAPACITY	TOXICITY
PARTITION COEFFICIENT	FLAMMABILITY
DENSITY	AVAILABILITY
MELTING/BOILING PTS	WASTE DISPOSAL
SURFACE TENSION	PRICE
VISCOSITY	

DOWNSTREAM PROCESSING

The development of two-liquid phase biocatalytic systems into industrial processes will require detailed understanding of downstream operations as well as the knowledge of biocatalyst kinetics and stability necessary for reactor design. The rationale behind such an approach, developing both reactor and downstream studies in parallel, lies in the understanding that many of the key reactor performance parameters (e.g. solvent selection and organic/aqueous phase ratio) are also key downstream performance parameters. An integrated approach therefore leads to a fuller understanding of the effect of such parameters on the overall process. This is a prerequisite for process design and ultimately optimisation. Fig. 5 represents a schematic two-liquid phase biocatalytic process.

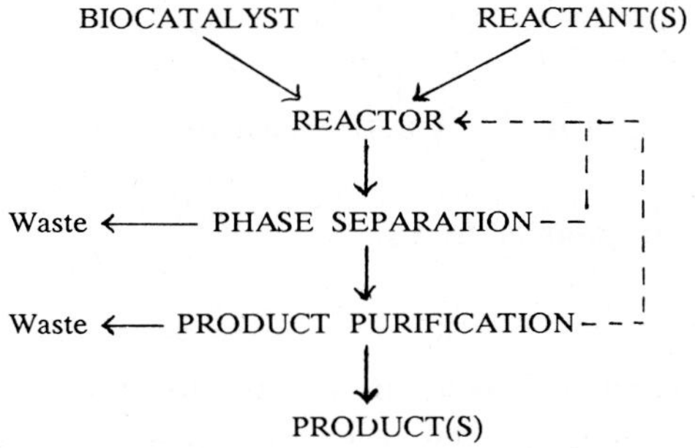

Fig. 5. Schematic diagram of a two-liquid phase biocatalytic process.

The plant downstream of the reactor will potentially have three types of output streams:

. waste stream(s), e.g. spent biocatalyst
. recycle stream(s), e.g. unreacted substrate, reusable biocatalyst
. product stream(s)

The downstream process can be divided into two main operations; first, a phase separation unit in which the multiphasic reactor effluent is split into its constituent phases, and secondly, a product purification unit in which the product is separated from its host phase, whether this be aqueous or organic, by means of standard operations, e.g. distillation and crystallisation. The more important of these two downstream processes, when considering an integrated design approach, is the phase separation unit since it is upon this unit in particular that the key reactor performance parameteres will have a direct influence. We consider here the potential design of such a phase separation unit.

In order to design the phase separation unit it is important to consider the input and output streams from the proposed unit. The input stream will contain a multiphasic mixture of product, and potentially un-reacted substrate, organic solvent, water and biocatalyst, unless the process involves a conversion in a reactor in which biocatalyst is retained, e.g. a packed bed reactor. Consideration of the product form reveals the nature of the effluent streams from the phase separation unit and the subsequent purification steps. Fig. 6 shows a simple classification of the possible product forms. Gaseous product

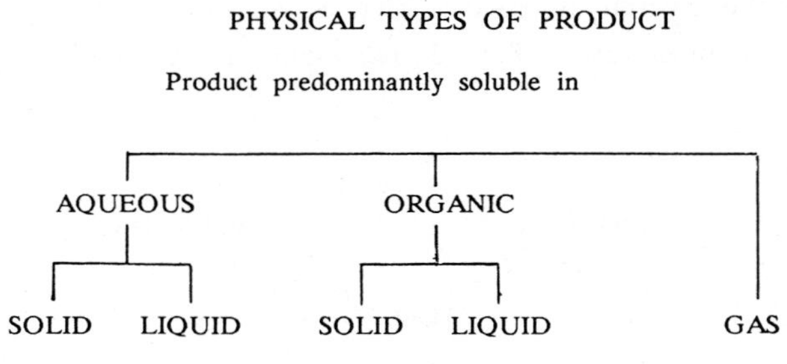

Fig. 6. Classification of product forms

would probably be separated directly in the reactor and the necessary product purification be effected from the flue gas downstream. However, for reactions producing a liquid or a solid product at reaction temperature, the product will be soluble in either the aqueous phase or organic phase. While the solid or liquid nature of the product determines the purification system used, it is the phase in which the product is predominantly soluble (aqueous or organic) that will determine the nature of the phase separation unit. Hence three types of reaction can be identified: (a) reactions producing product(s) predominantly soluble in the aqueous phase; (b) reactions productimg product(s) predominantly soluble in the organic phase; (c) reactions producing products at least one of which is soluble in the aqueous phase and one of which is soluble in the organic phase.

In order to separate the phases a property needs to be chosen which can be exploited such that a separation occurs (ref. 36) and for the following analysis the density has been chosen. One measure of the difficulty of the separation is the difference in densities of the phases to be separated. Specific density data about a reaction system can therefore lead to a relative measure of both the rate of separation and the resolution to be expected. Through the time course of a reaction the phase densities may change as the phase composition changes.

In a simple system in which biocatalyst either does not enter the downstream plant (because of retention in the reactor) or alternatively is in a soluble form, the phase separation unit is merely required to separate organic phase from aqueous phase. However, for those reaction systems in which insoluble biocatalyst enters the downstream plant it is possible to synthesise four separation schemes (Fig. 7) dependent upon the sequence and number of phase separations and assuming that the biocatalyst has a higher density than the aqueous phase. This may not always be the case, but is a reasonable generalisation from which to start.

The separation scheme I could be used for a reaction with an organic product, and scheme II with an aqueous and/or organic product. Scheme III is of no use since it cannot separate organic from aqueous phases. Scheme IV has the same potential to separate as II. In order to identify which scheme provides the most appropriate separation in a given case it is necessary to know what the relative densities of the three phases are. An analysis of the suitability of the four separation schemes is summarised in Table 4. It has been assumed that systems which separate such that biocatalyst becomes entrained within the organic phase are less favourable because they could result in serious damage to the catalyst.

Table 4 indicates that there is some flexibility in the design of the product recovery step. In this case the most versatile system would be scheme IV because it is appropriate to all cases but when $\rho_B \gtrsim \rho_A \gtrsim \rho_O$ then scheme II could also be used. However, the decision which to choose depends upon the number of operations involved and the relative difficulty of the to be achieved. Industrial process design implies industrial grade reaction

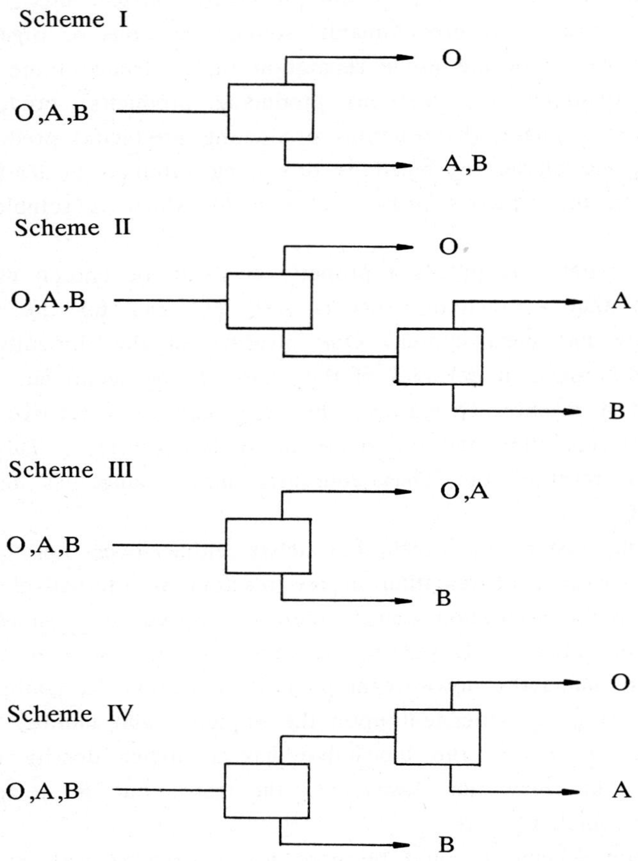

Fig. 7. The four possible phase separation schemes (insoluble biocatalyst, B; aqueous phase, A; organic phase, O) using density difference as separation criterion.

TABLE 4

PHASE SEPARATION HEURISTICS

SYSTEM	PRODUCT	I	II	III	IV
$\rho_B > \rho_A > \rho_O$	O	+	+	-	+
	A	-	+	-	+
	O & A	-	+	-	+
$\rho_B > \rho_O > \rho_A$.	O	-	O	-	+
	A	O	O	-	+
	O & A	-	O	-	+
$\rho_O > \rho_B > \rho_A$	O	-	O	+	+
	A	O	O	-	+
	O & A	-	O	-	+

\+ appropriate scheme
0 less appropriate as catalyst recovered with organic phase
\- not appropriate

components and dirty solvent, for example, may lead to much tougher separations than experienced in the laboratory.

There are several methods which could be employed to separate the phases: centrifugation which could damage the biocatalyst; electrostatic coalescence which would only separate the two-liquid phases and any biocatalyst present would require another method; membrane separations which are still at the developmental stage for liquid-liquid separations and have the problem that the organic phase may damage the membrane. Of these methods only the centrifugal separation relies on exploiting the density differences between the phases. The other methods exploit other property differences.

However, the philosophy of synthesising the possible separation schemes for each type of reaction system is fundamental and the same approach would be required regardless of the property being exploited. These downstream processing considerations together with detailed knowledge of microbial and enzymatic kinetics and stability are essential for the development of general process design rules which will ultimately form a framework from which specific reaction processes may be designed.

Acknowledgements

The authors wish to thank the Biotechnology Directorate of the Science and Engineering Research Council for its support of our work.

REFERENCES

1 M.D. Lilly and J.M. Woodley, Biocatalytic reactions involving water-insoluble organic compounds, in: J. Tramper, H.C. van der Plas, and P. Linko (Eds), Biocatalysts in Organic Syntheses, Elsevier, Amsterdam, 1985, pp 179–192.
2 S. Fukui, A. Tanaka and T. Iida, Immobilization of biocatalysts for bioprocesses in organic solvent media, this volume.
3 A. Freeman and M.D. Lilly, The effect of water-miscible solvents on the Δ^1-dehydrogenase activity of free and PAAH-entrapped *Arthrobacter simplex*, Appl. Microbiol. Biotechnol., in press.
4 F. Widmer, P. Thorbek, J. Lauridsen, G. Houen, S. Bayne, A.J. Andersen and J.T. Johansen, The strategy and preparative potential of enzymatic peptide synthesis exemplified on the synthesis of oxytocin, Ac-mEGF(3-20)-NH_2 and LH-RH, this volume.
5 B.C. Buckland, P. Dunnill and M.D. Lilly, The enzymatic transformation of water-insoluble reactants in nonaqueous solvents. Conversion of cholesterol to cholest-4-ene-3-one by *Nocardia* sp., Biotechnol. Bioeng., 17 (1975) 815–826.
6 J.M.C. Duarte and M.D. Lilly, The use of free and immobilised cells in the presence of organic solvents. The oxidation of cholesterol by *Nocardia rhodochrous*, in: H.H. Weetall and G.P. Royer (Eds), Enzyme Engineering, Vol. 5, Plenum Press, New York, 1980, pp. 363–367.
7 A. R. Macrae, Interesterification of fats and oils, in: J. Tramper, H.C. van der Plas, and P. Linko (Eds), Biocatalysts in Organic Syntheses, Elsevier, Amsterdam, 1985, pp. 195–208.
8 R.A. Wisdom, P. Dunnill, M.D. Lilly and A.R. Macrae, Enzymic inter-esterification of fats: factors influencing the choice of support for immobilized lipase, Enzyme Microb. Technol., 6 (1984) 443–446.
9 A. Zaks and A.M. Klibanov, Enzymatic catalysis in organic media at 100°C, Science, 224 (1984) 1249–1251.
10 Y. Inada, K. Takahashi, T. Toshimoto, A. Ajima, A. Matsushima and Y. Saito, Application of PEG-modified enzymes in biotechnological processes: organic solvent soluble enzymes, Trends Biotechnol., 4 (1986) 190–194.
11 M.D. Lilly and P. Dunnill, Immobilized-enzyme reactors, in: K. Mosbach (Ed.), Methods in Enzymology, Vol. XLIV, Academic Press, New York, 1976, 717–738.
12 M.D. Lilly, Immobilized enzyme reactors, Proc. 1st Europ. Congress Biotechnol., Dechema Monograph No. 82 (1979) 165–180.
13 M.M. Hoq, T. Yamane, S. Shimizu, T. Funada and S. Ishida, Continuous synthesis of glycerides by lipase in a microporous membrane bioreactor, JAOCS, 61 (1984) 776–781.
14 I.K. Brookes, M.D. Lilly and J.W. Drozd, Stereospecific hydrolysis of *d,l*-menthyl acetate by *Bacillus subtilis*: mass transfer-reaction interactions in a liquid-liquid system, Enzyme Microb. Technol., 8 (1986) 53–7.
15 A.C. Williams, J.M. Woodley, P.A. Ellis and M.D. Lilly, Denaturation and inhibition studies in a two-liquid phase biocatalytic reaction: the hydrolysis of menthyl acetate by pig liver esterase, this volume.
16 L.E.S. Brink and J. Tramper, Optimization of organic solvent in multiphase biocatalysis, Biotechnol. Bioeng., 27 (1985) 1258–69.

17 C. Laane, S. Boeren and K. Vos, On optimizing organic solvents in multi-liquid-phase biocatalysis, Trends Biotechnol., 3 (1985) 251-2.
18 M.D. Hocknull and M.D. Lilly, The Δ^1-dehydrogenation of hydrocortisone by *Arthrobacter simplex* in organic-aqueous two-liquid phase environments, this volume.
19 T. Yamane, H. Nakatani, E. Sada, T. Omata, T. Tanaka and S. Fukui, Steroid bioconversions in water insoluble organic solvents: Δ^1-dehydrogenation by microbial cells and by cells entrapped in hydrophilic and lipophilic gels, Biotechnol. Bioeng., 21 (1979) 1887-1903.
20 T. Omata, N. Iwamoto, T. Kimura, A. Tanaka and S. Fukui, Stereoselective hydrolysis of *d,l*-menthyl succinate by gel-entrapped *Rhodotorula minuta* var *texensis* cells in organic solvent, Eur. J. appl. Microbiol. Biotechnol., 11 (1981) 194-204.
21 J.M.C. Duarte and M.D. Lilly, Cholesterol degradation by polymer-entrapped *Nocardia* in organic solvents, in: A.I. Laskin, G.T. Tsao and L.B. Wingard, Jr (Eds), Enzyme Engineering 7, New York Academy of Sciences, New York, 1984, 573-6.
22 R.W. Jackson and J.A. De Moss, Effects of toluene on *Escherichia coli*, J. Bacteriol., 90 (1965) 1420-5.
23 M.J. De Smet, J. Kingma and B. Witholt, The effect of toluene on the structure and permeability of the outer cytoplasmic cell membranes of *E.coli*, Biochem. Biophys. Acta, 506 (1978) 64-80.
24 S. Silver and L. Wendt, Mechanism of action of phenethyl alcohol: breakdown of the cellular permeability barrier, J. Bacteriol., 93 (1967) 560-6.
25 L.O. Ingram and T.M. Buttke, Effects of alcohols on micro-organisms, Adv. Microbial Physiol., 25 (1982) 253-300.
26 J.S. Teh and K.H. Lee, Effects of *n*-alkanes on *Cladosporium resinae*, Can. J. Microbiol., 20 (1976) 971-6.
27 H. Felix, Permeabilized cells, Analyt. Biochem., 120, (1982) 211-234.
28 P.S. Cohen and H.L. Ennis, The requirement for potassium for bacteriophage T4 protein and deoxyribonucleic acid synthesis, Virology, 27 (1965) 282-9.
29 M. Lubin and H.L. Ennis, On the role of intracellular potassium in protein synthesis, Biochem. Biophys. Acta, 80 (1964) 614-31.
30 H. Galbraith, T.B. Miller, A.M. Paton and J.K. Thompson, Antibacterial activity of long chain fatty acids and the reversal with calcium, magnesium, ergocalciferol and cholesterol, J. Appl. Bacteriol., 34 (1971) 803-13.
31 C.W. Sheu, and E. Freese, Effects of fatty acids on growth, and envelope proteins of *Bacillus subtilis*, J. Bacteriol., 111 (1983) 516-24.
32 M.J. Playne and B.R. Smith, Toxicity of organic extraction reagents to anaerobic bacteria, Biotechnol. Bioeng., 25 (1983) 1251-1265.
33 G. Berrah and W.A. Konetzka, Selective and reversible inhibition of the synthesis of bacterial deoxyribonucleic acid by phenethyl alcohol, J. Bacteriol., 83 (1962) 738-44.
34 M.C. Laurerio-Dias and J.M. Peinado, Effect of ethanol and other alaknaols on the maltose transport system of *Saccharomyces cerevisiae*, Biotechnol. Lett., 4 (1982) 721-4.
35 M.J. Hampe, Selection of solvents in liquid/liquid extraction according to physico-chemical aspects, Ger. Chem. Eng., 9 (1986) 251-263.
36 D.F. Rudd, G.J. Powers and J.J. Siirola, Process Synthesis, Prentice-Hall Inc., New Jersey, 1973.

SESSION II

BIOCATALYST AND MEDIUM ENGINEERING

Chairman : E. Flaschel

Co-chairman : C. Veeger

C. Laane, J. Tramper and M.D. Lilly (Editors), *Biocatalysis in Organic Media*, Proceedings of an International Symposium held at Wageningen, The Netherlands, 7–10 December 1986.

IMMOBILIZATION OF BIOCATALYSTS FOR BIOPROCESSES IN ORGANIC SOLVENT MEDIA

S. Fukui[1], A. Tanaka[1] and T. Iida[2]

[1]Department of Industrial Chemistry, Faculty of Engineering, Kyoto University, Yoshida, Sakyo-ku, Kyoto 606, Japan

[2]Kansai Paint Co., Higashi Yawata, Hiratsuka 245, Japan

SUMMARY

Enzymes and microbial cells were immobilized by entrapment with prepolymers of photo-crosslinkable resins and urethane resins, respectively. The gels entrapping biocatalysts had desired hydrophilic or hydrophobic characters and net-work structures. The gel-entrapped biocatalysts were used for successful bioconversions of a variety of substrates with highly lipophilic or hardly water-soluble nature in reaction media composed of water-watermiscible organic cosolvents or water-containing organic solvents.

The experiments showed marked effects of the hydrophobicity-hydrophilicity balance and the net-work structure of the gels on the apparent activity as well as stability of the immobilized biocatalysts in such organic solvent media. Use of hydrophobic gels and less polar solvents is preferable for bioconversions of lipophilic compounds in general.

INTRODUCTION

In the case of bioconversion of highly lipophilic or hardly water-soluble compounds, it is desirable to carry out the enzymatic reaction in a mixture of water and a suitable organic cosolvent or in an adequate organic solvent (containing a little amount of water). The use of an organic solvent will improve the poor solubility of the substrate and other reaction components of hydrophobic nature in water. Moreover, for the utilization of hydrolytic enzymes for synthetic or group transfer reactions, the water fraction in the reaction mixture should be reduced by replacing water with appropriate organic solvents. Furthermore, along with the recent development of biotechnology, much wider applications of biocatalysts are demanded — for examples, bioconversions of xenobiotic compounds, such as aliphatic and aromatic compounds, having hydrophobic characters. In such bioprocesses, substrates and products are themselves organic solvents which are unconventional for biocatalysts.

Biocatalysts — enzymes, microbial cells *etc.* have been traditionally used in aqueous systems. It has been generally considered that biocatalysts are liable to be denatured in the presence of organic solvents, resulting in the loss of their catalytic abilities.

Attempts to render biocatalysts resistant to organic solvents have been made through different lines of approaches — chemical, biochemical and genetic techniques. Lately, site-specific mutations of enzyme molecules by gene manipulation attract world-wide interests. Of these approaches, immobilization of biocatalysts on or in suitable supports seem to be most general and promising.

The conformational structure of an immobilized enzyme will be more resistant to the distortion caused by organic solvents as compared with free native enzyme. In the cases of biotransformations of biological substances *in vivo*, many enzymes, particularly those catalyzing transformations of lipophilic biological compounds, function in membrane-bound states, and the stability of such membrane-associated enzymes is in general greater than that of enzymes released from the membrane.

Inclusion of enzymes or microbial cells within suitable gels would give an environment analogous to that *in vivo*. Multi-point interactions between entrapped enzymes (and microbial cells including enzymes) and gel matrices will give stabilizing effects. If the gels have desired hydrophobic characters and a suitable net-work structure, the environment around the biocatalysts entrapped in the gels will be more similar to that *in vivo*. However, the situation of immobilized biocatalysts *in vitro* is more complicated. For application of such gel-entrapped biocatalysts for conversions of lipophilic compounds in organic solvent systems, affinity of lipophilic substrates for the gels entrapping biocatalysts and diffusion of reactants through gel matrices are important factors. Low affinity of hydrophilic gels for lipophilic substrates will lower the apparent activity of the gel-entrapped biocatalysts. Thus, use of suitably hydrophobic gels with an adequate net-work structure will be preferable depending on hydrophobicity of substrates and polarity of solvents to be used.

In this article, we would like to report comprehensively on our experimental results using biocatalysts immobilized with prepolymers of photo-crosslinkable resins and urethane resins. The biocatalyst-entrapping gels had the desired hydrophobicity-hydrophilicity balances and net-work structures. They were used for bioconversions of a variety of highly lipophilic or hardly water-soluble substrates carried out in homogeneous reaction systems composed of water-watermiscible cosolvents or water-containing organic solvents. The effects of solvents and water content in reaction systems are also investigated.

METHODS AND MATERIALS

We have developed convenient methods to entrap biocatalysts inside gel matrices formed from synthetic prepolymers (ref. 1-5). Figures 1 and 2 show the structures of the prepolymers of photo-crosslinkable resins (ENT, hydrophilic and ENTP, hydrophobic) and the prepolymers of urethane resin (PU), respectively. Photo-crosslinkable resin prepolymers have photo-sensitive functional groups, such as acryloyl group at both terminals of the linear main chain. The chain length of prepolymers can be adjusted by using poly(ethylene glycol) or poly(propylen glycol) of optional chain length as the starting material for synthesis. Thus, ENT-4000, for instance, means that the prepolymer is formed with poly(ethylene glycol)-4000 (average Mw, *ca.* 4000; the chain length, *ca.* 40 nm). When the main skeleton consists of poly(ethylene oxide), the prepolymer and accordingly, the gels formed from the prepolymers, should have a hydrophilic character. On the other hand, the prepolymers containing poly(propylene oxide) in the main skeleton (ENTP) and the gels formed from ENTP should be hydrophobic. In the case of urethane prepolymers, the hydrophilic or hydrophobic character of prepolymers can be adjusted by changing the ratio of the poly(ethylene oxide) part and the poly(propylene oxide) part in the polyether moiety of the main skeleton. Thus PU-3 with a high content of poly(propylene oxide) gives more hydrophobic gels, while PU-6 and PU-9 with a high content of poly(ethylene oxide) hydrophilic gels (Figures 1 and 2, Tables 1 and 2).

Gelation of photo-crosslinkable resin prepolymers can be easily completed by illuminating the mixture of prepolymer solution, a small amount of photosensitizer, *e. g.* benzoin ethyl ether or benzoin isobutyl ether, and enzyme solution or microbial cell suspensions by near-ultraviolet ray for 3-5 min (refs. 1-5). Entrapment of biocatalysts with urethane prepolymers is much simpler. When liquid prepolymers are mixed with aqueous solutions of enzymes or aqueous suspensions of microbial cells, prepolymers react each other, being crosslinked by forming urea linkages with liberation of carbon dioxide (refs. 3-5).

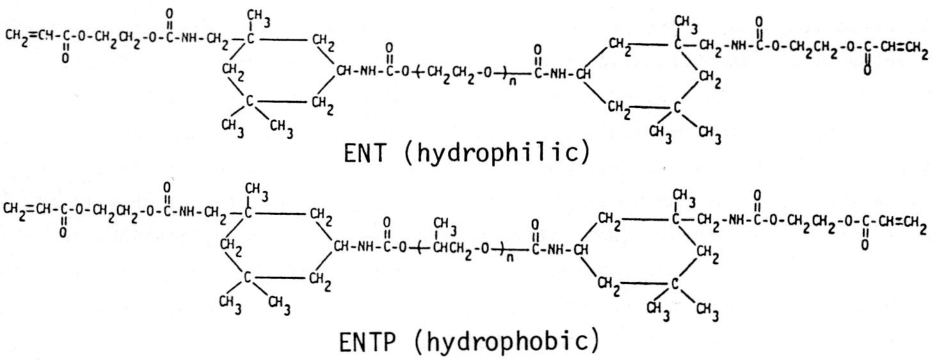

Fig. 1 Structures of Typical Photo-crosslinkable Resin Prepolymers

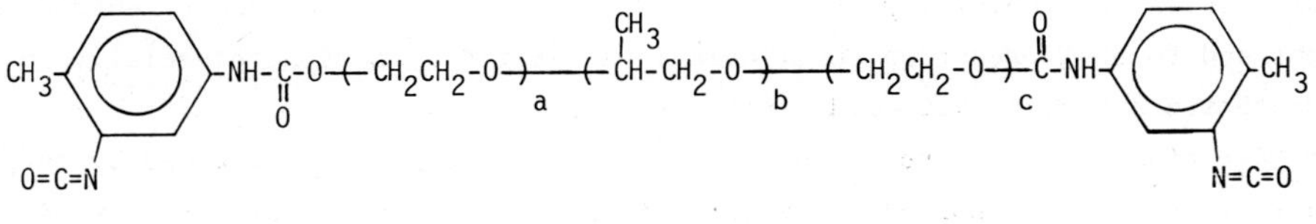

Fig. 2 Structure of Polyurethane Resin Prepolymers (PU)

Table 1. Properties of Typical Photo-crosslinkable Resin Prepolymers

Prepolymers	Component for Main Chain	Molecular Weight	Property
ENT-1000	Poly(ethylene glycol)	1000	Hydrophilic
-2000		2000	
-4000		4000	
-6000		6000	
ENTP-1000	Poly(propylene glycol)	1000	Hydrophobic
-2000		2000	
-3000		3000	
-4000		4000	

Table 2. Properties of Typical Polyurethane Resin Prepolymers

Prepolymers	$\bar{Mw}$ of Polyol	NCO Content (%)	Ethylene oxide Content (%)
PU-3	2529	4.2	57
PU-6	2627	4.0	91
PU-9	2616	4.0	100

Specific features of the prepolymer methods mentioned above are not only the simplicity of the immobilization processes but also the tailor-made net-work structure of gels with desired physical and chemical properties which can be obtained by the use of appropriate prepolymers. Another merit is that monomers which are liable to give bad effects on the activity of biocatalysts during immobilization processes and undesirable from sanitary viewpoints, if they remain in gels, are not involved. Especially, in the case of their applications to biocatalyses in organic solvent media, easy adjustment of the hydrophilicity-hydrophobicity balance and the net-work structure of the gels formed by the prepolymer methods is favourable as mentioned below.

RESULTS

[I] Bioconversion in Water-Organic Cosolvent Systems

Water-watermiscible organic cosolvent systems have been widely employed to dissolve water-insoluble, lipophilic compounds to prepare homogeneous reaction systems and to shift reaction equilibria in a desired direction, especially in the synthetic direction with hydrolyzing enzymes.

We have studied extensively bioconversions by biocatalysts entrapped by our prepolymer methods in appropriate water-watermiscible organic solvent systems for carrying out the reactions continuously. Table 3 shows our results on bioconversions of various lipophilic or hardly water-soluble compounds carried out in water-watermiscible organic solvent systems by the use of microbial cells entrapped with prepolymers.

Table 3. Bioconversions in Water-Organic cosolvent Systems by Microbial Cells Entrapped with Prepolymers

Microorganisms (Condition)	Organic Cosolvent	Application
Arthrobacter simplex (acetone-dried)	10% Methanol	Δ^1-Dehydrogenation of hydrocortisone
Curvularia lunata (living)	2.5% Dimethyl sulphoxide	11β-Hydroxylation of cortexolone
Rhizopus stolonifer (living)	2.5% Dimethyl sulphoxide	11α-Hydroxylation of progesterone
Sepedonium ampullosporum (living)	0.65% Dimethyl formamide	16α-Hydroxylation of estrone
Corynebacterium sp. (living)	15% Dimethyl sulphoxide	9α-Hydroxylation of 4-androstene-3,17-dione
Enterobacter aerogenes (thawed)	40% Dimethyl sulphoxide	Synthesis of adenine arabinoside

Acetone-dried cells of *Arthrobacter simplex*, entrapped with photo-crosslinkable resin prepolymers or urethane prepolymers, catalyzed Δ^1-dehydrogenation of hydrocortisone to form prednisolone. Introduction of an appropriate organic cosolvent (10%), such as methanol, ethylene glycol, propylene glycol *etc.*, stimulated significantly the reaction of the entrapped cells together with the solubilizing capacity of substrate and product. The entrapment markedly enhanced the stability of the cells at a certain high concentration level of the organic solvents and maintained the stability during long-term operations (ref. 6).

Hydroxylation of steroids by entrapped resting or living cells was also successfully carried out in the presence of organic cosolvents (refs. 7, 8). Cortexolone (Reichstein's Compound S) was hydroxylated at the 11β-position by photo-crosslinked gel-entrapped mycelia of *Curvularia lunata*, which were derived from spores germinated and developed inside the gel matrices in a similar way to that of Ohlson *et al.* (ref. 9). Dimethyl sulphoxide or methanol at 2.5% by volume was insufficient to dissolve the substrate (cortexolone) but was sufficient to dissolve the product (hydrocortisone). The hydroxylation activity of the entrapped mycelia could be reactivated by incubating the mycelia-entrapping gels in a nutrient medium containing cortexolone as an inducer. The hydroxylation activity was correlated to the growth of mycelia inside gel matrices, both of which were affected by the chemical and physical properties of gel materials. The photo-crosslinkable prepolymers with a suitably hydrophilic nature having an adequate net-work structure, *e. g.*, ENT-4000 (the chain length, *ca.* 40 nm) gave excellent gels for entrapping the spores of *Curvularia lunata* and for making well propagated mycelia having a strong 11β-hydroxylating activity. The entrapped mycelia were far more stable than their free counterparts and could be utilized repeatedly for at least 50 batches (total operational period, 100 days). The reaction system was connected sequentially with the Δ^1-dehydrogenation system of immobilized *A. simplex* cells to convert cortexolone to prednisolone *via* hydrocortisone (ref. 10). Spores of *Rhizopus stolonifer* were also entrapped with photo-crosslinkable prepolymers and allowed to germinate and develop inside gel matrices. The immobilized mycelia thus obtained hydroxylated progesterone at the 11α-position to form 11α-hydroxyprogesterone in the presence of 2.5% (by volume) of methanol, ethanol or dimethyl sulphoxide. In this case also, the enzyme system in the entrapped mycelia was reactivated by incubation in a nutrient medium (ref. 11).

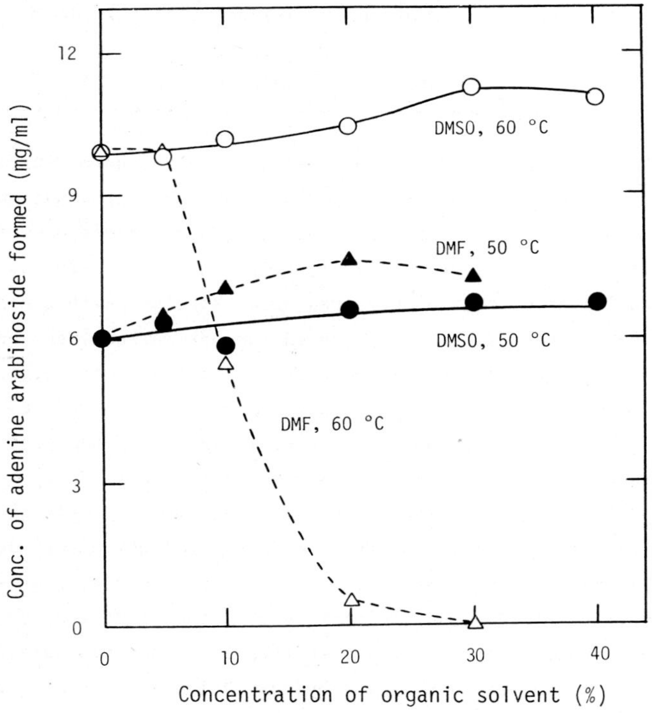

Fig. 3 Effect of organic cosolvents on adenine arabinoside synthesis by free cells of *Enterobacter aerogenes*.
Recations were carried out for 20 hr in dimethyl sulphoxide or in N,N-dimethylformamide at 50° C or 60°C.

Similarly, spores of *Sepedonium ampullosporum* were entrapped with ENT and the resulting gel-entrapped mycelia successfully catalyzed 16α-hydroxylation of estrone in the presence of a low concentration of N,N-dimethylformamide. Entrapped cells of *Corynebacterium* sp. were more torelant to organic solvent than entrapped fungal mycelia. Thus, 9α-hydroxylation of 4-androstene-3,17-dione to 9α-hydroxy-4-androstene-3,17-dione was performed in a nutrient medium containing 15% (by volume) of dimethyl sulphoxide (ref. 12).

The synthesis of adenine arabinoside, an antibiotic, from uracil arabinoside and adenine was almost quantitatively mediated by entrapped cells of *Enterobacter aerogenes*. Although the reaction had been carried out in an aqueous system, the productivity was fairly low due to the poor solubility of adenine and adenine arabinoside in water. By selecting an appropriate organic cosolvent on the criteria of stability of the enzyme system and solubility of adenine and adenine arabinoside, dimethyl sulphoxide at 40% (by volume) was chosen (see, Fig. 3).

Introduction of the organic solvent made it possible to increase the concentrations of the reactants 5 times higher than those in the previous aqueous system and thus brought about the high concentration of the product, adenine arabinoside. The biocatalysis should be carried out at 60°C to prevent

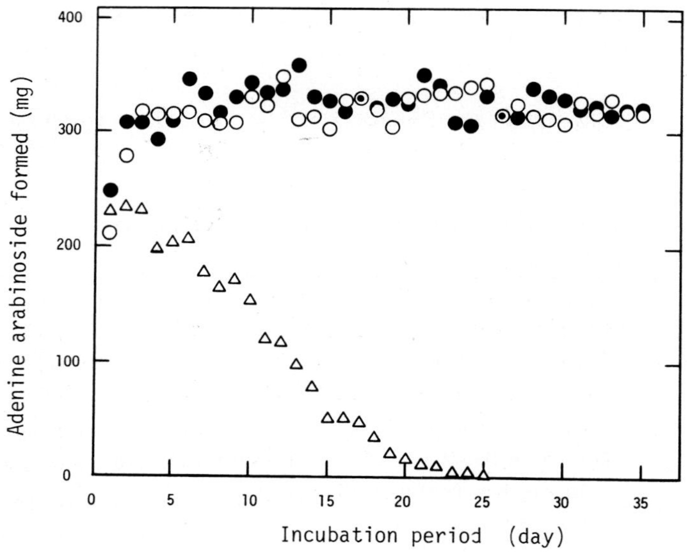

Fig. 4 Repeated use of *Enterobacter aerogenes* cells for production of adenine arabinoside.
Each reaction was carried out for 24 hr at 60°C in the presence of 40% dimethyl sulphoxide. (○), ENT-4000-entrapped cells; (●), PU-6-entrapped cells; (Δ), free cells.

side reactions occurring at lower temperature ranges. In such the reaction at the high temperature in the presence of dimethyl sulphoxide, the bacterial cells entrapped with photo-crosslinked gels or polyurethane gels showed a remarkable operational stability. The entrapped cells could be used repeatedly for at least 35 days without any loss of the catalytic activity (see, Fig. 4)(ref. 13).

Such enhanced stability of the gel-entrapped microbial cells was ascribed to multi-point interactions of the gels with cell membrane systems maintaining intracellular enzymes. Even if enzymes within cells would be liberated from the cells, interactions of gels and enzyme molecules will render the enzymes more stabilized than native enzymes in water-organic cosolvent systems.

[II] Bioconversion in Organic Solvent Systems

Although water-immiscible organic solvents have been used for bioreactions to increase the solubility of substrates and products of hydrophobic nature, most cases concern water-organic solvent two-phase systems. Only a limited number of papers have described the use of organic solvents alone (mostly, water-saturated organic solvents) in bioreactions.

Klibanov *et al.*(ref. 14) reported pioneering works on the use of organic solvents in the synthesis of N-acetyl-L-tryptophan ethyl ester from N-acetyl-L-tryptophan and ethanol by chymotrypsin covalently bound to porous glass. In this case, chloroform, benzene, or ether was used as the reaction solvent. Toluene or carbon tetrachloride was also employed in the dehydrogenation of cholesterol to cholestenone by DEAE-cellulose-adsorbed cells of *Nocardia erythrophilis* (ref. 15).

We have extensively investigated bioconversions of lipophilic compounds by immobilized biocatalysts in organic solvent systems (Table 4).

Table 4. Bioconversions in Organic Solvent Systems by Biocatalysts Entrapped with Prepolymers

Biocatalysts	Organic solvent (Water-saturated)	Application
Nocardia rhodocrous (cells)	Benzene-n-Heptane (1:1)[a]	Δ^1-Dehydrogenation of ADD[b]
	Benzene-n-Heptane (4:1)	Δ^1-Dehydrogenation of TS[c]
	Benzene-n-Heptane (1:1)	3β-Hydroxysteroid dehydrogenation
	Benzene-n-Heptane (4:1)	17β-Hydroxysteroid dehydrogenation
	Chloroform-n-Heptane (1:1)	3β-Hydroxysteroid dehydrogenation
Rhodotorula minuta (cells)	n-Heptane	Stereoselective hydrolysis of *dl*-menthyl ester
Candida cyclindracae (lipase)	Cyclohexane or Isooctane	Stereoselective esterification of *dl*-menthol
Rhizopus delemar (lipase)	n-Hexane	Interesterification of triglycerides

[a]Mixed ratio by volume, [b]4-Androstene-3,17-dione, [c]Testosterone

The activities of immobilized biocatalysts were found to be affected significantly by the hydrophilicity-hydrophobicity balance of gels, the hydrophobicity of substrates, and the polarity of reaction solvents.

Of the bioconversions summarized in Table 4, stereoselective esterification of *dl*-menthol and stereoselective hydrolysis of *dl*-menthyl esters will be mentioned in some more detail.

The stereoselective esterification of chemically synthesized *dl*-menthol was successfully carried out by gel-entrapped lipase preparations in water-saturated cyclohexane or isooctane. The lipases used were commercially available preparations obtained from *Candida cylindracea* (MY and OF 360) which were selected on the criteria of their ester-forming ability and stereospecificity toward the substrate. The acyl donor, 5-phenylvaleric acid, was chosen from various fatty acids of different chain length on the basis of both relative reactivity and the E. E. value of the esterification product. In the case of synthesis of *l*-menthyl 5-phenylvalerate from *dl*-menthol and 5-phenylvaleric acid, no significant difference was observed between the immobilized lipase preparations formed with hydrophilic urethane prepolymers (PU-6, PU-9) and those with hydrophobic prepolymer (PU-3) (Fig. 5).

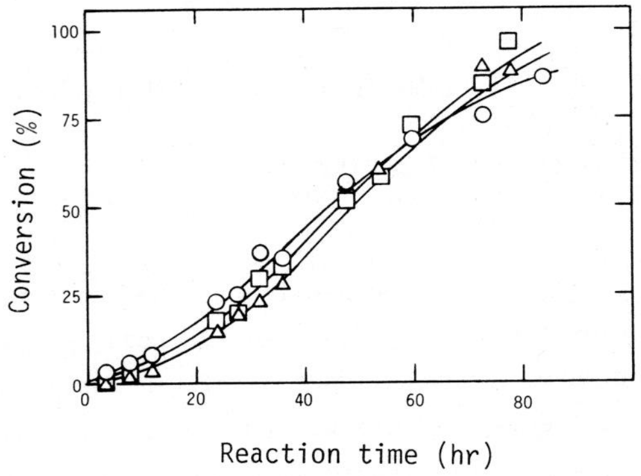

Fig. 5 Time-course of menthyl ester formation by *Candida cylindracea* lipase entrapped with different urethane prepolymers.

Candida cylindracea lipase OF 360 was entrapped with PU-3 (o), PU-6 (Δ) or PU-9 (□). The enzyme reaction was carried out in water-saturated sio-octane.

This is explained by the almost similar affinities of these gels for the substrates. The partition coefficient (the concentration ratio of substrate between gels and external solvent) of 5-phenylvaleric acid and *l*-menthol was *ca.* 5 or more with PU-3, PU-6 and PU-9, and 1.7, 1.3 and 1.0, respectively. When celite-adsorbed lipase was further entrapped in PU or ENT(P) gels, the operational stability of the immobilized lipase was markedly improved (ref. 16). In the case of transesterification of triglycerides by a regio-specific lipase of *Rhizopus delemar*, a similar improvement of the operational stability was observed together with the effects of gel hydrophobicity toward the lipophilic substrates, triglycerides and fatty acids (ref. 17).

The stereoselective hydrolysis of *dl*-menthyl succinate by gel-entrapped cells of *Rhodotorula minuta var. texensis* (Fig. 6) was performed successfully in water-saturated *n*-heptane (ref. 18). *l*-Menthol, a compound having a pepermint flavour and useful for the food and pharmaceutical industries, was obtained by stereospecific hydrolysis of an appropriate ester of chemically synthesized *dl*-menthol. The ammonium salt of *dl*-menthol succinate is water-soluble, and its stereoselective hydrolysis could be achieved in an aqueous system by using free cells of a yeast having esterase activity.

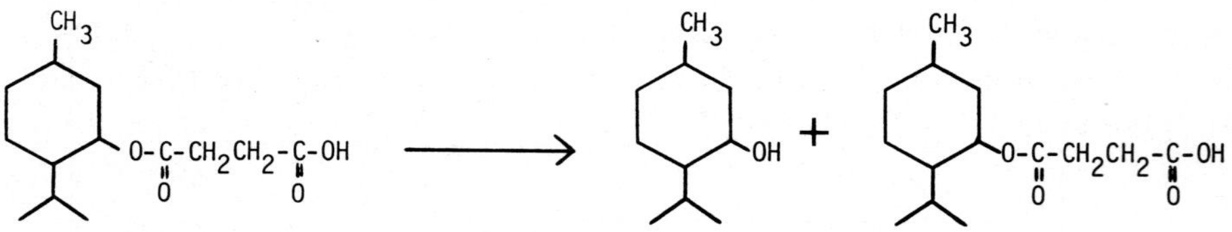

dl-Menthyl succinate *l*-Menthol *d*-Menthyl succinate

Fig. 6. Stereoselective hydrolysis of *dl*-menthyl succinate catalyzed by *Rhodotorula minuta var. texensis*

However, due to the poor solubility in aqueous buffers, the product, *l*-menthol accumulated on the surface of the yeast cells, thus decreasing the activity. To prevent this accumulation of *l*-menthol on the cell surface, various kinds of watermiscible organic solvents were tested as cosolvents. However, the hydrolytic activity of the yeast cells was reduced in water-organic cosolvent systems tested. After many trials using free cells in appropriate organic solvents and suitable two-phase systems consisting of potassium phosphate buffer and organic solvents, a combination of gel-entrapped cells and water-saturated *n*-heptane was finally employed to construct a homogeneous reaction system. Although the effects of gel hydrophobicity was not so remarkable as in the case of steroid transformation (see, the next Section), the activity of the gel-entrapped cells increased with increased hydrophobicity of the gels used.

The half-life of the free cells was 2 days, whereas that of PU-3 entrapped cells was *ca.* 63 days. Thus, immobilization greatly improved the operational stability of the hydrolytic enzyme in the yeast cells (Fig. 7).

The optical purity of the product was also constantly maintained at 100% after a long-term operation. In a semi-pilot scale production using immobilized *R. minuta* cells, the overall yield of *l*-menthol from the starting material, *dl*-menthol, was 86%.

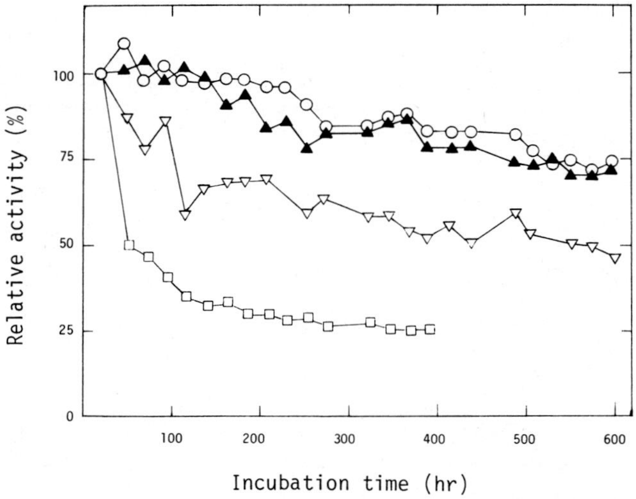

Fig. 7 Repeated use of free and entrapped *Rhodotorula minuta* cells in hydrolysis of *dl*-menthyl succinate. Each reaction was carried out for 24 hr. (□), Free cells; (○), PU-3 entrapped cells; (▲), PU-6-entrapped cells; (▽), ENT-4000-entrapped cells.

[III] Correlative Effects of Gel Hydrophobicity, Substrate Hydrophobicity and Water Concentration on Biocatalysis in Organic Solvent Media with Gel-entrapped Biocatalysts

As shown in Table 4, *Nocardia rhodocrous* cells mediate different types of bioconversions of a variety of steroids: Δ^1-dehydrogenation, 3β-hydroxysteroid dehydrogenation, and 17β-hydroxysteroid dehydrogenation. In order to study the effect of gel hydrophobicity — that is, the influence of the affinity between hydrophobic substrates and gels entrapping biocatalysts — a very simple parameter, the partition coefficient (the ratio of substrate concentration between gels and external solvent) was employed. The partition coefficient (P) was estimated according to the following equation:

$$P = \left(\frac{C_0 - C}{C}\right)\left(\frac{V_0}{V - V_0}\right)$$

where C_0 is the initial substrate concentration in the solvent, C is the final substrate concentration in the solvent, V_0 is the initial volume of the system without gels, and V is the final volume of the system with gels.

The effect of gel hydrophobicity was investigated by using the conversion of 3β-hydroxy-Δ^5-steroids to 3-keto-Δ^4-steroids (Fig. 8) in a water-saturated mixture of benzene and n-heptane (1:1 by volume). The solvent system was chosen on the basis of the following criteria: substrates and products are adequately soluble; the enzyme system is not damaged; and the solvent does not cause the gels to swell.

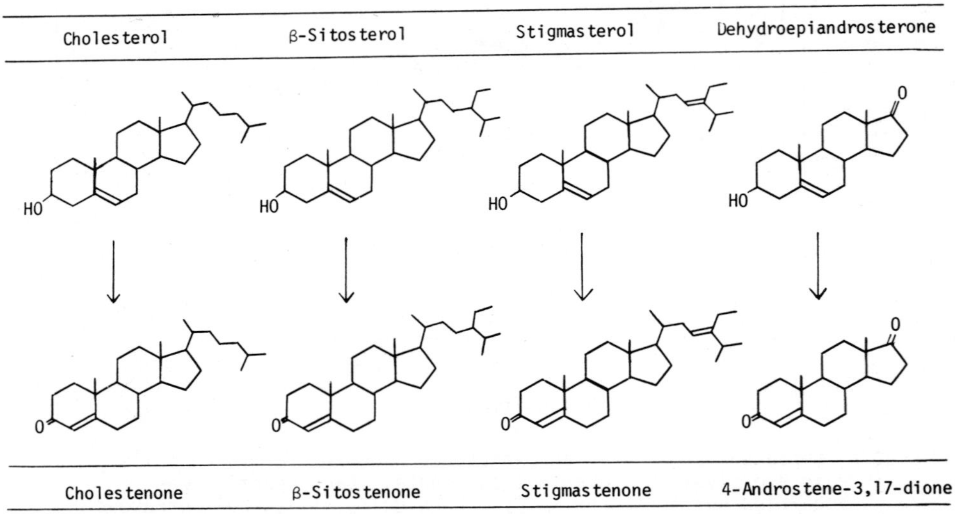

Fig. 8 Bioconversion of 3β-Hydroxy-Δ^5-steroids to the Corresponding 3-Keto-Δ^4-steroids Catalyzed by *Nocardia rhodocrous*

Nocardia rhodocrous cells entrapped in hydrophobic gels, such as ENTP=2000 and PU-3, converted dehydroepiandrosterone (DHEA) to 4-androstene-3,17-dione (4-AD) with a high reaction rate comparable to that of the free cells. On the other hand, the cells entrapped in hydrophilic gels, ENT-4000 and PU-6, were less active. Figure 9b shows a close relationship between the relative activity of the gel-entrapped cells and the partition coefficient of DHEA, both of which changed according to the hydrophobicity of gels. The abscissa is the mixed ratio of hydrophobic urethane prepolymer PU-3 and hydrophilic urethane prepolymer PU-6. In the case of transformation of cholesterol, a more lipophilic substrate than DHEA, to cholestenone, a much more clear-cut interrelationship was observed between gel hydrophobicity and the activity of the gel-entrapped cells. In the latter case (Fig. 9a), only the cells entrapped with hydrophobic prepolymers exhibited the catalytic activity. This phenomenon was well confirmed when the hydrophobicity of gels was adjusted by changing the mixing ratio of PU-3 and PU-6. In accordance with the low partition coefficient of the substrate, cholesterol, no conversion was observed with a low proportion of PU-3 (ref. 19).

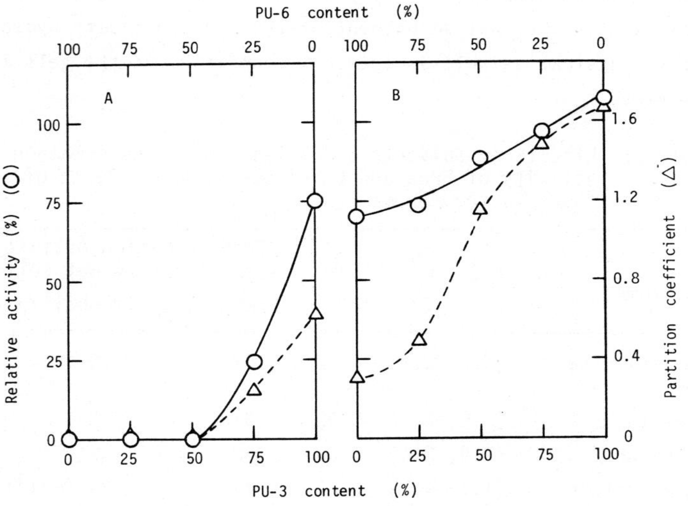

Figure 9. Effect of Hydrophobicity of Polyurethane Resin Gels on Relative Activity of Steroid Transformation and Partition Coefficient. (a) Cholesterol; (b) Dehydroepiandrosterone. (o), Relative activity; (Δ), Partition Coefficient.

In addition to gel hydrophobicity and substrate hydrophobicity, the polarity of the reaction solvent also had a marked effect on the conversion of steroids (ref. 20). *N. rhodocrous* cells entrapped in hydrophilic gels could not transform cholesterol in a non-polar solvent, such as a mixture of benzene and *n*-heptane (1:1 by volume). However substitution of benzene by chloroform made the hydrophilic gel-entrapped cells active, in harmony with the increased partition coefficient of the substrate between the gels and the external solvent. Increase in the solvent polarity lowered the activity and stability of the free cells, and subsequently, those of the entrapped cells. In the water-saturated mixture of chloroform and *n*-heptane (1:1 by volume), little effect of gel hydrophobicity was observed in the conversion of cholesterol and DHEA, differing from the results in the non-polar solvent system. The hydrophobic gel-entrapped cells had a rather low activity in the conversion of pregnenolone, a less hydrophobic substrate which was not soluble in the mixture of benzene and *n*-heptane. In spite of these facts, the usefulness of hydrophobic gels in the transformation of highly lipophilic compounds is clear because the transformation activity of the gel-entrapped cells is usually high and stable in less polar solvents. This leads to the conclusion that it is very important for successful bioconversions of lipophilic substrates to select gel materials with suitable hydrophobicity and reaction solvents with appropriate polarities, both of which should be selected according to substrate hydrophobicity. For the entrapment of biocatalysts, our prepolymer methods will be very useful, because selection of appropriate prepolymers enables optional adjustment of the hydrophobicity-hydrophilicity balance and other physical as well as chemical properties of the gels formed from the prepolymers.

Table 5. Effects of Solvents on Cholesterol Transformation Activity of Free and ENT-4000-entrapped Cells of *Nocardia rhodocrous*

Solvent (1:1 by volume)	(D_m^a, D_c^b)	P^c	Transformation Activity (μmol·hr-1·g wet cell-1) Free cells	Immobilized cells
Carbon tetrachloride - n-Heptane	(0.0, 2.2)	0.02	68	0
Benzene-n-Heptane	(0.0, 2.3)	0.02	57	0
Toluene-n-Heptane	(0.4, 2.4)	0.06	68	0
Chloroform-n-Heptane	(1.1, 4.7)	0.82	42	29 (69%[d])
Methylene chloride -n-Heptane	(1.5, 8.9)	0.81	19	15 (80%)
Ethyl acetate-n-Heptane	(1.9, 6.0)	0.25	27	0
Acetone-n-Heptane	(2.7, 20.7)	——	trace	0
Ethanol-n-Heptane	(1.7, 24.3)	——	0	0
Methanol-n-Heptane	(1.7, 32.6)	——	0	0

[a]Dipolar moment of organic solvent excluding n-heptane.

[b]Dielectric constant of organic solvent excluding n-heptane.

[c]Partition coefficient of cholesterol between ENT-4000 gel and external solvent

[d]Activity of free cells in each solvent was expressed as 100%.

It has been observed that gel hydrophobicity affected the conversion routes from testosterone (TS) to 1,4-androstadiene-3,17-dione (ADD) (ref. 21). Figure 11 illustrates the diverse transformation pathways of TS into ADD, mediated by *N. rhodocrous* in a water-saturated mixture of benzene and *n*-heptane (4:1 by volume). In the presence of an electron acceptor, such as phenazine methosulphate (PMS), the free bacterial cells converted TS to ADD *via* two different pathways. In these two routes, the 17β-dehydrogenation product of TS, 4-androstene-3,17-dione (4-AD) and the Δ^1-dehydrogenation product, Δ^1-dehydrotestosterone (DTS), appeared as the intermediates, respectively.

In these reactions, Δ^1-dehydrogenation absolutely required PMS, whereas 17β-dehydrogenation could proceed without the exogenous electron acceptor although PMS stimulated the reaction. When the cells were entrapped in gels of different hydrophilicity or hydrophobicity, the properties of the gels had striking effects on the conversion routes. With hydrophobic gel-entrapped cells, 4-AD was formed as the major reaction product. On the other hand, DTS was the main product in the reaction with hydrophilic gel-entrapped cells. This different pattern of dehydrogenation products can be explained by a marked difference in the affinity of PMS, a hydrophilic compound, for the hydrophilic and hydrophobic gels. With hydrophilic gel-entrapped cells, Δ^1-dehydrogenation of TS to yield DTS is stimulated by PMS taken up inside the gels; DTS so accumulated inhibits 17β-hydroxysteroid dehydrogenase converting DTS to ADD even at a low concentration. On the other hand,PMS was hardly taken up in the cells entrapped within hydrophobic gels and hence 17β-dehydrogenation of TS becomes predominant.

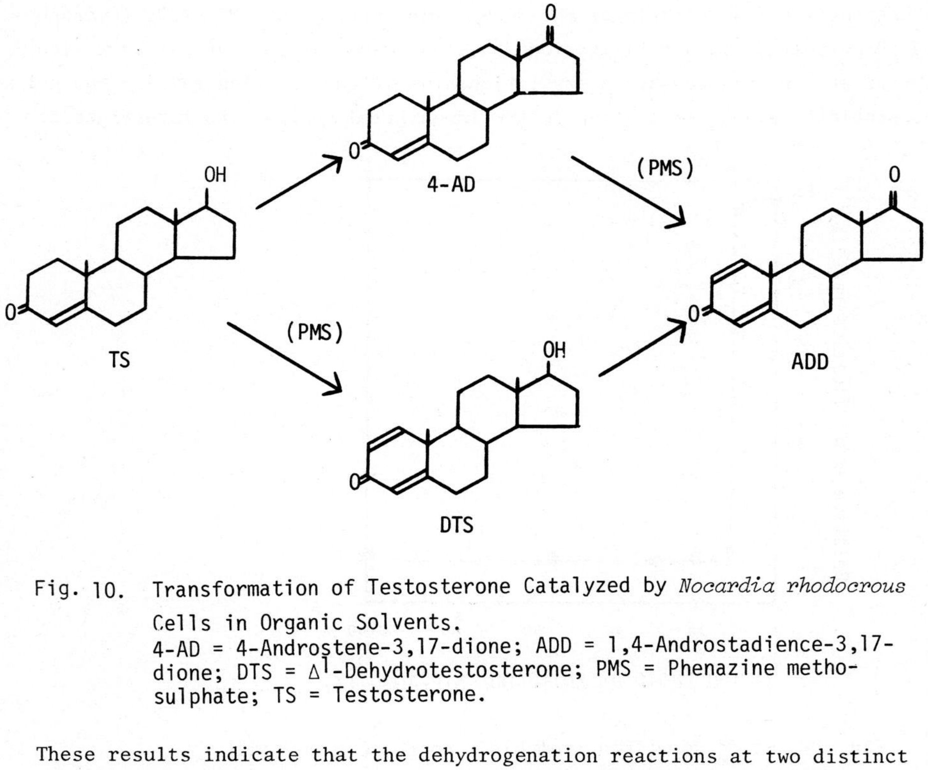

Fig. 10. Transformation of Testosterone Catalyzed by *Nocardia rhodocrous* Cells in Organic Solvents.
4-AD = 4-Androstene-3,17-dione; ADD = 1,4-Androstadience-3,17-dione; DTS = Δ^1-Dehydrotestosterone; PMS = Phenazine methosulphate; TS = Testosterone.

These results indicate that the dehydrogenation reactions at two distinct positions of TS can be controlled by selecting hydrophilic or hydrophobic gels for entrapping *N. rhodocrous* cells. Thus, selective formation of a desired product among diverse products from a single substrate has been achieved by appropriate use of hydrophilic or hydrophobic gels to entrap the cells. This principle would be applicable to the bioconversions of some other compounds having analogous diverse transformation routes.

The water content is one of the most important factors affecting the reaction rate, reaction equilibrium, product yield and stability as well as activity of biocatalysts in bioprocesses in organic media. Recently much attention is being denoted to basic and practical studies on the effects of water content on biocatalyses in organic media.

We have studied the effects of water concentrations on our immobilization systems. The model reactions are the syntheses of the ester from oleic acid and *n*-heptyl alcohol and the glycerides from oleic acid and glycerol. The substrates were mixed with a little amount of water (graded proportions) in the absence of organic solvents, and the esterification reactions were mediated by lipase (*Candida cylindracea* lipase OF-360, Meito Sangyo Co., Japan) which was homogeneously mixed with the reaction components and then entrapped within photo-crosslinked gels of different hydrophobicity. The hydrophobicity of the gels was adjusted by changing the mixing ratio of ENT-3400 (hydrophilic) and ENTP-4000 (hydrophobic) prepolymers. The esterification reactions were carried out at 30°C with stirring at 600 rpm and the interrelationships of the reaction rate to the gel hydrophobicity and water content in the immobilized system were investigated.

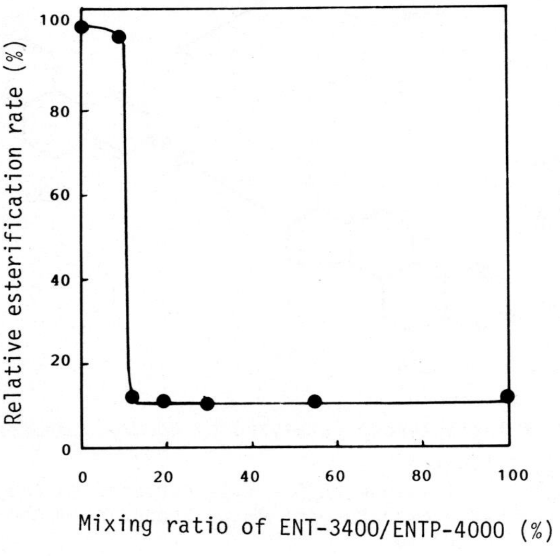

Fig. 11 Ester Formation from Oleic Acid and *n*-Heptanol with Lipase Entrapped within Photo-crosslinked Gels of Different Hydrophobicity

Figures 11 and 12 show the effects of gel hydrophobicity on the syntheses of the ester from oleic acid and *n*-heptanol and the glyceride formation from oleic acid and glycerol, respectively. As seen in Fig. 11, the ester formation from both lipophilic substrates was only mediated by the lipase entrapped in highly hydrophobic gels (the proportion of ENTP-4000, 90-100%), whereas the glyceride formation from oleic acid and glycerol was catalyzed by lipase entrapped with gels of a little more hydrophilic nature (the optimal mixing ratio of ENTP-4000 and ENT-3400; *ca.* 83:17). Thus, the effect of the gel hydrophobicity was again observed as the case of steroid transformation.

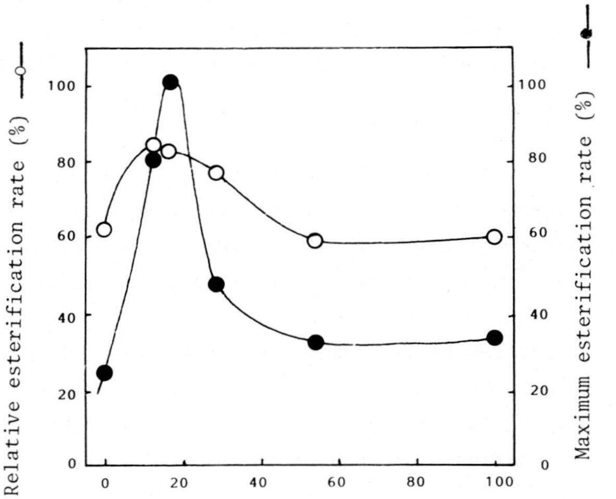

Fig. 12. Glyceride Formation from Oleic Acid and Glycerol with Lipase Entrapped within Photo-crosslinked Gels of Different Hydrophobicity

In the case of formation of ester and glyceride(s) by the use of lipase, the water concentration in the reaction system, especially within the gels entrapping the enzyme, will give significant effects on the reaction rate and product yield. In our experiments, the water concentration in the gels entrapping lipase and reaction components was *ca.* 3-5%. The DSC melting endotherms of the water in the gels showed the presence of two different states of water molecules. Although a quantitative determination of the water molecules in the vicinity of the enzyme molecules has not been carried out, the behaviour of water molecules in gels will be different between highly hydrophobic gels and less hydrophobic gels. Namely, physical and chemical properties of gels, particularly the hydrophobicity-hydrophilicity balance, will have dominant influences on the synthetic reactions and group transfer reactions catalyzed by hydrolytic enzymes through the interaction of gels with both substrates and water molecules.

DISCUSSION

As mentioned above, it is essential to introduce appropriate organic solvents into reaction systems for effective bioconversions of highly lipophilic or hardly water-soluble compounds. Applications of hydrolytic enzymes for synthetic or group transfer reactions also should be carried out in reduced concentrations of water.

Various ideas and devices have been proposed for the purposes mentioned above: water and water-immiscible organic solvent two-phase systems (refs. 22, 23), reaction of extremely high concentrations of substrates with free or modified enzymes in the presence of a small amount of water, reversed micell systems, and immobilized biocatalyst systems.

As for the catalytic activity and stability of some enzymes in a 99% organic medium, Zaks and Klibanov have published a stimulative paper (ref. 24). Uses of polyethylene glycol-modified enzymes — organic solvent-soluble enzymes — for biocatalyses in organic solvent media were reported by Inada and coworkers (ref. 25).

Concerning the design for immobilized enzyme reactors applicable to bioconversions of lipophilic compounds or reverse reactions of hydrolytic enzymes, Yamane and his collaborators have published the application of microporous hydrophobic membrane bioreactors for continuous synthesis of glycerides and hydrolysis as well as glycerolysis of fat and oil. The microporous membrane with a highly hydrophobic nature provides for spontaneous immobilization of lipase by hydrophobic interactions and thus gives hydrophobic environments for the immobilized lipase (refs. 26, 27, 28).

A significant superiority of continuous stirred tank reactors (CSTR) over plug flow-type reactors (PFR) was demonstrated for continuous syntheses of biologically important oligopeptides by Nakanishi, Matsuno *et al.* (refs. 29, 30). They have done detailed kinetic studies and established optimal conditions using a CSTR for the syntheses of N-(benzoylcarbonyl)-L-aspartyl-L-phenylalanine methyl ester and des-Tyr-leucine enkehalin mediated by immobilized thermolysin in ethyl acetate media (ref. 31).

Concerning the effects given by hydrophobicity and hydrophilicity around gel-entrapped enzymes, Suzuki and his collaborators have made interesting experiments using carrier-bound spiropyran derivatives. Spiropyran derivatives covalently bound to gels entrapping enzymes show photoisomerism — the open ring structure of hydrophilic character in the dark and the closed ring structure of hydrophobic nature in the light. By using this photochroism of gel-bound spiropyran, they controlled the hydrophilicity and hydrophobicity of the environment around immobilized hydrolytic enzymes such as lipase and chymotrypsin *etc.* Clear-cut influences of the adjustment of hydrophobicity and hydrophilicity of gels have been demonstrated on the directions of reactions mediated by the immobilized hydrolytic enzymes. For examples, syntheses of esters, ethers and glycerides were catalyzed by the spiropyran-gel entrapped hydrolytic enzymes in the light, while hydrolytic reactions by the gel-entrapped enzymes in the dark (ref. 32 and unpublished data).

We have studied the effects of the hydrophobicity and hydrophilicity and the net-wrok structure of gels entrapping biocatalysts for bioprocesses in organic solvent media. Our apporaches are the optional adjustment of these characters of gels by employing our new immobilization methods with synthetic resin prepolymers. Some of the fundamental studies and practical applications are mentioned in this article.

The advantages of the prepolymer methods presented here are not only the simplicity of immobilization procedures but also the ease in adjustment of the hydrophobicity-hydrophilicty balance and net-work structure of the gels entrapping biocatalysts. Biocatalysts immobilized by the methods have been successfully used for a variety of bioconversions in homogeneous reaction systems composed of water-watermiscible organic solvents and water-containing organic solvents. The effects of the hydrophobicity and net-work structure of gels entrapping biocatalysts have been conveniently studied by the use of such systems.

Acknowledgement

We wish to express our sincere gratitude to Professor Malcolm D. Lilly for his valuable help in preparing our manuscript.

REFERENCES

1 S. Fukui, A. Tanaka, T. Iida and F. Hasegawa, Application of photo-crosslinkable resin to immobilization of an enzyme, FEBS Letters, 66 (1976) 179-182.
2 S. Fukui, K. Sonomoto, N. Itoh and A. Tanaka, Several novel methods for immobilization of enzymes, microbial cells and organelles, Biochimie, 62 (1980) 381-386.
3 S. Fukushima, T. Nagai, K. Fujita, A. Tanaka and S. Fukui, Hydrophilic urethane prepolymers: Convenient materials for enzyme entrapment, Biotechnol. Bioengin., 20 (1978) 1465-1469.
4 S. Fukui and A. Tanaka, Application of biocatalysts immobilized by prepolymer methods, in: A. Fiechter (Ed.), Adv. Biochem. Engin./Biotechnol.Springer, Berlin-Heidelberg, Vol. 29, 1984, pp. 1-33.
5 S. Fukui and A. Tanaka, Immobilized microbial cells, in: A. Ornston (Ed.), Annual Rev. Microbiol., Annual Rev. Inc., Palo-Alto, Vol. 36, 1982, pp. 145-172.
6 K. Sonomoto, A. Tanaka, T. Omata, T. Yamane and S. Fukui, Application of photo-crosslinkable resin prepolymers to entrap microbial cells. Effects of increased cell-entrapping gel hydrophobicity on the hydrocortisone Δ^1-dehydrogenation, Eur. J. Appl. Microbiol. Biotechnol., 6 (1979) 325-334.
7 A. Tanaka, K. Sonomoto, M. M. Hoq, N. Usui, K. Nomura and S. Fukui, Hydroxylation of steroids by immobilized microbial cells, in: I. Chibata, S. Fukui and L. B. Wingard, Jr. (Ed.), Enzyme Engineering, Vol. 6, Plenum Press, New York-London, 1982, pp. 131-133.
8 K. Sonomoto, M. M. Hoq, A. Tanaka and S. Fukui, 11β-Hydroxylation of cortexolone (Reichstein's Compound S) to hydrocortisone by *Curvularia lunata* entrapped in photo-crosslinked resin gels, Appl. Environ. Microbiol., 45 (1983) 436-443.
9 S. Ohlson, S. Flygare, P. O. Larsson and K. Mosbach, Steroid hydroxylation using immobilized spores of *Curvularia lunata* germinated *in situ*, Eur. J. Appl. Microbiol. Biotechnol., 10 (1980) 1-9.
10 T. K. Mazmder, K. Sonomoto, A. Tanaka and S. Fukui, Sequential conversion of cortexolone to prednisolone by immobilized mycelia of *Curvularia lunata* and immobilized cells of *Arthrobacter simplex*, Appl. Microbiol. Biotechnol., 21 (1985) 154-161.
11 K. Sonomoto, K. Nomura, A. Tanaka and S. Fukui, 11α-Hydroxylation of progesterone by gel-entrapped living *Rhizopus stolonifer* mycelia, Eur. J. Appl. Microbiol. Biotechnol., 16 (1982) 57-62.
12 K. Sonomoto, N. Usui, A. Tanaka and S. Fukui, 9α-Hydroxylation of 4-androstene-3,17-dione by gel-entrapped *Corynebacterium* sp. cells, Eur. J. Appl. Microbiol. Biotechnol., 17 (1983) 203-210.
13 K. Yokozeki, S. Yamanaka, T. Utagawa, K. Takinami, Y. Hirose, A. Tanaka, K. Sonomoto and S. Fukui, Production of adenine arabinoside by gel-entrapped cells of *Enterobacter aerogenes* in water-organic cosolvent system, Eur. J. Appl. Microbiol. Biotechnol., 14 (1982) 225-231.
14 A. M. Klibanov, G. P. Samokhin, K. Martinek and I. V. Berezin, A new approach to preparative enzymatic synthesis, Biotechnol. Bioengin., 19 (1977) 1351-1361.
15 P. Atrat, E. Hüller and C. Hörhold, Steroid transformation mit immobilierten Mikroorganismen: Transformation von cholesterol zu cholestenone im organischen lösungsmittel, Z. Allg. Mikrobiol., 20 (1980) 79-84.
16 S. Koshiro, K. Sonomoto, A. Tanaka and S. Fukui, Stereoselective esterification of *dl*-menthol by polyurethane-entrapped lipase in organic solvent, J. Biotechnol., 2 (1985) 47-57.
17 K. Yokozeki, S. Yamanaka, K. Takinami, Y. Hirose, A. Tanaka, K. Sonomoto and S. Fukui, Application of immobilized lipase to regio-specific interesterification of triglyceride in organic solvent, Eur. J. Appl. Microbiol. Biotechnol. 14 (1982) 1-5.
18 T. Omata, N. Iwamoto, T. Kimura, A. Tanaka and S. Fukui, Stereoselective hydrolysis of *dl*-menthol succinate by gel-entrapped *Rhodotorula minuta var. texensis* cells in organic solvent, Eur. J. Appl. Microbiol. Biotechnol., 11 (1981) 199-204.

19 T. Omata, T. Iida, A. Tanaka and S. Fukui, Transformation of steroids by gel-entrapped *Nocardia rhodocrous* cells in organic solvent, Eur. J. Appl. Microbiol. Biotechnol., 8 (1979) 143-155.
20 T. Omata, A. Tanaka and S. Fukui, Bioconversion under hydrophobic conditions: Effects of solvent polarity on steroid transformations by gel-entrapped *Nocardia rhodocrous* cells, J. Ferment. Technol., 58 (1980) 339-343.
21 S. Fukui, S. A. Ahmed, T. Omata and A. Tanaka, Bioconversion of lipophilic compounds in non-aqueous solvent. Effect of gel hydrophobicity on diverse conversions of testosterone by gel-entrapped *Nocardia rhodocrous* cells, Eur. J. Appl. Microbiol. Biotechnol., 10 (1980) 289-301.
22 M. D. Lilly, Two-liquid-phase biocatalytic reactions, J. Chem. Technol. Biotechnol., 32 (1982) 162-169.
23 G. Carrea, Biocatalysis in water-organic solvent two-phase systems, Trends in Biotechnol., 2 (1984) 102-106.
24 A. Zaks and A. M. Klibanov, Enzymatic catalysis in organic media at 100°C, Science, 224 (1984)1249-1251.
25 Y. Inada, K. Takahashi, T. Yoshimoto, A. Ajima, A. Matsushima and Y. Saito, Application of polyethylene glycol-modified enzymes in biotechnological processes: Organic solvent-soluble enzymes, Trends in Biotechnol., 4, (1986) 190-194.
26 M. M. Hoq, T. Yamane and S. Shimizu, Continuous synthesis of glycerides by lipase in a microporous membrane bioreactor, J. Am. Oil Chem. Soc., 61 (1984) 776-781.
27 M. M. Hoq, T. Yamane and S. Shimizu, Continuous hydrolysis of olive oil by lipase in microporous hydrophobic membrane bioreactor, J. Am. Oil Chem. Soc., 62 (1985) 1016-1021.
28 T. Yamane, M. M. Hoq, S. Itoh and S. Shimizu, Continuous glycerolysis of fat by lipase in microporous hydrophobic membrane bioreactor, J. Jap. Oil Chem. Soc., 35 (1986) 632-636.
29 K. Nakanishi, Y. Kimura and R. Matsuno, Design of proteinase-catalyzed synthesis of oligopeptides in an aqueous-organic biphasic system, Biotechnology, 4 (1986) 452-454.
30 K. Nakanishi, T. Kamikubo and R. Matsuno, Continuous synthesis of N-(benzyloxycarbonyl)-L-aspartyl-L-phenylalanine methyl ester with immobilized thermolysin in an organic solvent, Biotechnology, 3 (1985) 459-464.
31 K. Nakanishi, Y. Kimura and R. Matsuno, Kinetics and equilibrium of enzymatic synthesis of peptides in aqueous/organic biphasic systems, Eur. J. Biochem., in press.
32 I. Karube, Y. Ishimori and S. Suzuki, Photocontrolled binding of cytochrome c to immobilized spiropyran, J. Soild-phase Biochem., 2 (1977) 9-17.

C. Laane, J. Tramper and M.D. Lilly (Editors), *Biocatalysis in Organic Media*,
Proceedings of an International Symposium held at Wageningen,
The Netherlands, 7–10 December 1986.

ENANTIOSELECTIVE REACTIONS IN AQUEOUS AND IN ORGANIC MEDIA USING CARBOXYL ESTERASE FRACTIONS OBTAINED FROM CRUDE PORCINE PANCREAS LIPASE PREPARATIONS

G.M. RAMOS TOMBO, H.-P. SCHAER, X. FERNANDEZ I BUSQUETS and O. GHISALBA
Central Research Laboratories of CIBA-GEIGY AG, CH-4002 Basel (Switzerland)

SUMMARY

A direct entry to both enantiomeric forms of chiral diol monoacetates is described. The method is based on enzyme catalyzed transformations of prochiral precursors in aqueous and in organic media. The catalysts used are carboxyl esterase preparations obtained from crude porcine pancreas lipase.

INTRODUCTION

The advantages of introducing chirality by means of enantioselective enzymatic transformations of prochiral starting materials are well established (ref. 1). We were interested in the enzymatic production of compounds of type b starting from a prochiral precursor a in which the substituent R^1 should be widely variable.

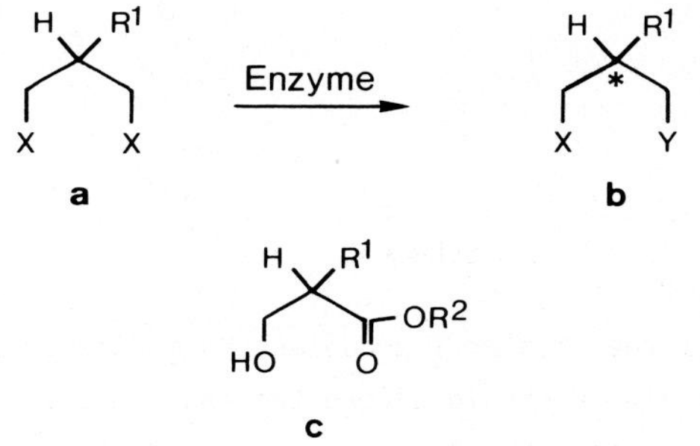

Scheme 1

The synthetic utility of this type of chiral building blocks is illustrated by the extended use of chiral ß-hydroxy isobutyric acid derivatives c ($R^1 = CH_3$) in the synthesis of optically active compounds (ref. 2) as well as by the efforts of many research groups in developing enantioselective synthesis of related 2-substituted hydroxy acids (c) (ref. 3).

Looking for a method of general applicability in the average organic chemistry laboratory, we focused our interest on the use of hydrolytic enzymes for the preparation of optically active 2-substituted 1,3-propanediol derivatives. Although it is possible to chemically interconvert the enantiomeric monoesters g and ent-g, in practice this would require the tedious manipulation of protecting groups. Thus, we felt that a direct entry to both enantiomeric esters would be very advantageous.

Given the fact that most of the hydrolytic enzymes used for the enantioselective hydrolysis of esters also catalyze acyl transfer reactions in organic media (ref. 4), we chose the following approach for the synthesis of both series g and ent-g (Scheme 2).

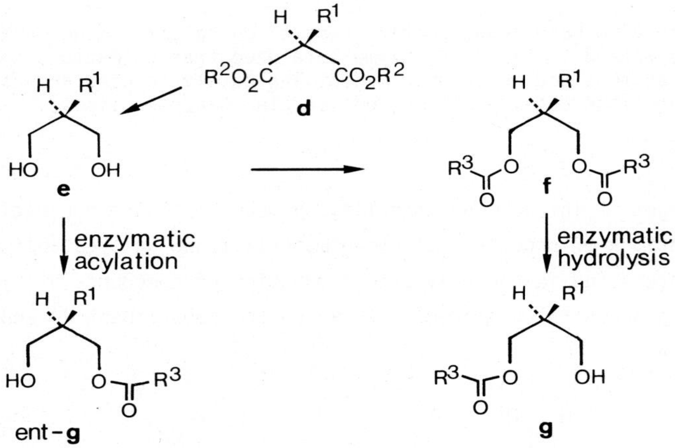

Scheme 2

Enzymatic hydrolysis of the diester f produces the monoester g, while the enzymatic acylation of the diol e should afford the enantiomeric monoacetate ent-g, provided that both processes take place on the same enantiotopic side of the substrates.

SELECTION OF THE ENZYME

A number of commercially available hydrolases were tested for the enantioselective hydrolysis of diesters of type f (R^1 = $CH(CH_3)_2$, R^3 = $CH_3CH_2CH_2$, CH_3CH_2 or CH_3) to the corresponding chiral monoesters of type g. The best results were obtained for the hydrolysis of diacetate 1 to the monoacetate 7 and are summarized in Table 1. Although the observed e.e.-values were quite low, various reasons induced us to investigate the reaction with crude porcine pancreas lipase (PPL) in more detail (see also ref. 5).

TABLE 1

Enzyme screening for the enantioselective hydrolysis of diacetate 1 (→7)

Enzyme (Sigma)	e.e.%[a)]	(Monoacetate 7)[b)]
Esterase from wheat germ	39	(+)
Esterase from orange peel	0	
Esterase from pig liver (PLE)	26	(+)
Cholesterol esterase from pig pancreas	22	(-)
Lipase from pig pancreas (PPL) "crude"	37	(-)
Lipase from pig pancreas "pure" (EC 3.1.1.3)	very slow reaction	
Lipase from *Candida cylindracea*	42	(-)
Lipase from *Rhizopus arrhizus*	no reaction	
Lipase from *Chromobacterium viscosum*	0	
Papain	no reaction	
Subtilisin	0	
α-Chymotrypsin	no reaction	

a) e.e.-values determined by 360 MHz ^{1}H-NMR spectroscopy of the corresponding (-)MTPA-derivatives.

b) See (ref. 7)

The commercially available crude PPL preparation (Sigma Nr. L3126), which was shown by FPLC (ref. 6) to be a mixture of different hydrolytic activities, produced the monoacetate (-)-7 (ref. 7) with 37% e.e. . On the other hand, the reaction catalyzed by the pure PPL (triacylglycerol acyl hydrolase EC 3.1.1.3, Sigma Nr. L0382) was extremely slow and showed a poor selectivity. These observations indicated that the reaction 1 → (-)-7 was not catalyzed by the lipase EC 3.1.1.3 itself but by other carboxyl esterase(s) present in the crude PPL preparation. We therefore undertook the isolation of this esterase activity(ies).

FRACTIONATION OF CRUDE PPL

A sample of crude PPL (Sigma Nr. L3126) was chromatographed on Sephacryl S 300 sf and the fractions were tested for hydrolytic activity on tributyrin, triacetin and diacetate 1, respectively. As shown in Fig. 1 the hydrolytic activity is concentrated in the high molecular weight fractions. There are two distinct activity peaks for tributyrin and triacetin. The main activity for the hydrolysis of 1 to (-)-7 coincides with the main activity for triacetin. This fraction (Pool 1) showed also the highest enantioselectivity for the hydrolysis of 1 (Fig. 2).

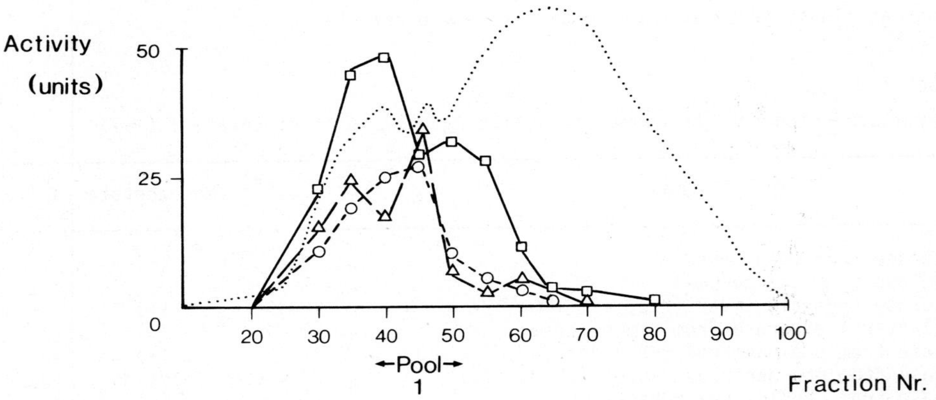

Fig. 1. Gel filtration of crude PPL (Sigma Nr. L3126) on Sephacryl S 300 sf. Elution with 18 ml/h of 69 mM sodium phosphate buffer pH 7, fraction size 4.5 ml. (□): activity for tributyrin (reduced by a factor of 5). (○): activity for triacetin. (△): activity for the hydrolysis of 1 to (-)-7. (···): protein detected by monitoring the absorbance at 280 nm.

The e.e.-value of 75% for the monoacetate (-)-7 obtained with the fraction Pool 1 (esterase P1) represented a marked increase with respect to the e.e. obtained with the crude PPL preparation for the same reaction (e.e. 37%).

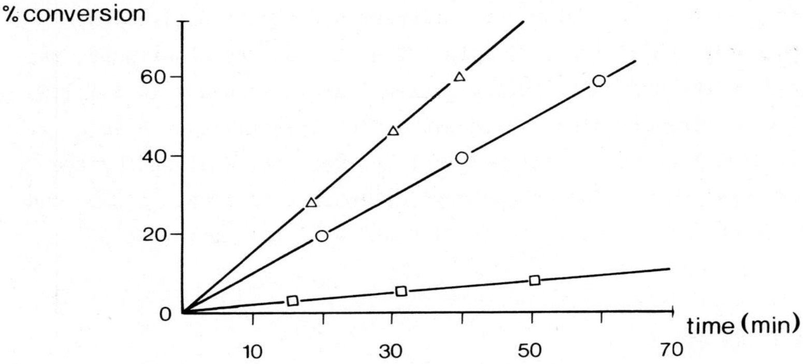

Fig. 2. Initial velocities (v_0) for the reaction 1 to (-)-7 in 69 mM sodium phosphate buffer pH 7 at 25°C. (△): Pool 1, v_0 = 1,50 μmol/min mg protein, e.e. 75%. (○): crude PPL, v_0 = 1.27 μmol/min mg protein, e.e. 37%. (□): pure lipase from porcine pancreas (EC 3.1.1.3), v_0 = 0.09 μmol/min mg protein, e.e. 5-15% depending on the preparation.

The carboxyl esterase P1 was found suitable as a catalyst for the enantioselective hydrolysis of various 2-substituted 1,3-propanediol diacetates. For practical use, the esterase P1 was immobilized on Eupergit C (ref. 6). The specific activity of the covalently immobilized material on triacetin was 70%

of that of the soluble enzyme.

HYDROLYSIS OF 2-SUBSTITUTED 1,3-PROPANEDIOL DIACETATES

The enantioselective hydrolysis of the diacetates 1 - 6 (Scheme 3) with esterase Pl immobilized on Eupergit C was investigated.

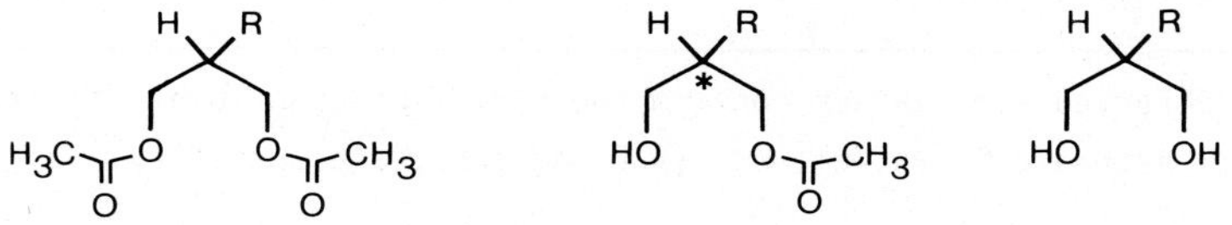

R			
$CH(CH_3)_2$	1	7	13
$CH_2CH_2CH{=}CH_2$	2	8	14
C_6H_5	3	9	15
$CH_2C_6H_5$	4	10	16
C_6H_{11}	5	11	17
$CH_2C_6H_{11}$	6	12	18

Scheme 3

The preparative experiments were performed in 10 - 100 mmol scale. The diacetate was suspended in 69 mM sodium phosphate buffer pH 7 and stirred at 4°C with 25 mg (0.15 mg protein) of immobilized esterase Pl per mmol of starting material. The products were purified by flash chromatography and subsequent bulb to bulb distillation (ref 6.). The absolute configuration is not yet determined for all monoacetates (ref 7.). The hydrolysis results are summarized in Table 2.

TABLE 2

Enantioselective hydrolysis of diacetates (f) by esterase Pl on Eupergit C

Diacetate (f)	Monoacetate (g)[a]	$[\alpha]^{20}_{365}$ (c in $CHCl_3$)		e.e.%[b]
1 R : $CH(CH_3)_2$	(-)-7 (91%)	- 24.8°	(0.46)	75
2 R : $CH_2CH_2CH{=}CH_2$	(-)-8 (80%)	- 38.7°	(1.00)	95
3 R : C_6H_5	(-)-9 (91%)	- 64.3°	(0.57)	95
4 R : $CH_2C_6H_5$	(+)-10 (65%)	+ 65.3°	(0.70)	61
5 R : C_6H_{11}	(-)-11 (96%)	- 32.4°	(0.50)	60

(continued)

TABLE 2 (cont.)

6 R : $CH_2C_6H_{11}$	(+/-)-12

a) Yield of isolated material in brackets.

b) e.e.-values determined by 360 MHz ^{1}H-NMR spectroscopy of the corresponding (-)MTPA-derivatives.

The observed e.e.-values confirm the excellent enantioselectivity of the carboxyl esterase P1 as well as its broad substrate specifity.

PREPARATION OF THE ACYLATION CATALYST

Enzyme catalyzed acyl exchange reactions in organic solvents are very sensitive to the concentration of water in the reaction system (refs. 4 and 9). For this reason the used enzyme should be in a form which easily allows the control of its water content. This form should also be preparable in a reproducible manner and should have good mechanical properties. Adsorption of the carboxyl esterase P1 on Hyflo Supercel by acetone precipitation (ref. 10) gave a catalyst meeting these requirements. For this purpose the enzyme (ref. 11) was dissolved in 18 mM sodium phosphate buffer pH 7 at 0°C and precipitated on suspended Hyflo Supercel by the addition of chilled acetone. The adsorbed enzyme was filtered off and then dried in vacuum (ref. 6).

Various acetates of lower alcohols were tested for the acylation of diol 13 (→ 7), both as solvents and as acyl donors. In all cases we observed the formation of the desired (+)-ester. The best chemical and optical yields were obtained by performing the reaction in methyl acetate.

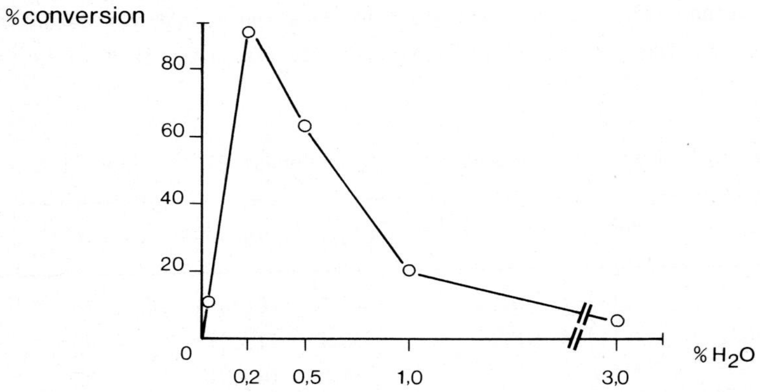

Fig. 3. Optimization of the water content (%, w/w) in the adsorbed carboxyl esterase P1 on Hyflo Supercel for the acylation of 13 to (+)-7 in dry methyl acetate.

The optimization of the concentration of water in the reaction mixture was done by using dry methyl acetate (less than 0.05% of water after drying for 24 h on 4 Å molecular sieves) and carrying out the reaction 13 → (+)-7 with enzyme preparations of different water content (i.e. 0.05% - 3%, w/w). As shown in Fig. 3, the best results were obtained with adsorbed esterase containing 0.2% water.

ENZYME CATALYZED ACYLATION OF 2-SUBSTITUTED 1,3-PROPANEDIOLS

The preparative monoacylation of the diols 13 - 18 (Scheme 3) was performed in 10 - 100 mmol scale in dry methyl acetate (less than 0.05% water) using 0.5 g (2.5 mg protein) of esterase P1 adsorbed on Hyflo Supercel (0.2% water content) per mmol of starting material (ref. 6). The results are summarized in Table 3.

TABLE 3

Enantioselective acylation of diols (e) in methyl acetate catalyzed by esterase P1 adsorbed on Hyflo Supercel

Diol (e)	Monoacetate (ent-g)[a)]	$[\alpha]^{20}_{365}$	(c in $CHCl_3$)	e.e.%[b)]
13 R : $CH(CH_3)_2$	(+)-7 (91%)	+ 21.5°	(0.95)	65
14 R : $CH_2CH_2CH=CH_2$	(+)-8 (70%)	+ 33.0°	(0.75)	90
15 R : C_6H_5	(+)-9 (98%)	+ 61.0°	(0.80)	92
16 R : $CH_2C_6H_5$	(-)-10 (90%)	- 13.9°	(0.45)	13
17 R : C_6H_{11}	(+)-11 (90%)	+ 31.0°	(1.40)	58
18 R : $CH_2C_6H_{11}$	(+)-12 (90%)	+ 12.8°	(1.70)	10

a) Yield of isolated material in brackets.

b) e.e.-values determined by 360 MHz ^{1}H-NMR spectroscopy of the corresponding (-)MTPA-derivatives.

Comparison of the experimental data presented in Tab. 2 and 3 shows that the carboxyl esterase P1 catalyzed acetate hydrolysis, as well as the corresponding acyl transfer, take place in the same enantiotopic side of the diacetates f and the diols e, respectively. The e.e.-values of the monoacetates obtained by enzyme catalyzed acylation of the diols in dry methyl acetate are somewhat smaller than the corresponding values obtained by hydrolysis of the diacetates. This is due to a very slow enzyme catalyzed racemization of the monoacetates, which can be minimized by terminating the acylation reaction at approximately 90% conversion.

CONCLUDING REMARKS

It has been shown that it is possible to switch the absolute configuration of the product of an enzyme reaction by choosing the suitable prochiral starting material and performing the reaction in the appropriate modus. This approach, presented for 2-substituted 1,3-propanediol monoacetates, is of general applicability. As shown in Scheme 4 for the meso compounds 20 and 21 (ref. 12), the described enzyme preparations can efficiently be used for the direct synthesis of both enantiomers of other types of chiral diol monoacetates.

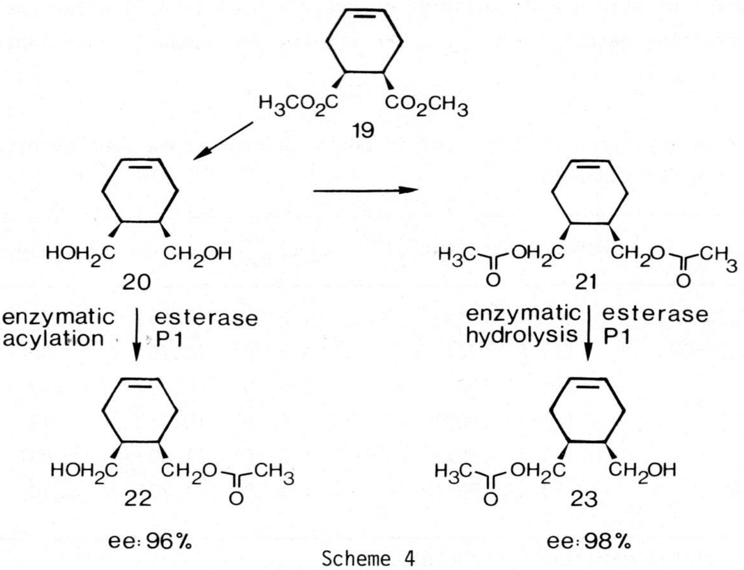

Scheme 4

REFERENCES

1 For a review see J.B. Jones, Tetrahedron 42 (1986) 351.
2 A. Fischli in: R. Scheffold (Ed.), Modern Synthetic Methods, Salle + Sauerländer, Aarau, 1980, p.269.
3 M.F. Züger, F. Giovannini and D. Seebach, Angew. Chem. 95 (1983) 1024. P.K. Matzinger and H.G.W. Leuenberger, Appl. Microbiol. Biotechnol. 22 (1985) 208.
4 For a review see S. Fukui and A. Tanaka, Endeavour 9 (1985) 10.
5 Y.-F. Wang and C.J. Sih, Tetrahedron Lett. 25 (1984) 4999.
6 G.M. Ramos Tombo, H.-P. Schär, X. Fernândez i Busquets and O. Ghisalba, Tetrahedron Lett. 27 (1986) 5707.
7 The absolute configuration of (-)-7 was shown to be S by correlation with the corresponding ß-hydroxy ester c ($R1$: $CH(CH_3)_2$, R2: CH_3CH_2) (compare ref. 8).
8 B. Sonnleitner, F. Giovannini and A. Fiechter, J. Biotechnol. 3 (1985) 33.
9 A. Zaks and A. Klibanov, Science 224 (1984) 1249.
10 R.A. Wisdom, P. Dunnill, M.D. Lilly and A.Macrae, Enzyme Microb. Technol. 6 (1984) 443.
11 Also crude PPL can be utilized for this purpose. See (ref. 6).
12 Compare K. Laumen and M. Schneider, Tetrahedron Lett. 26 (1985) 2073.

C. Laane, J. Tramper and M.D. Lilly (Editors), *Biocatalysis in Organic Media*, Proceedings of an International Symposium held at Wageningen, The Netherlands, 7–10 December 1986.

STEROID SIDE CHAIN CLEAVAGE WITH IMMOBILIZED LIVING CELLS IN ORGANIC SOLVENTS

H.-J. STEINERT[1], K.D. VORLOP[1] and J. KLEIN[2]
[1]Technische Universität Braunschweig, Institut für Technische Chemie, Hans-Sommer-Straße 10, D-3300 Braunschweig, F.R.G.
[2]Gesellschaft für Biotechnologische Forschung m.b.H., Mascheroder Weg 1, D-3300 Braunschweig-Stöckheim, F.R.G.

SUMMARY

Water insoluble substances like ß-sitosterol were transformed to ADD (1,4-androstadiene-3,17-dione) and AD (4-androsten-3,17-dione) with immobilized living cells by mycobacteria in organic media. For this reaction several matrices with hydrophilic or hydrophobic character were tested in aqueous or nonaqueous solutions. The hydrophilic or hydrophobic character of the matrices was determined by the distribution coefficients of indole between water and the matrices. Furthermore the distribution coefficients of AD and ADD between organic solvents and the carriers were investigated. The side chain cleavage rate of ß-sitosterol was tested with free and immobilized cells in an aqueous solution and organic media.

INTRODUCTION

Recently several authors like Fukui (ref. 1,2), Lilly (ref. 3) and Mattiasson (ref. 4) reported on steroid conversion (one-enzyme reactions) with immobilized cells in organic solvents or two-phase systems (organic solvent/ water). Fukui used hydrophobic polyurethane gels in organic solvents for the 1-2 dehydrogenation of steroids. We studied, the side chain cleavage of ß-sitosterol a water insoluble substance, with free and immobilized living cells in organic solvents like soybean oil, cyclohexane, chloroform and in an aqueous solution.The substrate was transformed to ADD and AD. For this reaction we prepared several matrices with hydrophilic or hydrophobic character and these matrices were tested in water media and organic solutions. The hydrophilic or hydrophobic character of the matrices was determined by the distribution coefficients of indole between water and the matrix phase. Furthermore the distribution coefficients of the reaction products AD and ADD between organic solvents and the biocatalysts were determined.

MATERIALS AND METHODS

Analytics

AD and ADD were estimated by HPLC (Si 60 250 mm column, E. Merck, 6100 Darmstadt 1). A mobile phase of n-hexane : i-propanol , 8.5 : 1.5 ratio was used at a flow rate of 1.0 ml / min., when organic media were employed. AD and ADD were detected at 254 nm UV. A typical chromatogram of this analysis is shown in Figure 1.

A C_{18} reverse-phase 250 mm column (E. Merck, 6100 Darmstadt 1) was used when side chain cleavage was performed in aqueous media used for side chain cleavage. A mobile phase of methanol : water; 1 : 1 ratio was used at a flow rate of 0.6 ml / min. The products were detected at 254 nm UV.

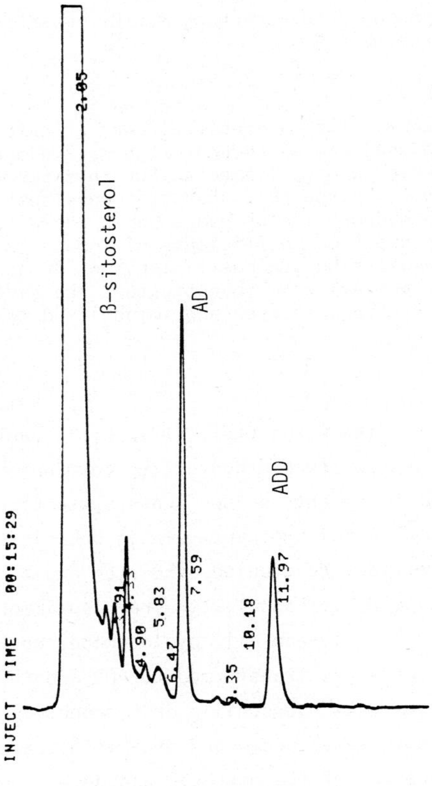

Fig. 1. Chromatogram shows the separation of ß-sitosterol, AD and ADD

Cell culture

The strain Mycobacterium fortuitum (DSM 1134) (ref. 5) was fermented in a 10 - litre Biostat E fermenter (Braun Melsungen, 3508 Melsungen) under the following conditions: The fermantation broth contained 8 g/l nutrient broth, 5 g/l glycerol, and 1 g/l yeastextract. After 65 hours at 28°C, pH 7.2 and 200 rpm the cells were harvested and washed with 0.9 % NaCl-solution. The wet cells were directly used for immobilization.

Side chain cleavage conditions

The side chain cleavage of ß-sitosterol was tested with free and immobilized cells in an aqueous solution, chloroform, cyclohexane and soybean oil. 8 g wet cells or immobilized cells containing the same weight of cells were added to 100 ml medium. The reaction was carried out at 28°C at 160 rpm in shaking flasks. The substrate and product structures are given in Fig. 2.

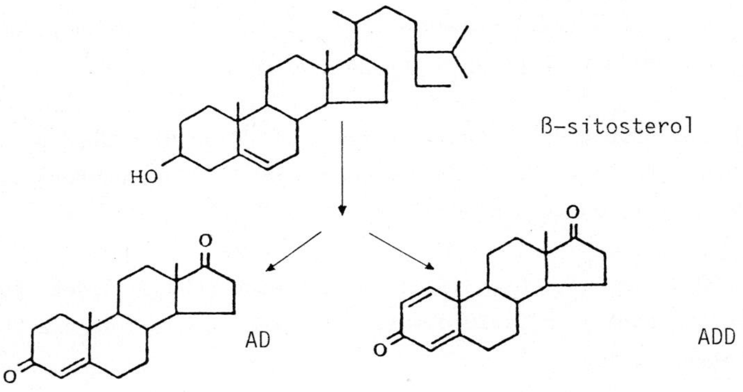

Fig. 2. Degradation of ß-sitosterol catalyzed by Mycobacterium fortuitum

Cell immobilization

Matrices based on ionotropic gels like Ca-alginate, hydrophobic alginate (Ca-alginate/PA 18), chitosan, hydrophobic chitosan (chitosan/E 100, chitosan/laurylsulfate) and on the other hand silicone and polyurethane matrices were used. The formation of biocatalysts by ionotropic gelation was described by Vorlop und Klein (ref. 6). As could be shown a hydrophobic gel can be formed in two ways: a) by the addition of a hydrophobic compound (PA 18, E 100), b) or by using a hydrophobic counterion.

The polyelectrolyte cell suspension was dropped through a capillary tube into a crosslinking solution. A 2 % $CaCl_2$ solution was used for formation of alginate beads and a 2 % Na-tripolyphosphate solution (pH 8.0) for chitosan beads. In the case of chitosan/ laurylsulfate matrix, the hydrophobic substance (Na-laurylsulfate) was directly used as crosslinking counterion. The chitosan / cell suspension was dropped into a 2 % Na-laurylsulfate crosslinking solution. After hardening of the beads, the biocatalysts were washed, sieved and used for ß-sitosterol side chain cleavage.

To form a silicone biocatalyst, the wet cells were mixed with a silicone prepolymer and the crosslinking catalyst (ref. 7). After hardening the gel was cut into pieces (each approximately 2 x 2 x 2 mm) and used for transformation of ß-sitosterol.

In forming the polyurethane matrix the wet cells were suspended in a soybean oil/water emulsion. This suspension was mixed strongly in a beaker with the prepolymer (ref. 8). The gel was cut into pieces in the same way described above.

Matrices

Alginate: 100 g 3.5 % (w/v) Protanal LF 20/60 (Hellmut Carroux, 2000 Hamburg 11) plus 20 g wet cells were dropped into a 2 % $CaCl_2$ solution.

Alginate/PA 18: 66.6 g 3.5 % (w/v) Protanal LF 20/60 plus 33.3 g 20 % (w/v) polyanhydrid resin PA 18 (Fig. 3) (SERVA GmbH & Co, 6900 Heidelberg 1) plus 20.0 g wet cells (ref. 9) were dropped into a 2 % $CaCl_2$ solution.

Chitosan: 200 g 1 % (w/v) high viscosity chitosan (Chugai Boyeki Europe Office, 4000 Düsseldorf 1) plus 20 g wet cells were dropped into a 2 % Na-tripolyphosphate-solution.

Chitosan / Na-laurysulfate: 200g 1 % (w/v) h. v. chitosan plus 20g wet cells were dropped into a 2 % (w/v) Na-laurysulfate solution (SIGMA GmbH, 8020 Taufkirchen).

Chitosan / E 100: 200 g 1% (w/v) h. v. chitosan were 20 g (w/v) Eudragit E 100 (Fig. 4) (Röhm GmbH, 6100 Darmstadt 1) (ref. 9) plus 20 g wet cells were dropped into a 2 % Na-tripolyphosphatesolution.

Silicone-matrix: 20 g wet cells plus 80 g silicone RTV -M 457; 2.4 g of catalyst T 40 (Waker-Chemie GmbH, 8000 München 70).

Polyurethane-matrix: 20 g wet cells plus 10 g soybean oil plus 10 g water plus 20 g PU 2002 (Industrial Chemicals Group, W.R. Grace & Co, Lexington Mass. 02173).

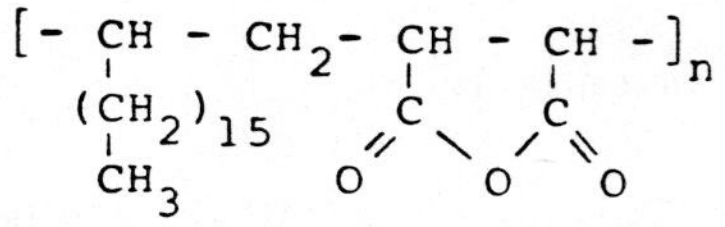

Fig. 3. Structure of polyanhydrid resin PA 18

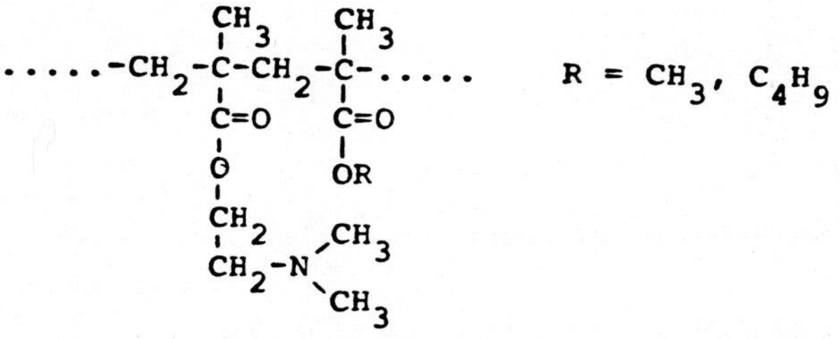

Fig. 4. Structure of Eudragit E 100

MEASUREMENT OF DISTRIBUTION COEFFICIENTS

The distribution coefficients of indole between water and matrices and the cell containing matrices were determined under the following conditions.

10 g wet matrices were shaken (300 rpm) at room temperature in 100 ml indole solution (0.024 mg indole / ml buffer solution). The indole concentration was estimated by measuring its absorbance at 270 nm after attaining equilibrium. The distribution coefficient was determined using the equation

$$K_{indole} = \frac{C_M}{C_L} \qquad (1)$$

C_M = indole concentration in the matrix (mol / l)

C_L = indole concentration in the buffer solution (mol / l) and the results are summarized in table 1.

Furthermore the distribution coefficients of AD and ADD between the outer soybean oil phase and the carrier aqueous phase (without and with cells) were determined.

$$K_{AD/ADD} = \frac{C_M}{C_L} \qquad (2)$$

C_M = AD or ADD concentration in the matrix (mol / l)

C_L = AD or ADD concentration in the buffer solution (mol / l)

The results are given in table 2.

RESULTS

The separation of ß-sitosterol, AD and ADD was achieved by HPLC. The chromatogram (Fig. 1) shows three main peaks. The first peak appearing is ß-sitosterol followed by AD and ADD. The small peaks are intermediate products. Using soybean oil as medium the ß-sitosterol peak was covered by the UV-active substances which are contained in the soybean oil.

The different matrices were characterized by the distribution coefficients of indole between water and the carriers. A matrix with a low coefficient was defined as a hydropilic one. Consequently, a high coefficient characterized a carrier with hydrophobic properties. The alginate and chitosan matrices showed distribution coefficients like K = 1.0 (alginate 3.5 %), K = 2.8 (alginate 7 %) and K = 1.9 (chitosan 1.5%). The cell containing carriers somewhat exhibit higher coefficients for alginate 7 % (w/v) K = 3.7, due to the weakly hydrophobic nature of the cells.

The hydrophobic matrices like Alginate/PA 18 exhibit a coefficient of K = 21.4 but the cell-containing matrix showed a lower one (Tab. 1). On the other hand we determined the distribution coefficients of AD and ADD between soybean oil and the different matrices. These coefficients are more directly related to our specific reaction. In this case chitosan 1.5 % (w/v) K_{AD} = 0.002 characterize a hydrophilic carrier. An example is given for a hydrophobic matrix like Chitosan/Laurylsulfate K_{AD} = 0.0417. Other results are shown in table 2.

TABLE 1

Distribution coefficients of indole between water and the matrices (without and with cells)

Matrices	$K_{indole} = C_M/C_L$	carrier dry weight (%)	cell concentration (%) (wet weight)
Alginate 3.5 %	1.0	3.8	
Alginate 7.0 %	2.8	5.0	
Alginate/PA 18	21.4	16.1	
Chitosan 1.5 %	1.9	11.4	
Chitosan (1.5%)/Lauryl.	12.2	8.3	
Chitosan (1.5%)/E 100	9.3	11.8	
Matrices with cells			
Alginate 7.0 %	3.7	7.1	19.6
Alginate/PA 18	19.4	17.1	18.6
Chitosan 1.5 %	5.9	10.2	27.8
Chitosan (1.5%)/Lauryl.	10.1	7.6	23.9
Chitosan (1.5%)/E 100	7.6	10.4	27.6

TABLE 2

Distribution coefficients of AD and ADD between soybean oil and the matrices without and with cells

Matrices	$K_{AD} = C_M/C_L$ ($*10^{-3}$)	$K_{ADD} = C_M/C_L$ ($*10^{-3}$)	carrier dry weight (%)	cell concentration (%) (wet weight)
Silicone	0.0	0.0	98.9	
Polyurethane	0.0	93.9	46.0	
Alginate	0.0	0.0	3.8	
Alginate/PA 18	12.8	0.0	9.6	
Chitosan	2.0	21.7	12.2	
Chitosan/E 100	0.0	10.0	13.0	
Chitosan/Lauryl.	41.7	49.7	8.5	
Matrices with cells				
Silicone	0.0	23.4	81.2	14.9
Polyurethane	0.0	88.1	50.9	21.5
Alginate	8.1	19.2	8.0	21.8
Alginate/PA 18	4.7	74.8	12.8	20.0
Chitosan	20.5	81.6	11.7	27.2
Chitosan/E 100	3.8	134.6	10.0	20.6
Chitosan/Lauryl.	30.6	321.7	19.6	27.1

The matrices were then used for the side chain cleavage of ß-sitosterol as a water insoluble substrate. The degradation rate using free cells in water is shown in figure 5. A very low transformation activity was observed by entrapped cells in an aqueous solution, too (Tab. 3). Furthermore there were very low degradation activities with free and immobilized cells in cyclohexane and chloroform. The best results were obtained with soybean oil and cells entrapped in silicone and Chitosan/E 100 (Fig. 7,8).

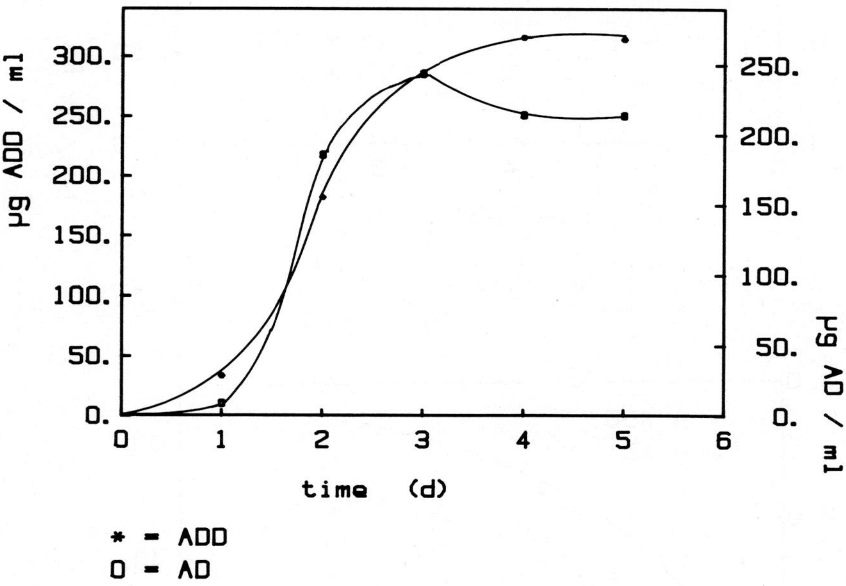

Fig. 5. ß-sitosterol side chain cleavage with free cells in an aqueous solution to ADD and AD; 200 mg ß-sitosterol / 100 ml; 8 g free cells / 100 ml; 28°C; 160 rpm; glycerol 10.0 g/l, K_2HPO_4 0.5 g/l, NH_4Cl 1.0 g/l, $MgSO_4$ * 7 H_2O 0.5 g/l, $FeCl_3$ * 6 H_2O 0.05 g/l, Tween 20 1 ml/l, N,NDimethylformamide 2 ml/100 ml;

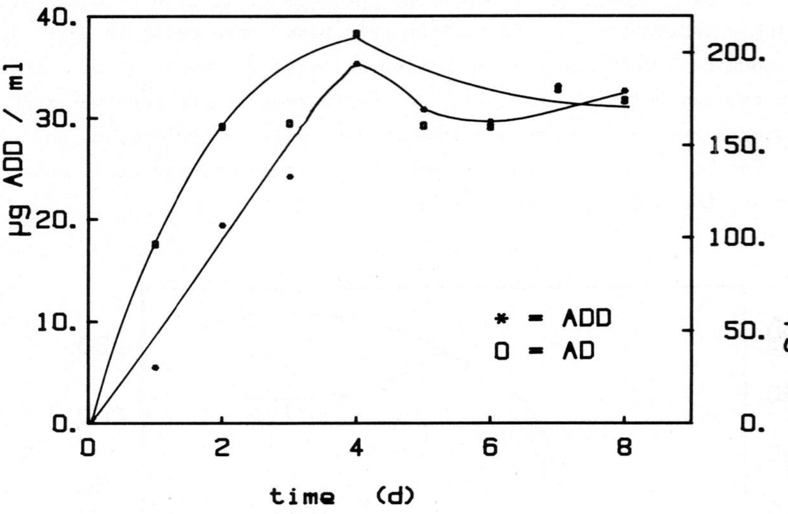

Fig. 6. ß-sitosterol side chain cleavage with free cells in vegetable oil to ADD and AD; 200 mg ß-sitosterol / 100 ml; 8 g free cells / 100 ml; 28°C; 160 rpm;

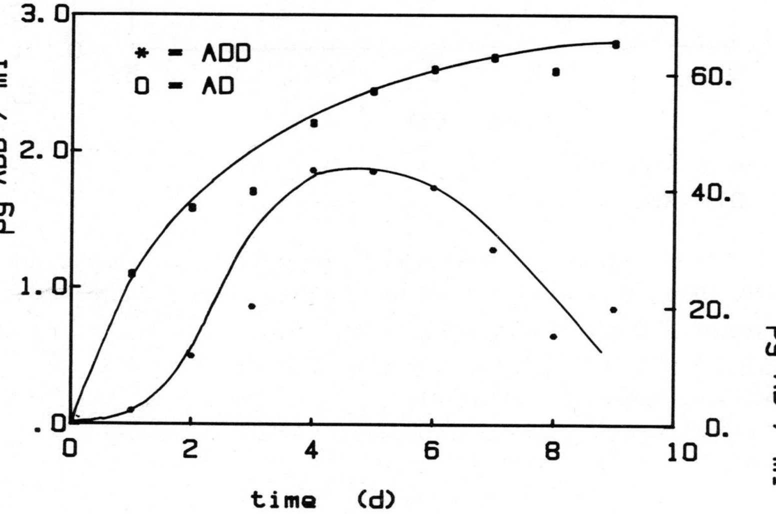

Fig. 7. ß-sitosterol side chain cleavage with immobilized cells in vegetable oil to ADD and AD; silicone matrix corresponding 8 g wet cells / 100 ml; 200 m ß-sitosterol / 100 ml; 28°C; 160 rpm;

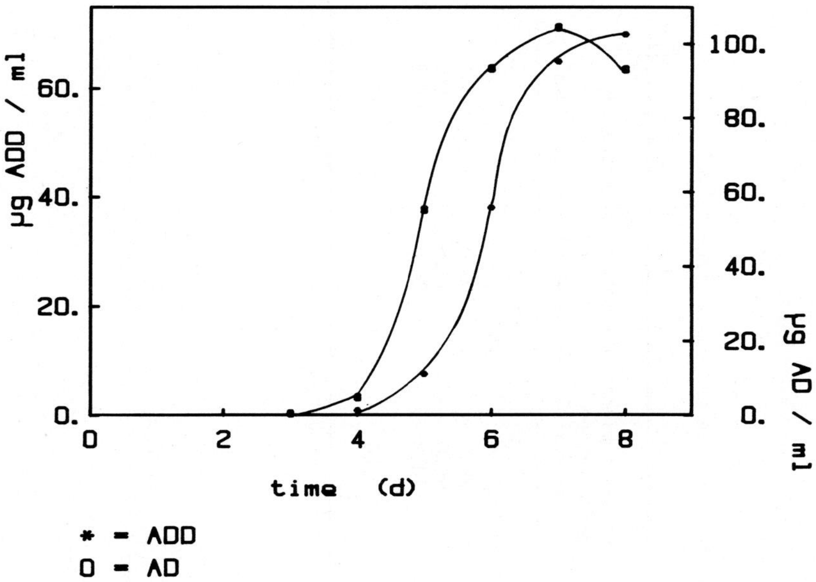

Fig. 8. ß-sitosterol side chain cleavage with immobilized cells in vegetable oil to ADD and AD; Chitosan/E 100 matrix corresponding 8 g wet cells / 100 ml; 200 mg ß-sitosterol / 100 ml; 28°C; 160 rpm;

DISCUSSION

In this paper we reported about methods to characterize hydrophobic and hydrophilic matrices. The indole method was used to determine the matrix character and to delivere a quantitative value independent of the under investigation reaction. High distribution coefficents correspond with a hydrophobic character of the matrix (Tab. 1). Carriers like alginate and chitosan are examples for hydrophilic matrices. In comparing different matrices, the dry weight content of the carrier materials was to be taken into account. Higher coefficients for the hydrophilic carriers were appeared by the cell containing carriers through the hydrophobic character of the cell wall. On the other hand K dropped with cell addition to the hydrophobic carriers because the cell wall was of lower hydrophobicity than the matrix.

The determination of the distribution coefficients of AD and ADD between soybean oil and the matrices are more characteristic for the studied reaction. The results demonstrate the different behaviour of AD and ADD against the matrices. Clearly the higher matrix adsorption of ADD that was supported through the entrapped hydrophobic cells can be seen. Another reason for the good adsorption of ADD could be the complex-forming effect of chitosan and the

TABLE 3

ß-sitosterol side chain cleavage with free and immobilized cells in oil and aqueous solution

Matrices	reaction time (d)	initial substrate concentration µg ß-sito./ ml	product concectration µg AD/ml	µg ADD/ml
(soybean oil)				
Silicone	4	2000	103.4	3.4
Chitosan	7	1000	71.9	5.3
Chitosan/E 100	7	1000	104.4	64.8
free cells	4	2000	207.4	37.7
(aqueous solution)				
Chitosan	4	2000	--	--
Chitosan/E 100	7	2000	1.2	0.6
free cells	4	2000	316.0	214.0

quasi-aromatic structure of the A-ring of ADD. In this comparison the higher dry weights of the silicone and PU- matrices should be observed. The addition of cells had the same effects as described above.

The degradation activity was tested with free and immobilized cells in water and organic media. The entrapped cells showed no or very low conversion rates, if water was used as the bulk phase (Tab. 3). The reason for the low activity of this system is the insolubility of ß-sitosterol in water although containing N,NDimethylformamide and Tween 20.

Also very low degradation activity was observed in the case of organic bulk phase media like chloroform and cyclohexane. The reason for the limited conversion rate is interpreted by the toxical effect of the organic media to the immobilized living cells. The highest side chain cleavage rate was obtained with entrapped cells in vegetable oil as bulk phase media. The cells immobilized in chitosan, chitosan/E 100 and silicone showed the best activity in this organic system. As a result of this study it can be seen, that in a two phase system of a bulk phase and an aqueous catalyst phase the solution power and the low toxicity of the bulk phase solvent is more importent than the matrix influence on the distribution coefficient.

ACKNOWLEDGMENT

The support of this work by Deutsche Forschungs Gemeinschaft (DFG) is gratfully acknowledged.

REFERENCES

1 T. Omata, T. Iida, A. Tanaka and S. Fukui, Transformation of steroids by gel-entrapped Nocardia rhodocrous cells in organic solvent, European J. Appl. Micraobiol. Biotechnol. 8, 143 - 155 (1979)
2 S. Fukui, S.A. Ahmed, T. Omata, and A. Tanaka, Bioconversion of lipophilic compounds in non-aqueous solvent. Effect of gel hydrophobicity on diverse conversions of testosterone by gel-entrapped Nocardia rhodocrous cells. European J. Appl. Microbiol. Biotechnol. 10, 289 -301 (1980)
3 M.D. Lilly, Two-liquid-phase biocatalytic reactions, J. Chem. Tech. Bio-technol. 32, 162 - 189 (1982)
4 B. Mattiasson and B. Hahn-Hägerdal, Utilization of aqueous twophase systems for generation soluble immobilized preparations of biocataysts, Immobilized cells and organelles Volume I, CRC PRESS 121 - 134 (1983)
5 Upjohn Co, Deutsche Offenlegungschrift 2746383 (1978)
6 K.D. Vorlop and J. Klein, New developments in the field of cell immobilization formation of biocatalysts by ionotropic gelation, Enzyme Technology, Edited by R.M. Lafferty, Springer Verlag, Berlin, Heidelberg, New York, Tokyo, 219 - 235 (1983)
7 B. Kressdorf (1986) unpublished results
8 J.W. Becke (1986) unpublished results
9 H.-J. Steinert, Entwicklung hydrophober Matrizes zum Seitenkettenabbau an ß-Sitosterol mit immobilisierten Zellen, Diplom Thesis, TU Braunschweig FRG (1983)

C. Laane, J. Tramper and M.D. Lilly (Editors), *Biocatalysis in Organic Media*,
Proceedings of an International Symposium held at Wageningen,
The Netherlands, 7–10 December 1986.

OPTIMIZATION OF BIOCATALYSIS IN ORGANIC MEDIA

Colja LAANE*, Sjef BOEREN, Riet HILHORST, and Cees VEEGER

Department of Biochemistry, Agricultural University, De Dreijen 11, 6703 BC Wageningen (The Netherlands)

*Unilever Research Laboratorium, P.O. Box 114, 3130 AC Vlaardingen (The Netherlands)

SUMMARY

Solvent hydrophobicity is related to various biocatalytic activities. As indicators of solvent hydrophobicity different parameters are taken viz. the Hildebrand solubility parameter, the dye E_T, the dielectric constant, the dipole moment, and log P. Best correlations between hydrophobicity and biocatalytic activity are found using log P. In general biocatalysis in organic solvents is low in relatively hydrophilic solvents having a log P<2, is moderate in solvents having a log P between 2 and 4, and is high in hydrophobic solvents having a log P>4. It is argued from literature data that high biocatalytic rates are also feasible in solvents having a log P<4 when the essential water layer around a biocatalyst is stabilized by a hydrophilic support.
Further optimization of biocatalysis in organic solvents is achieved when the hydrophobicity of the microenvironment of the biocatalyst and the continuous organic phase is tuned to the hydrophobicities of both the substrate and the product according to simple optimization rules.
Furthermore it is discussed that solvent effects on biocatalysts can only be understood completely when in addition to activity, solvent effects are studied on viability and stability of biocatalysts.

INTRODUCTION

This book contains many examples showing the advantages of organic solvents over water as biocatalytic reaction medium. In this paper we will use some of these examples, together with literature data and own results to try to answer a few basic questions in the field of biocatalysis in organic media; viz. "which of the about 200 commonly used organic solvents can be applied for general biosynthetic purposes and why?". Not so long ago the answer to the first question would be none, making the second question irrelevant. In the last few years, however, it became apparent that biocatalytic systems can thrive in various organic media. This was especially true for rather hydrophobic solvents. To date the general consensus seems to be that high biocatalytic activity is favoured in relatively hydrophobic solvents and that none or low activity is observed in relatively hydrophilic solvents. This trend, which is depicted in Figure 1, seems to be biocatalyst and system independent, since it holds for whole (bacterial) cells, free- or immobilized enzymes, as well as for two-liquid-phase systems, pure (dry) organic solvents, or aqueous media saturated with organic solvents (refs. 1,2).

HIGH ACTIVITY ⟷ HYDROPHOBIC SOLVENTS

LOW ACTIVITY ⟷ HYDROPHILIC SOLVENTS

Fig. 1. Trend between biocatalytic activity and solvent hydrophobicity derived from literature.

In the next part of this paper we will try to quantify this trend with particular attention for the quantification of hydrophobicity. Later on we will discuss the effects of solvent hydrophobicity on viability, stability and activity of biocatalysts.

QUANTIFICATION OF HYDROPHOBICITY

In the literature many different ways have been described to quantify hydrophobicity (ref. 3). A list of the most important ones is given in Table 1.

TABLE 1

Indicators of hydrophobicity

δ	:	Hildebrand solubility parameter
E_T	:	Solvatochromism of dye
ε	:	Dielectric onstant
μ	:	Dipole moment
log P	:	logarithm of partition coefficient

The first parameter in this list, the Hildebrand solubility parameter (δ), has been used recently by Brink and Tramper (ref. 4) to relate biocatalytic activity and solvent hydrophobicity. They studied the influence of various water-immiscible organic solvents on biocatalysis in general and on the microbial epoxidation of propene and 1-butene in particular. For every organic solvent the δ-value was obtained either from the literature (refs. 3,5,6), or by means of the following formula:

$$\delta = \left(\frac{\rho(\Delta H_v - RT)}{M} \right)^{\frac{1}{2}} \qquad (1)$$

where ΔH_v is the heat of vaporization, M the molecular weight of the solvent, ρ the specific gravity of the solvent, R the gas constant, and T the temperature. The only term in this expression that has to be determined experimentally is ΔH_v; the others are usually known.

By plotting these δ-values against the rates of epoxidation, expressed as percentage activity after a few hours relative to the activity observed in water, the results presented in Figure 2 were obtained. The qualitative trend depicted in Figure 1 can now be quantified, viz. high activities can be found over the complete δ-range, but no low activities were observed below $\delta < 8$.

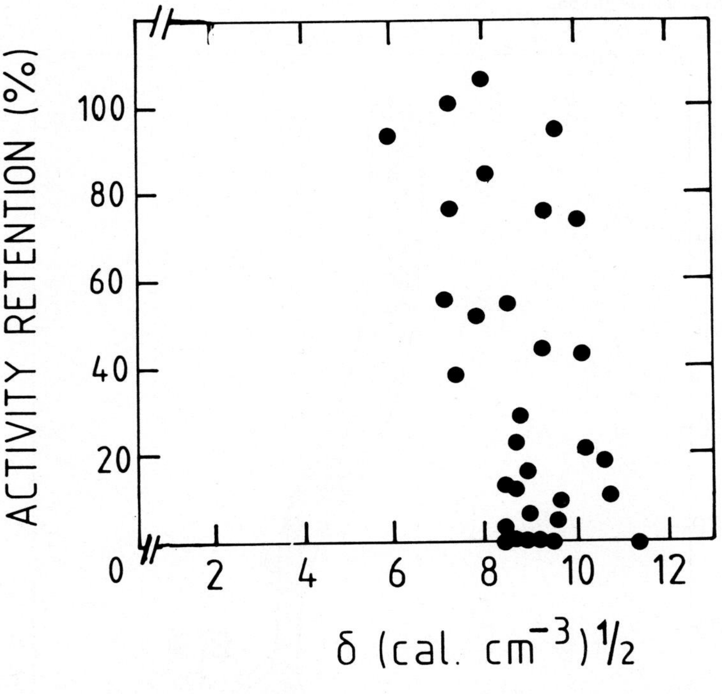

Fig. 2. Activity retentions of epoxidizing cells exposed to different organic solvents versus the Hildebrand solubility parameter δ . Data were taken from Brink and Tramper (ref. 4). The more hydrophobic the solvent the lower δ .

Further quantification was achieved by taking the molecular weight of the organic solvent into account, reasoning that high molecular weight solvents might be less harmful to the epoxidizing microbes due to the fact that they cannot penetrate into the cell. The results are shown in Figure 3, where the molecular weight of the organic solvent is plotted against δ . Every box represents an organic solvent which has been numbered by Brink and Tramper in their original paper (ref. 4). The number closest to a box is the activity retention found for that particular solvent. For example, solvent number 3 is toluene: its molecular weight is 92; its δ-value 8.9; and the activity found in toluene is zero. In this plot we shaded the area where activity retentions are > 75%. By doing so the final conclusion of the pioneering work of Brink and Tramper becomes visible immediately: High biocatalytic activity is favoured in solvents having δ < 8 and a molecular weight > 150. On the other hand the shape of the shaded area also shows that there are still numorous exceptions which do not fit into this conception.

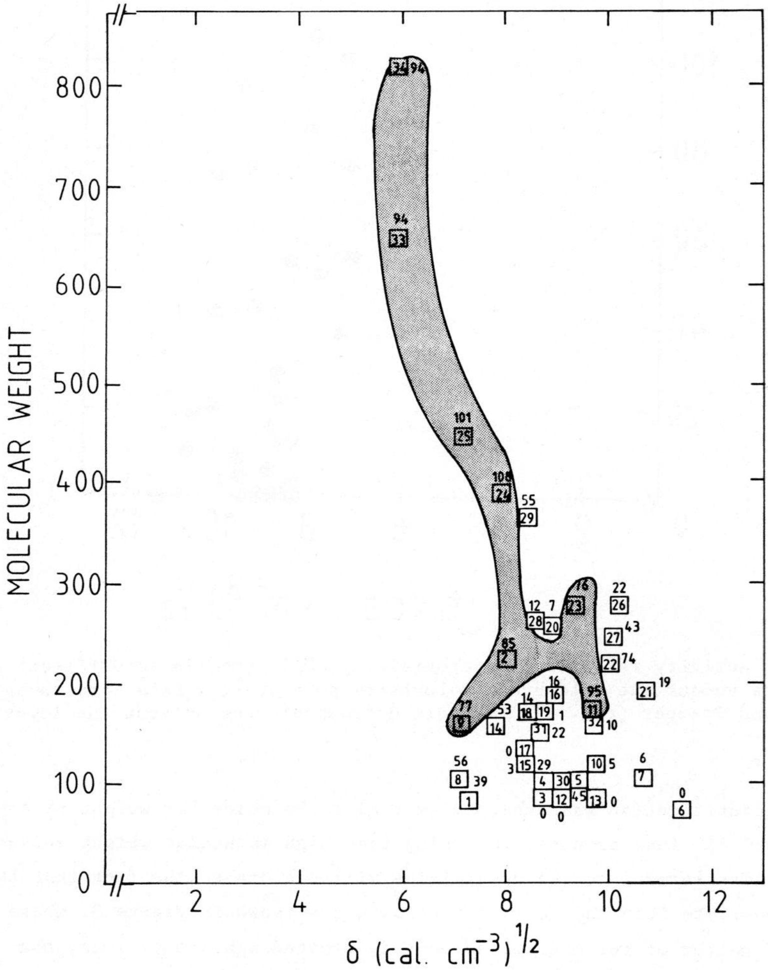

Fig. 3. Activity retentions of epoxidizing cells exposed to organic solvents as a function of molecular weight and δ . For explanation see text.

Furthermore, this figure demonstrates that δ hardly senses differences in hydrophobicity for solvents having a δ -value around 7. This is not surprising since the key parameter in the expression for δ (equation 1), the heat of vaporization depends most strongly on polar interactions between solvent molecules and to a lesser extent to attraction forces of apolar nature (ref. 7). This prompted us to test the applicability of other well-known indicators of hydrophobicity.

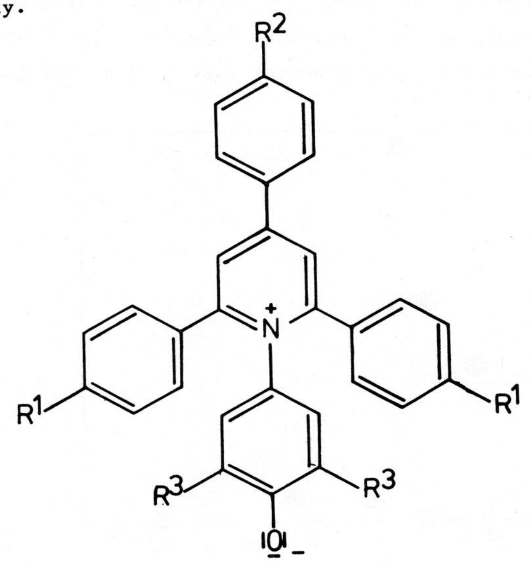

$E_T(30)$: $R^1=R^2=H$: $R^3=C_6H_5$

Fig. 4. Structure of dye E_T. Taken from ref. 3.

One such indicator is E_T. E_T stands for the difference in transition energy between the ground state and first excited state of a probe, whose structure is depicted in Figure 4. Among physical-organic chemist E_T is a rather popular indicator of polarity. This dye is a remarkable one, because its colour depends strongly on the hydrophobicity of the solvent in which it is dissolved. For example, the solution colour of E_T is red in methanol, violet in ethanol, green in acetone, blue in isoamylalcohol, and greenish-yellow in anisole (ref. 3). Hence with this cameleon-like compound nearly every colour of the visible spectrum can be obtained by applying suitable binary mixtures of solvents of different hydrophobicity. In Figure 5 we have plotted the data of Brink and Tramper against E_T. Although few E_T-values could be traced in the literature, the amount is sufficient to illustrate that there is a poor correlation between bio-epoxidation and E_T. The problem associated with E_T, like δ , is that it is a sensitive hydrophobicity-indicator for relatively polar (water-miscible) organic solvents, but quite insensitive for the more hydrophobic solvents. For example, for solvents more hydrophobic than hexane the hypochromic shift of E_T is close to zero. Since such solvents are of particular importance for biocatalysis, the use of E_T as an indicator for this purpose has lost most of its glance.

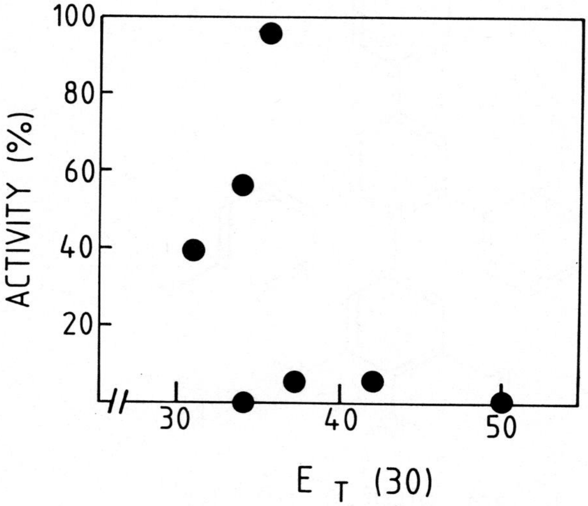

Fig. 5. Activity retentions of epoxidizing cells exposed to organic solvents versus E_T. For explanation see text.The more hydrophobic the solvent the smaller its E_T-value. For example, E_T (hexane) = 30.

In the list of Table 1 the next parameter of hydrophobicity is the dielectric constant (ε). Although ε is not a direct measure of hydrophobicity it reflects to some extent differences in hydrophobicity. Figure 6 shows again the data of Brink and Tramper now plotted against ε. The proper ε-values for the various solvents were taken from ref. 6. Again poor correlations are found and problems arise with relatively hydrophobic solvents ($\varepsilon < 6$). Like δ and E_T, ε-values between polar solvents differ markedly, but hardly or not between apolar solvents. For example, the ε-value for hexane, octane, tetradecane, hexadecane etc. are all zero, while it is common knowledge that they do differ in hydrophobicity. Nevertheless these indicators might be applied successfully in case biocatalytic activities are studied in rather polar, usually water-

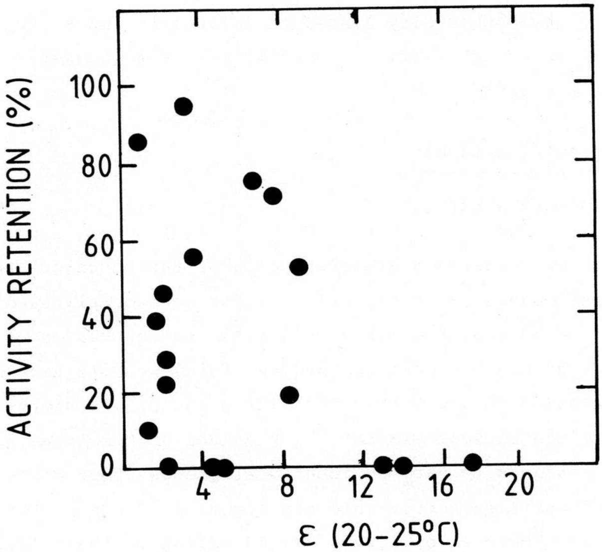

Fig. 6. Activity retentions of epoxidizing cells exposed to organic solvents versus ε . For explanation see text. The more hydrophobic the solvent the smaller ε.

miscible, solvents. The work of Rodgers et al. in this proceedings seems to substantiate this contention. They studied the effect of various, hydroxyl-containing, water-miscible organic solvents on the hydrolytic and synthetic activity of thermolysin. In case of hydrolysis a correlation was found between activity and ε . On the other hand there was no simple correlation between solvent dielectric constant and effect on synthetic activity.

Another measure of hydrophobicity that should be tried is the dipole moment, denoted as μ . To avoid falling into repetitions we will not show the results. Again correlations, if any, are poor. Good correlations can probably only be obtained with the parameters described so far, when indicators of low hydrophobicity are combined with indicators of high hydrophobicity. For example by combining δ with parameters that reflect more specific the attraction forces of apolar nature i.e. μ and the capacity to form hydrogen bridges γ (ref. 8). However, this so-called three-dimensional approach to solubility is in our view too complicated for practical purposes.

The final sensor of hydrophobicity listed in Table 1 is log P. Log P is defined as the logarithm of the partition coefficient in a standard octanol-water two-phase system:

$$P = \frac{[\text{solute}]\ \text{octanol}}{[\text{solute}]\ \text{water}} \qquad (2)$$

For simple molecules log P-values are either known or can be calculated from hydrophobic fragmental constants (hfc) (ref. 9). For more complicated structures log P can be determined by measuring the equilibrium concentration of the solute in both phases by conventional techniques. Below we will give an example of the calculation of log P from hfc. The principle is simple: by comparing ample experimentally-determined log P-values of homoloque series of compounds, hfc were assigned to various functional groups. Some common functional groups and their respective hfc are listed in Table 2. For a more extensive overview see Rekker et al. (ref. 9). By adding up the proper hfc log P-values can be calculated for any relatively simple compound. For example, the log P-value of dipropylether (CH_3-CH_2-CH_2-O-CH_2-CH_2-CH_3) is: 2 * CH_3 + 4 * CH_2 + O = 2 * 0.701 + 4 * 0.519 + (-1.595) = 1.88, which is close to the experimentally determined value of 1.90.

TABLE 2

List of some functional groups with their respective hydrophobic fragmental constant. Taken from ref. 9. For explanation see text.

Functional group	Hydrophobic fragmental constant
CH	0.337
CH_2	0.519
CH_3	0.701
C_6H_4	1.658
CH_2=CH	0.856
(al) O	-1.595
(al) OH	-1.470
(al) CO	-1.643
(al) COO	-1.251
(al) CHO	-1.172
(al) Br	0.249
(al) Cl	0.057

It should be noted that for conjugated structures calculations are slightly more complicated, since corrections are required to yield the proper log P-value.

In analogy with other indicators we have plotted the activity-retention data of Brink and Tramper against calculated log P-values (Figure 7). As indicated previously (ref. 1) a clear, sigmoidal-shaped correlation exists between epoxidation-activity retention and log P. Activities are low in relatively hydrophilic solvents having a log P < 2, is quite variable in solvents having a log P between 2 and 4, and is high in hydrophobic solvents having a log P > 4. Similar correlations were found for completely different biocatalytic systems in organic media viz. anaerobic cells suspended in aqueous media saturated with water-immiscible organic solvents (ref. 10), and two different lipases in dry solvents (ref. 11). From own experience we know that the activity of enzymes such as cholesterol oxidase, xanthine oxidase, and enoate reductase follow the same trend.

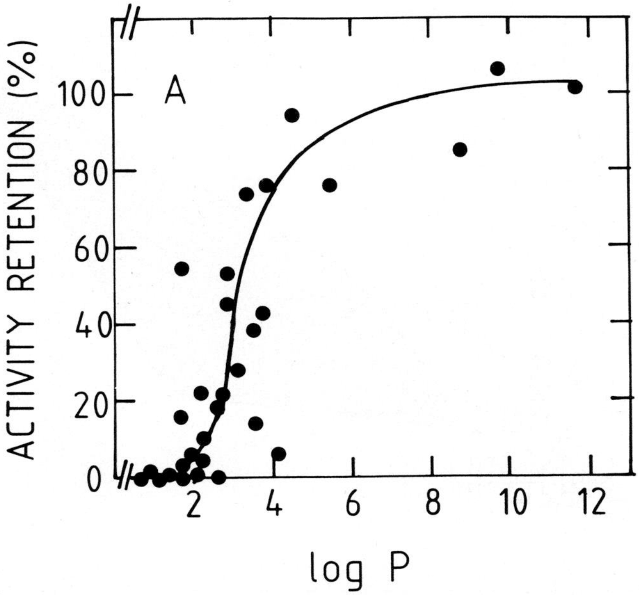

Fig. 7. Activity retentions of epoxidizing cells exposed to organic solvents versus log P. For explanation see text.

In this proceedings a few other examples are given which are in agreement with our observations:

(i) the Δ^1-dehydrogenation of hydrocortisone by <u>A. simplex</u> in organic-aqueous two-liquid-phase environments (Hocknull <u>et al</u>.);

(ii) the α-chymotrypsin-catalyzed esterification of N-acetyl-L-phenylalanine with ethanol (Reslow <u>et al</u>.);

(iii) the production of L-tryptophan in a two-liquid-phase system from indole and L-serine by free- and immobilized cells of <u>E. coli</u> (Ribeiro <u>et al</u>.);

(iv) the oxidation of various 3α-hydroxysteroids by cholesterol oxidase in organic solvents (Snijder-Lambers <u>et al</u>.).

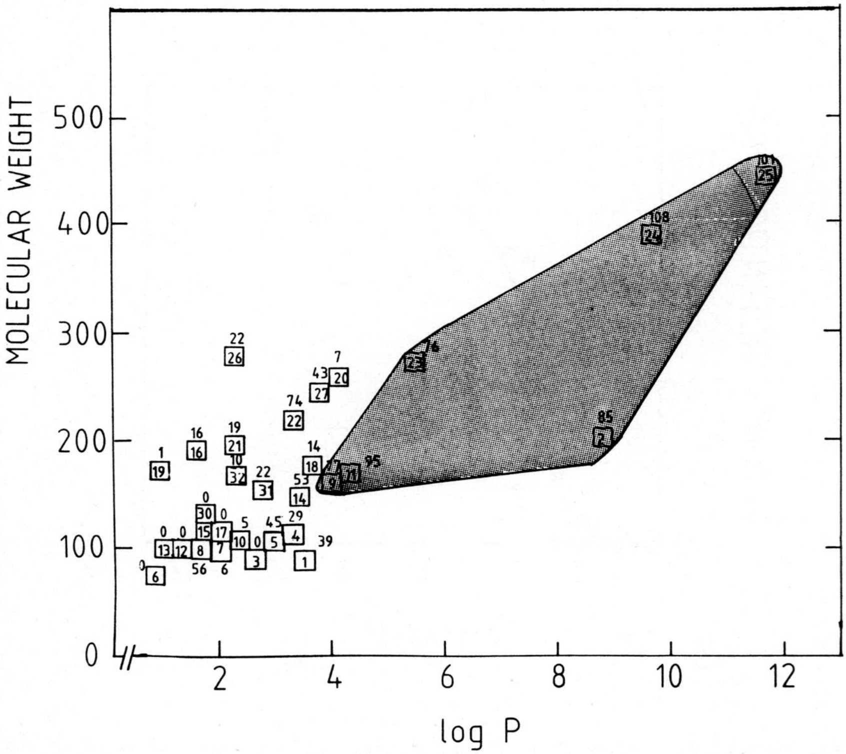

Fig. 8. Activity retentions of epoxidizing cells exposed to organic solvents as a function of molecular weight and log P. For explanation see text.

As we will show later on these examples not only support our generalisations but also refine them. All these results together clearly illustrate the superiority of log P as a measure of hydrophobicity above all other parameters, and furthermore that the log P-activity correlations are biocatalyst and system independent. In our opinion, the parameter log P meets all the requirements for a good indicator. Firstly, log P is a complete measure of polarity and not an incomplete one such as δ, ε, μ and to a certain extent also E_T, since all interactions between molecules are represented in log P. Secondly, log P-values can be determined easily by a standard method, or calculated simply from hfc. Thridly, in contrast to all other indicators log P is very sensitive for hydrophobicity differences in a quite broad range. Fourthly, log P as such is sufficient to describe the trend. In contrast to δ correlations do not improve by considering the molecular weight of the solvent. The reason is simple: all solvents with a log P > 4 have a M > 150 (see Figure 8).

TABLE 3

Log P values of commonly used organic solvents. Log P values were calculated from hydrophobic fragmental constants according to Rekker (ref. 9). Since the specific place of a fragment is not significant for the log P-value, Table 3 includes only one compound for every fragmental pattern (e.g. the log P-values of pentyl-methylether, isopentyl-methylether, sec. pentyl-methylether, butyl-ethylether, diisopropylether etc. are identical).

Solvents		logP
1.	dimethylsulfoxide	-1.3
2.	dioxane	-1.1
3.	N,N-dimethylformamide	-1.0
4.	methanol	-0.76
5.	acetonitrile	-0.33
6.	ethanol	-0.24
7.	acetone	-0.23
8.	acetic acid	-0.23
9.	ethoxyethanol	-0.22
10.	methylacetate	0.16
11.	propanol	0.28
12.	propionic acid	0.29
13.	butanone	0.29
14.	hydroxybenzylethanol	0.40
15.	tetrahydrofuran	0.49
16.	diethylamine	0.64
17.	ethylacetate	0.68
18.	pyridine	0.71
19.	butanol	0.80
20.	pentanone	0.80
21.	butyric acid	0.81
22.	diethylether	0.85
23.	benzylethanol	0.90
24.	cyclohexanone	0.96
25.	methylpropionate	0.97

(continued)

26.	dihydroxybenzene	1.0
27.	methylbutylamine	1.2
28.	propylacetate	1.2
29.	ethylchloride	1.3
30.	pentanol	1.3
31.	hexanone	1.3
32.	benzylformate	1.3
33.	phenylethanol	1.4
34.	cyclohexanol	1.5
35.	methylcyclohexanone	1.5
36.	phenol	1.5
37.	m-phthalic acid	1.5
38.	triethylamine	1.6
39.	benzylacetate	1.6
40.	butylacetate	1.7
41.	chloropropane	1.8
42.	acetophenone	1.8
43.	hexanol	1.8
44.	nitrobenzene	1.8
45.	heptanone	1.8
46.	benzoic acid	1.9
47.	dipropylether	1.9
48.	hexanoic acid	1.9
49.	chloroform	2.0
50.	benzene	2.0
51.	methylcyclohexanol	2.0
52.	methoxybenzene	2.1
53.	methylbenzoate	2.2
54.	propylbutylamine	2.2
55.	pentylacetate	2.2
56.	dimethylphthalate	2.3
57.	octanone	2.4
58.	heptanol	2.4
59.	toluene	2.5
60.	ethylbenzoate	2.6
61.	ethoxybenzene	2.6
62.	dibutylamine	2.7
63.	pentylpropionate	2.7
64.	chlorobenzene	2.8
65.	octanol	2.9
66.	nonanone	2.9
67.	dibutylether	2.9
68.	styrene	3.0
69.	tetrachloromethane	3.0
70.	pentane	3.0
71.	ethylbenzene	3.1
72.	xylene	3.1
73.	cyclohexane	3.2
74.	benzophenone	3.2
75.	propoxybenzene	3.2
76.	diethylphtalate	3.3
77.	nonanol	3.4
78.	decanone	3.4
79.	hexane	3.5
80.	propylbenzene	3.6
81.	butylbenzoate	3.7
82.	methylcyclohexane	3.7
83.	ethyloctanoate	3.8
84.	dipentylether	3.9
85.	benzylbenzoate	3.9
86.	decanol	4.0
87.	heptane	4.0
88.	cymene	4.1
89.	pentylbenzoate	4.2
90.	diphenylether	4.3
91.	octane	4.5
92.	undecanol	4.5
93.	ehtyldecanoate	4.9
94.	dodecanol	5.0
95.	nonane	5.1
96.	dibutylphthalate	5.4
97.	decane	5.6
98.	undecane	6.1
99.	dipentylphthalate	6.5
100.	dodecane	6.6
101.	dihexylphthalate	7.5
102.	tetradecane	7.6
103.	hexadecane	8.8
104.	dioctylphthalate	9.6
105.	butyloleate	9.8
106.	didecylphthalate	11.7
107.	dilaurylphthalate	

What is so special about log P > 4 solvents? As discussed earlier (ref. 1,2 and 12) a reasonable answer to this question is that log P > 4 solvents do not distort the essential water layer around biocatalysts, thereby leaving the biocatalyst in an active state. On the other hand solvents having a log P < 2 are in general not suitable in biocatalytic systems since they strongly distort the essential water-biocatalyst interactions, thereby inactivating or denaturating the biocatalyst. Between these extremes there are solvents (2 < log P < 4) which are weak water-distorters, and will affect biological activity to an extent which is yet rather unpredictable. In Table 3 the most commonly used basic organic solvents are listed and their corresponding log P-value. It can be seen that many solvents (± 50%) have a log P < 2 and are therefore not very suitable for bioorganic synthesis. Only about 20% of the solvents listed are applicable for this purpose.

The next question we asked ourselves is: "is it possible to shift the log P-activity curve to lower log P-values, so that we can apply more solvents for general biocatalytic reactions?" (see Figure 9). We think this is possible by binding biocatalysts to relatively hydrophilic supports. The underlying idea, is that these supports will bind water very tightly, and hence will help to stabilize the essential water layer around the biocatalyst. So far we have no proof for this contention, but in the literature as well as in this book data are presented which support this idea:

(i) Thermolysin bound to amberlite or other hydrophilic supports is more stable in low log P solvents than the free enzyme (ref. 13 and Oyama in this book);

(ii) Liver alcohol dehydrogenase adsorbed to glass beads has still some activity in isopropylether (log P = 1.9), while the free enzyme is less active (ref. 14);

(iii) Mushroom polyphenol oxidase precipitated onto glass powder is more active and stable in chloroform (log P = 2.0) than the free, suspended enzyme (ref. 15);

(iv) Lipases attached to Celite is still capable of interesterification in trichloroethane (log P = 3.0) while the free enzyme is not (Critchley in this book);

(v) Carboxyesterase immobilized onto the polar matrix Hyflow supercel is more productive than the free enzyme (Ramos Tombo in this book).

As shown by Hocknull and Lilly in this proceedings the log P-curve can be shifted also in the undesired direction by increasing the stirring speed of the reaction mixture. It was concluded that the stability of free

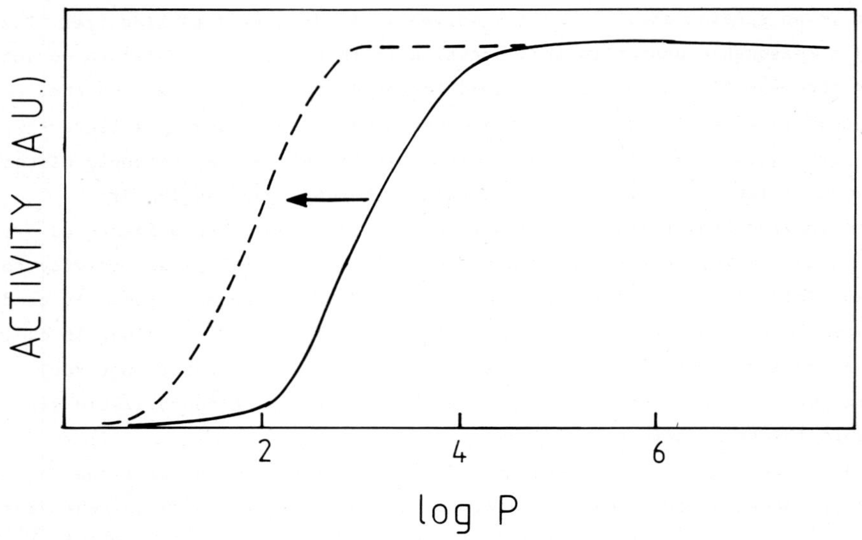

Fig. 9. Schematic plot between biocatalytic activity and log P. Expected influence of hydrophylic supports on the position of the curve.

cells in two-liquid phase environments is not only a function of the organic solvent log P, but also of the agitation rate in the reaction vessel. Another parameter that might cause a shift in the log P-activity correlations is the temperature. At elevated temperatures we expect the curve to shift towards higher log P values. To date we do not have evidence to test this contention.

HYDROPHOBICITY VERSUS ACTIVITY, STABILITY AND VIABILITY

So far we described the effect of hydrophobicity on activity as such. However, the term activity needs further specification, since in some instances initial activities are ment, while in other cases activities over several hours are measured. In the former cases we are dealing with activity pur sang, while in the latter cases activities are probably influenced by changes in the stability of the biocatalyst. Hence, the overall term activity is sufficient in a phenomenological, descriptive sense, but unfortunately does not give insight in the mechanism of action of solvents on biological materials. For that purpose

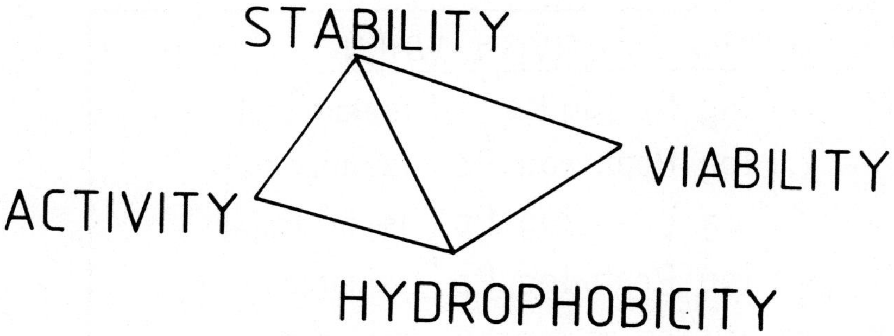

Fig. 10. Schematic view of cross-effects between hydrophobicity, activity, stability and viability. For explanation see text.

the effects of solvents on activity as such and stability have to be studied separately and in detail. Furthermore, in case of whole cells solvent effects on cell-viability have to be included as well. The coherence between these separate effects is depicted in Figure 10. Meanwhile we have studied the effects of solvent hydrophobicity on the viability of Flavobacterium cells in a two-liquid-phase system. Some preliminary data are presented by Boeren et al. in this proceedings.

Till now only the effects of the solvent on the biocatalyst has been discussed. However, for optimal catalysis also the properties of the substrate(s) and product(s) should be taken into account. In the final part of this paper we would like to stress the importance of substrate- and product hydrophobicity in determining biocatalytic rates. It can be expected that high reaction rates are favoured when the biocatalyst is amply supplied by substrate molecules i.e. when the substrate concentration in the microenvironment of the biocatalyst is high. Translated into log P-values this means that the hydrophobicity of this so-called interfacial region (log P_i) should approach as much as possible the fixed hydrophobicity of the substrate (log P_s), since small differences between log P-values indicate good mutual solubilities (see rule number 1 in Figure 11).

OPTIMAL ACTIVITY WHEN :

1. | log Pi - log Ps | is minimal
2. | log Pcph - log Ps | is maximal
3. | log Pi - log Pp | is maximal
4. | log Pcph - log Pp | is minimal

EXCEPT IN CASE OF SUBSTRATE INHIBITION THEN

5. log Pi should be optimized with respect to log Ps.

Fig. 11. Rules for the optimization of biocatalysis in organic media.

Reasoning along these lines it can be expected that by increasing the difference in hydrophobicity between the continuous phase (log P_{cph}) and the substrate, according to rule number 2, substrate molecules will be pushed in the interphase, thereby increasing reaction rates even further. To test these simple optimization rules we needed a biocatalytic model system in which the hydrophobicity of the interphase and the continuous phase could be varied easily and separately from each other. A system that meets these requirements consists of an enzyme-containing reversed micellar solution. Figure 12B depicts schematically the situation of our model system. The enzymes we have used for these type of studies were 20β-hydroxysteroid dehydrogenase (ref. 16), cholesterol oxidase (ref. 17), and enoate reductase (ref. 18). Typically our reversed micellar solution contained the cationic surfactant cetyltrimethylammonium bromide (CTAB) in an organic solvent having a log P > 2.5, a few percent of a buffered aqueous solution and hexanol as cosurfactant. CTAB is localised preferentially in the interphase, whereas hexanol partitions between the interphase and the continuous phase. Hence, the interphase consists of a mixture of CTAB and cosurfactant, whereas the continuous phase contains cosurfactant as well as organic solvent. To calculate the hydrophobicity of mixtures the following semi-empirical formula was defined by us (ref. 16).

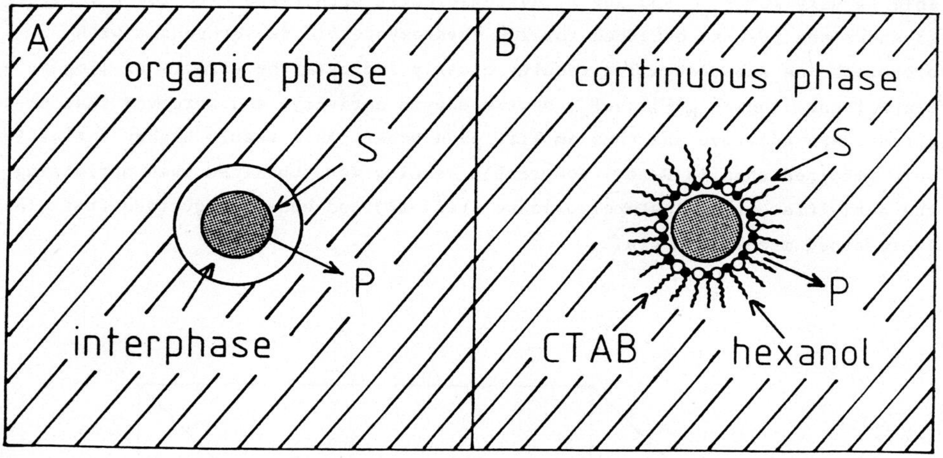

Fig. 12. Schematic representation of a biocatalyst in an organic medium (A) and in a reversed micellar medium (B). The shaded area mimics the biocatalyst. S, P and CTAB are abbreviations of substrate, product and the surfactant cetyltrimethylammonium bromide, respectively.

$$\log P_{mixture} = X_1 \log P_1 + X_2 \log P_2 \qquad (3)$$

where X_1 and X_2 are the mole fraction of the constituents. Using this general formula $\log P_i$ and $\log P_{cph}$ become:

$$\log P_i = \frac{a_o}{a_o + 1} \log P_{cosurfactant} + \frac{1}{a_o + 1} \log P_{CTAB} \qquad (4)$$

where a_o is the molar ratio of cosurfactant to CTAB in the interphase, and;

$$\log P_{cph} = c_o \log P_{cosurfactant} + (1 - c_o) \log P_{org.\ solvent} \qquad (5)$$

where c_o is the mole fraction of the cosurfactant in the continuous phase. Both a_o and c_o can be determined by phase-boundary titrations as described in (ref. 16,19). By indepently varying the hexanol content in the reversed micellar medium, the water concentration, or the type of organic solvent we were able to vary log P_{cph} and log P_i systematically with respect to log P_s. Figure 13 shows the results obtained for 20β -hydroxysteroid dehydrogenase with prednisone as substrate. The results clearly indicate the effects of |log P_i - log P_s| and |log P_{cph} - log P_s| on the enzyme activity, and moreover that these effects are additive and thus amplify each other. As a result higher activities were obtained than in water. Comparable results were obtained when optimizing the activities of cholesterol oxidase (ref. 17) and enoate reductase (ref. 18) in reversed micellar media.

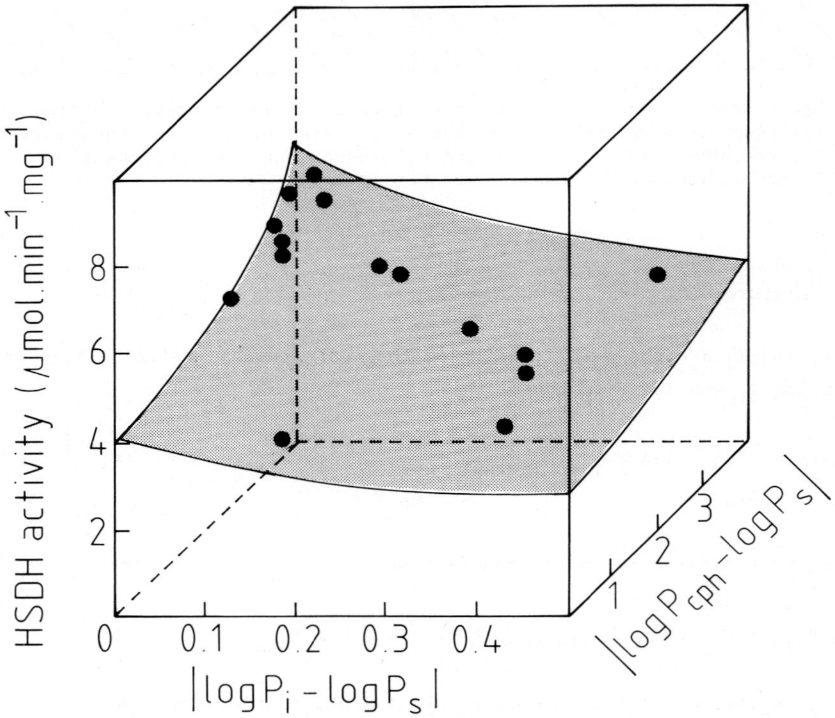

Fig. 13. Optimization of 20 β -hydroxysteroid dehydrogenase activity in reversed micellar media. Data taken from ref. 16.

It can be expected that optimal activities are reached when the concentration of product in the microenvironment is low. This can be achieved when $|\log P_i - \log P_p|$ is maximal and $|\log P_{cph} - \log P_p|$ is minimal (see rules number 3 and 4 in Figure 11). Under these circumstances the product is scavenged by the continuous phase immediately after production thereby shifting the equilibrium of the reaction into the desired direction, and avoiding possible product inhibition. Following this line of reasoning rule number 5 can be formulated for reactions subjected to substrate inhibition. In that case log P_i should be optimized with respect to log P_s. Yet the validity of these latter rules (3-5) has not been tested.

It should be noted, however, that optimization with respect to the substrate might yield conditions which are in conflict with those obtained by product optimization. Then other means of removing the inhibitory product during catalysis, such as continuous precipitation, chromatography, or by membranes should be incorporated in the system.

We hypothesize that perhaps all biocatalytic systems in organic solvents containing an interphase can be optimized in a similar manner as we have demonstrated for enzymes in reversed micellar media. It should be possible to estimate log P_i and log P_{cph} for every biocatalytic system, even those containing a polymer. In that case the hydrophobicity of the polymer might be estimated by calculating the log P-value of the monomeric unit. For those systems which do not contain a real interphase, such as for example pure enzymes in nearly anhydrous organic solvents, log P_i is the same as log P_{cph}, and hence the medium should be optimized with respect to log P_s and log P_p.

In this proceedings Reslow et al. show that the log $P_{mixture}$-formula can be used to improve log P-activity correlations for solvents in the range of log P = 0.5-2.5. These solvents do contain respectable amounts of water (0.5 - 10%) (unpublished results). Consequently log $P_{org.\ solvent}$ does not reflect anymore the real hydrophobicity of the medium.

CONCLUSIONS

1. Log P is a good indicator of solvent hydrophobicity. The reason for this is that unlike all other tested parameters (δ, E_T, ε and μ) log P senses differences in hydrophobicity between all commonly used organic solvents. Furthermore it can be used to estimate the hydrophobicity of mixtures of solvents, surfactant interphases, and maybe polymeric supports.
2. Hydrophilic supports might be helpful in stabilizing biocatalysts in solvents having a log P < 4, thereby broadening the range of useful organic solvents.
3. Optimization of activity can be achieved when log P_i and log P_{cph} are tuned with respect to log P_p and log P_s according to simple optimization rules.
4. Further studies are needed to understand the effect of solvents on individual parameters such as viability, activity and stability of biocatalysts.

ACKNOWLEDGEMENTS

We wish to thank Mrs N. Zaal and Mrs. D. v.d. Hoeven for typing the manuscript and Mr.M.M. Bouwmans for drawing the figures.

REFERENCES

1. C. Laane, S. Boeren and K. Vos, Trends in Biotechnology, 3 (1985) 251-252.
2. C. Laane, S. Boeren, K. Vos and C. Veeger, Biotechnol. Bioeng. (1987) in press.
3. C. Reichardt, Solvent Effects in Organic Chemistry. Monographs in Modern Chemistry, 3 (1979) Verlag Chemie, Weinheim-New York.
4. L.E.S. Brink and J. Tramper, Biotechnol. Bioeng., 27 (1985) 1258-1269.
5. K. Shinoda and P. Becher, Principles of Solution and Solubility (1978) Marcel Dekker, New York.
6. R.C. Weast (Ed.), Handbook of Chemistry and Physics, 57th edition (1976) CRC Press, Cleveland, Ohio.
7. J.H. Hildebrand, J.M. Prausnitz and R.L. Scott, Regular and Related Solutions (1970) Van Nostrand Reinhold Company.
8. J.D. Crowley, G.S. Teague and J.W. Lowe, J. Paint Technol., 38 (1966) 269-280.
9. R.F. Rekker and H.M. de Kort, Eur. J. Med. Chim. Therapeutica, 14 (1979) 479-488.
10. M.J. Playne and B.R. Smith, Biotechnol. Bioeng., 25 (1983) 1251.
11. A. Zaks and A.M. Klibanov, Proc. Natl. Acad. Sci. U.S.A., 82 (1985) 3192-3196.
12. C. Laane, Biocatalysis (1987) in press.
13. K. Nakanishi, T. Kamikubo and R. Matsuno, Biotechnology, 3 (1985) 459-464.
14. J. Grundwald, B. Wirtz, M.P. Scollar and A.M. Klibanov, J. Am. Chem. Soc. 108 (1986), 6732-6734.
15. R.Z. Kazandjian and A.M. Klibanov, J. Am. Chem. Soc., 107 (1985) 5448-5440.
16. R. Hilhorst, R. Spruijt, C. Laane and C. Veeger, Eur. J. Biochem., 144 (1984) 459-466.
17. C. Laane and R. Spruijt, In: J.M.C. Duarte (Ed.), Proc. Recent Developments in Biotechnology, Troia, March 17-29, 1985, Reidel Publ. in press.
18. C. Laane and M. Dekker, In: Surfactants in Solution. Proc. VIIIth Int. Symp. on Surfactants in Solution. New Delhi, August 22-28, 1986.
19. C. Laane and C. Veeger, In: Meth. Enzym, 136 (1987) in press.

SESSION III

CONVERSIONS I

Chairman : B. Mattiasson

Co-chairman : H.J. Grande

C. Laane, J. Tramper and M.D. Lilly (Editors), *Biocatalysis in Organic Media*,
Proceedings of an International Symposium held at Wageningen,
The Netherlands, 7–10 December 1986.

LOG P AS A HYDROPHOBICITY INDEX FOR BIOCATALYSIS; COFACTOR REGENERATION DURING ENZYMATIC STEROID OXIDATION IN ORGANIC SOLVENTS

A.M. Snijder-Lambers, H.J. Doddema, H.J. Grande and P.H. van Lelyveld; TNO, Institute of Applied Chemistry, P.O. Box 108, 3700 AC ZEIST

Key words: biocatalysis in organic solvents, cholesterol oxidase, cofactor regeneration, log P

SUMMARY

Cholesterol oxidase from *Nocardia rhodochrous* was used in organic solvents to study redox reactions in non-aqueous environments. Various 3β-hydroxy steroids were oxidized in a series of solvents with increasing hydrophobicity. Both the activity and the stability of the enzyme are highest when the enzyme operates in the most hydrophobic solvent used (heptane).

The hydrophobicity of the solvent is best expressed as log P. The log P of a compound is the logarithm of its partition coefficient over a standard octanol-water system.

Efficient regeneration of the enzyme-bound cofactor could be achieved with several artificial hydrophobic electron acceptors. The hydrophobicity (expressed as log P) of a number of these, the quinones, appears to be a good measure for their effectiveness as electron acceptors.

INTRODUCTION

During the last few years increasing attention has been paid to biocatalysis in organic solvents. Many industrially interesting compounds are hydrophobic and almost insoluble in water. These hydrophobic substrates and their reaction products dissolve well in most organic solvents. In many cases enzyme activity is enhanced in such an environment, as described in recent reviews (1 - 4).

Steroids are compounds of considerable pharmaceutical interest. Industrial steroid-converting pathways usually include both chemical and biocatalytic steps. Most of the bioconversions are performed by fermentation in aqueous environments, and they all suffer from the limited solubility of the reactants (see, for instance, ref. 5). Enzymatic conversions, on the other hand, can be carried out in organic solvents. However, many of these reactions are redox reactions, requiring expensive cofactors. In this paper, a number of relatively cheap, artificial electron acceptors have been used for the regeneration of the enzyme-cofactor in an organic solvent.

Also, this paper deals with some other problems of enzymatic steroid oxidation in organic solvents, such as the effect of solvent polarity on enzyme activity and stability and the influence of water content on enzyme activity.

To study redox reactions in organic solvents, cholesterol oxidase (E.C. 1. 1.3.6.) was chosen, since it is readily available from different sources and well described in the literature (6). The reaction catalyzed is:

$$\text{cholesterol} + O_2 \dashrightarrow \text{cholestenone} + H_2O_2$$

The hydrogen peroxide formed in this reaction generally inactivates the enzyme (7). Replacing oxygen by another electron acceptor will prevent the formation of H_2O_2.

Cholesterol oxidase also oxidizes the 3β-hydroxy steroids pregnenolone and androstenediol. *Nocardia rhodochrous* cholesterol oxidase is stable in organic solvents, and contrary to the bioconversions in water, no byproducts were detected using the biocatalyst in non-aqueous solvents (8).

In biochemical and biotechnological literature many parameters are used to describe the polarity or hydrophobicity of solvents, carrier materials, membranes, substrates and products. Usually the terms hydrophobic and hydrophilic are used very loosely, without quantitation or reference values (9, 10). Numerical polarity or hydrophobicity indices would be a great help in a more systematical approach of biocatalysis in organic solvents.

In the present paper, we describe the use of log P as an easily obtainable numerical hydrophobicity measure for biocatalysis in organic solvents: the hydrophobicity of both the solvent and the electron acceptor for cholesterol oxidase proved to be important. P is defined as the partition coefficient of a compound in a standard octanol-water system (11-13).

MATERIALS AND METHODS

Organism: *Nocardia rhodochrous* (NCIB 10554) cells were grown in a 10 l fermentor according to Buckland et al. (14) with minor modifications (8). After 20 h, cells were harvested by centrifugation, washed twice with butylacetate to remove steroids and freeze dried. The cells were then stored at -20°C. Assay of enzyme activity: 25 mg rehydrated cells were suspended in 1 ml solvent saturated with buffer (20 mM potassium phosphate; pH=7) and incubated at 30°C under vigourous shaking. 0,1% (w/v) Cholest-ene-3β-ol, pregn-5-ene-3β-ol 20-on, androst-5-ene-3β, 17β-diol were used as substrates. All steroids were purchased from Janssen Chimica (Belgium). Solvents used were of highest grade available. The determination of the enzyme stability in butylacetate was performed as follows: 0,1 g rehydrated *N. rhodochrous* cells and 5 ml 0,1% (w/v) cholesterol in buffer-saturated butylacetate were incubated at 30°C and shaken vigourously. At 50 h and 100 h, additional cholesterol (5 mg in 1 ml of buffer-saturated butylacetate) was added. Cofactor regeneration: cholesterol oxidation in the presence of various artificial electron acceptors in the absence (<1 ppm O_2) of oxygen was performed at 30°C according to Doddema et al (15). Cholesterol and electron acceptor were added in equimolar quantities. Steroids were analysed with GC and/or HPLC: Packed GC conditions: 1 m column OV 101 5% (Chrompack) SS 1/8 inch. Oven temperatures were 290°C (cholesterol/ cholestenone), 270°C (pregnenolone/progesterone), or 250° C (androstenediol/testosterone). FID detection at 300°C, injection temperature was 300°C, nitrogen was used as the carriergas (15 ml/min). Capillary GC conditions: 25 m column CP Sil 5 CB d=0.22 mm (Chrompack). The following temperature programme was used: initial 120°C (2 minutes), raise 30°C/min (3 min), raise 2°C/min (45 min), final temperature 300°C. FID detection at 300°C, injection temperature was 300°C, carriergas was N_2. HPLC conditions: column CP Spher 250 x 4,7 (Chrompack). Heptane, containing 5% (cholesterol/ cholestenone), 10% (pregnenolene/progesterone) or 15% (androstenediol/testosterone) dioxane was used as eluent. Detection of the steroids was performed at 240 nm with a Spectrometer (Zeiss PM DLC) at room temperature. Calculation of log P: both calculated and measured log P values are listed in the literature (11-13). Unlisted log P values were calculated using the hydrophobic fragmental constants and factors (13).

RESULTS

Effect of water content on cholesterol oxidase activity in heptane.

Cholesterol oxidase from *Nocardia rhodochrous* is active in organic solvents. The enzyme activity depends on the amount of water present in the system: freeze-dried cells containing cholesterol oxidase were not active when no water was added. Maximal activity was obtained by cells containing amounts of water equalling approximately four times their original dry weight (Figure 1). No further increase of activity was measured above this value, which is close to the water content of freshly harvested cells. Thus, rehydrated, freeze-dried cells were used in further experiments.

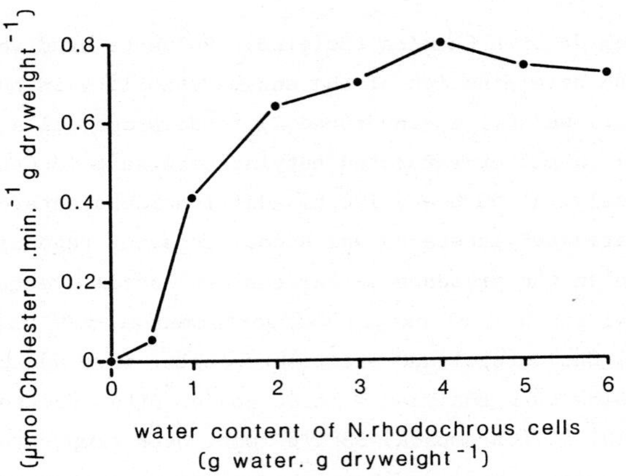

Fig. 1. Dependence of cholesterol oxidase activity on the amount of water that was present in the reaction mixture. Initial rates are shown. These rates were constant up to 4 hours of incubation. Solvent: heptane. [Cholesterol] = 2.6 mM.

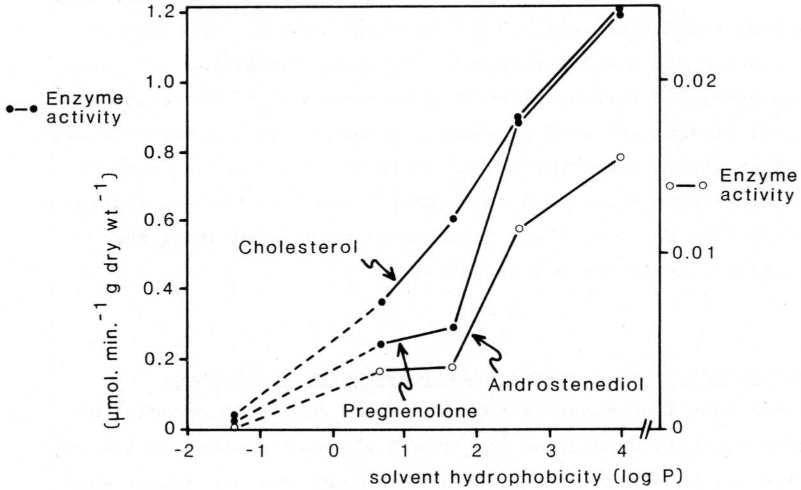

Fig. 2. Cholesterol oxidase activity using various substrates, and solvents of increasing hydrophobicity. The solvents used were: water (log P = -1,38), ethylacetate (0.68), butylacetate (1.7) toluene (2.6), and heptane (4.0). Oxygen served as electron acceptor. Androstenediol (log P = 3.0), pregnenolone (3.5) and cholesterol (4.4) were used as substrates. All substrates were present at 0.1% (w/v).

Effect of solvent polarity on enzyme activity

Besides water, four industrially useful solvents of low price and low volatility were tested as a medium for enzymatic steroid oxidation. For all three substrates tested, enzyme activity was higher in the more hydrophobic solvents (Fig. 2). The enzyme activity in water (log P = -1,38) was very low, mainly because of poor solubility of the substrates in water (approximately 0.002% for cholesterol and 0.005% for androstenediol, pregnenolone).

Effect of solvent polarity on enzyme stability

Most of the data on enzyme activity in organic solvents reported in the literature describes either initial conversion rates, or final conversion percentages. Data about enzyme stability are rarely given. Only a few examples (4, 10, 16, 17) are given in the literature where both activity and stability of the enzyme are described. The half lifes of cholesterol oxidase in whole cells of *N. rhodochrous* suspended in solvents of decreasing hydrophobicity are listed in Table 1.

TABLE 1
The half lifes of cholesterol oxidase in solvents of various hydrophobicity.

solvent	log P	t1/2	(ref.)
heptane	4.0	>14 d	(8)
carbon tetrachloride	2.8	69 h	(14)
butylacetate	1.7	50 h	(this paper)

Thus, not only the activity of *N.rhodochrous* cholesterol oxidase (initial rates) but also the stability of this enzyme increases with increasing hydrophobicity of the solvent.

Anaerobic cofactor regeneration

As artificial electron acceptors to be tested several quinones, which are known to be able to reoxidize FAD in aqueous environments, NAD analogues, redoxdyes, and redox mediators were chosen, Twenty-six electron acceptors were tested (18). The rate of cholesterol oxidation in the absence of oxygen was taken as the regenerative power of these acceptors. Nine compounds were able to sustain cholesterol oxidase activity in butylacetate (Table 2). Butylacetate was the solvent of choice for these experiments, because the various electron acceptors and their reduced forms dissolved best in this solvent.

With the following electron acceptors no conversion was obtained in butylacetate: benzoquinone, naphtoquinone, tetrachloro-o-benzoquinone, tetrachloro-p-benzoquinone, anthraquinone-2-sulphonic acid hydrate, anthroquinone-2,6- disulphonic acid hydrate, vitamine K_1 , methyl red, copperphthalocyanine, phthalazon, phenazine-meto-sulphate, phenazine-eto-sulphate, resazurin, methylene blue, dichlorophenol indophenol, ferrocene, 2,3,5-triphenyl tetrazolium chloride.

There is no correlation between the redox potential $E_0^{\prime}$ of the electron acceptors used and the enzyme activity of cholesteroloxidase (Table 2) This may be due to the fact that the actual redox potential in an organic solvent differs from the standard redox potential in aqueous solutions (C. van Dijk, personal communication). Attempts to determine the redoxpotential of some electron acceptors by cyclovoltammetry in butylacetate were not successful because of the extremely low dielectric constant (ε=4) of this solvent. Also, no correlation exists between the electron acceptors' log P values and their efficiency to regenerate the enzyme cofactor (Table 2). However, when a group of related compounds were taken together (the quinones), a log P 'window' can be readily seen (Fig. 3): only quinones with log P values in the range 2-3 are efficient electron acceptors for this enzyme.

The enzyme activities in the presence of the various electron acceptors cannot be compared with each other in a quantitative way. The solubilities of the various electron acceptors (and their reduced forms) in butylacetate vary, and Km and Vmax values may also vary considerably. No further studies were undertaken to investigate these parameters.

DISCUSSION

The enzyme activity in an apolar organic solvent usually depends on the amount of water in the reaction mixture. In the case of cholesteroloxidase in freeze dried *N.rhodochrous*, the cells have to be rehydrated for the enzyme to be fully active. Dry enzyme (no water added) showed no activity. The optimal amount of water is the same as the amount of water in the wet packed cells used by Buckland et al. (14). Higher amounts of water did not further increase the cholesterol oxidase activity (see also ref. 10). Similar results concerning the effect of water were reported by others for cells (10,14,20, 21), and for purified enzymes such as 20β-hydroxy steroid dehydrogenase (4, 22) and lipase (28), which also show an optimal enzyme activity at a specific water concentration. Our results and those of others (4,11,16,23-25), concerning enzyme activity and enzyme stability show that in many cases enzymes

TABLE 2

Enzyme activity in butylacetate in the presence of various electron acceptors. O_2 was excluded from the reactions.

Electron acceptor	Enzyme activity (μmol. min.$^{-1}$ g dry weight^{-1})	log P	$E_0'^{\,d}$ (mV)
None	0.00	-	-
Menadione	0.17	2.2[a]	+5
Anthraquinone	0.17	2.7[a]	-53
Meldola's Blue	0.21	-1.8[b]	-
Duroquinone	0.27	2.3[a]	+120
Procion Blue HB	0.33	-2.6	-
Azure A	0.33	-3.0[b]	-
9,10-Phenanthrene quinone	0.40	2.6[a]	-22
Nickel phthalocyanine	0.44	1.6[b]	-
N,N,N',N'-tetramethyl-phenylenediamine (TMPD)	0.47	1.9[a]	+240
O_2	1.33	0.7[c]	+815

(a) calculated, (b) measured and (c) log P values listed in ref. 13, (d) E_0' values listed in ref. 19 are shown.

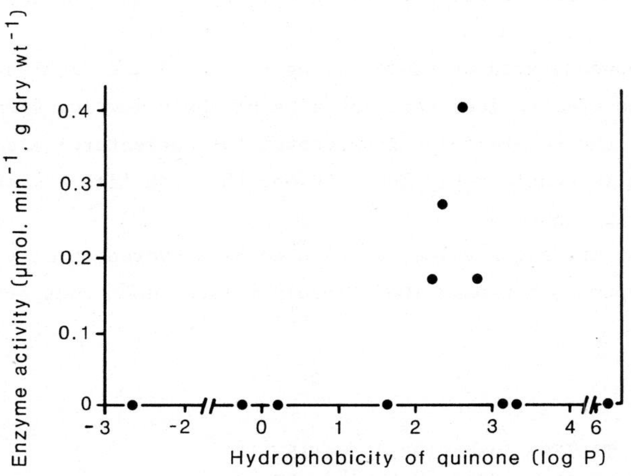

Fig.3. Effectivity of various quinones in cofactor regeneration. The solvent was butylacetate. Oxygen was excluded from the reactions. The log P values of several compounds are listed in Table 2. Additional data: anthraquinone -2,6-disulfonate hydrate (-2.7), anthraquinone-2-sulfonate hydrate (0.25), benzoquinone (0.22), naphtoquinone (1.64), tetrachloro-o-quinone (3.1), Tetrachloro-p-quinone (3.3) and vitamin K_1 (6.3).

display both their highest activity and highest stability in hydrophobic solvents. However, some exceptions to this general rule have been described (3, 4,16,17,22,23,25). Other solvent properties are also important: the solvent should not compete with one of the reactants (25), or be toxic for the enzyme.

Enzymes whose natural substrates are hydrophilic, like horseradish peroxidase (16) and naringinase (23) also show good enzyme activity in moderately hydrophobic (log P>0.6) solvents. However these enzymes show higher (initial) enzyme activity in hydrophilic solvents like DMSO (log P=-1,4) or ethanol (log P=-0.24). No stability measurements have been reported for these enzymes.

The ability of electron acceptors to accomplish cofactor regeneration in apolar organic solvents is not related to their standard redox potential E_0' (Table 2). When the hydrophobicities of the electron-accepting compounds used in this study were expressed as their δ-values according to Hoy (26,27) no correlation with their activity was found (18). However, the calculated log P values of the quinones are a great help in predicting the usefulness of the electron acceptor (Fig.3): quinones which have a log P value between 1.7 and 3 were able to reoxidize the enzyme-bound cofactor in butylacetate quite efficiently, in sharp contrast to the more polar (log P<1.7) and the apolar (log P>3.0) quinones.

However, other compounds, such as Azure A (log P = -3,0) and Procion Blue (log P = -2,6) do act as electronacceptors in spite of their low log P values (Table 2). Also oxygen, the natural electronacceptor for cholesteroloxidase, has a log P value which is considerably lower (0.65) than the 'log P window' for the quinones shown in Figure 3.

It is concluded that the log P value can be used as a hydrophobicity index for electron acceptors, but only when closely related compounds are compared.

REFERENCES

1 P. Luisi, C. Laane, TIBTECH June (1986), pp 153-161
2 H.J. Grande, PT Procestechnologie 10 (1985), pp 55-58
3 A.M. Klibanov, Chemtech. 16 (1986) pp 354-359
4 G. Carrea, Trends in Biotechnol. 2 (1984) pp 102-106
5 J. Kloosterman, M.D. Lilly; Biotechn. Bioeng. (1986), 28 pp 1390-1395
6 A.G. Smith, C.J.W. Brooks; Biochem. J. 167 (1977) pp 121-129
7 P. Adlercreutz, B. Matthiasson; Appl.Microbiol.Biotechnol. 20 (1984) pp 296-302

8 H.J. Doddema, J.P. v.d. Lugt, A.M. Lambers, J.K. Liou, H.J. Grande, C. Laane; Proc. Enzyme Eng. Conf. 8, 1986, in the press
9 S. Fukui, A. Tanaka; Endeavour, New Series 9 (1) (1985) pp 10-17
10 M.D. Lilly; J.Chem. Techn. Biotechnol. 32 (1982) pp 162-169
11 C. Laane, S. Boeren, K. Vos, C. Veeger; accepted for publication in Biotechn.Bioeng. (1986)
12 R.F. Rekker, H.M. de Kort, The hydrophobic fragmental constant, a compilation of 1054 data. Mentioned in: R.F. Rekker, H.M. de Kort; Eur. J. Med. Chem. Chimica Therapeutica, 14 (6) (1979) pp 483
13 C. Hansch, A. Leo; Substituent Constants for Correlation Analysis in Chemistry and Biology. Wiley, New York (1979)
14 B.C. Buckland, P. Dunill, M.D. Lilly; Biotechn. Bioeng. 17 (1975) pp 815-826
15 H.J. Doddema, A.M. Lambers, H.J. Grande, in D. de Nettancourt, A. Goffeau, P. Reiniger (eds) Second Generation Bioreactors, BEP CEC, Brussel (1986) pp 23-25
16 R.Z. Kazandjian, J.S. Dordick, A.M. Klibanov, Biotechnol. Bioeng. 28 (1986) pp 594-599
17 G.A.Homandberg, J.A. Mattis, M. Lawskowski, Biochemistry 17 (1978) pp 5220-5227
18 H.J. Doddema, A.M. Lambers, M.A. Visser, H.J. Grande, in E Magnien (ed.) Biomolecular Engineering in the European Community, M. Nijhoff, Dordrecht (1986) pp 159-171
19 G.D. Fasman (ed) Handbook of Biochemistry and Molecular Biology, CRC Press Inc., Florida.
20 T. Tanaka, E. Ono, M. Ishihar, S. Yamaraka, K. Takinami; Agric. Biol. Chem. 45 (1981) pp 2387-2389
21 M. Ueda, S. Mukataka, S. Sato, J. Takahashi; Agric. Biol. Chem. 50 (6) (1986) pp 1533-1537
22 M. Hilhorst, R. Spruyt, C. Laane, C. Veeger; Eur. J. Biochem. 144 (1984) pp 459-466
23 P. Turecek, F. Pittner; Appl. Biochem. Biotechnol. 13 (1986) pp 1-14
24 B.C. Buckland, M.D. Lilly, P. Dunill; Biotechn. Bioeng. 18 (1976) pp 601-621
25 A. Zaks, A.M. Klibanov; PNAS USA 82 (1985) pp 3192-3196
26 H.F. Mart, D.F. Othmer C.G. Overberger, G.T. Seaborg (eds); Kirk Othmer Encyclopedia of Chemical Technology 21 3rd edn. Wiley, New York (1983) pp 337-401
27 K.L. Hoy; J.Paint Technol. 42 (1970) pp 71-118
28 A. Zaks, A.M. Klibanov; Science 224 (1984) pp 1249-1251

8 [illegible] Goodman, [illegible], [illegible] Cameron, [illegible] London, [illegible] (1980) in the press.
9 [illegible], A. [illegible], Kawashima, [illegible] 17 [illegible] (1977) pp 10-17.
10 R.H. [illegible] (1966) pp [illegible]-100.
11 [illegible], [illegible], [illegible] submitted for publication in [illegible] (1980).
12 [illegible], [illegible] de [illegible], The hydrophobic fragmental constant, a compilation of [illegible] data, [illegible], [illegible] (Eds), [illegible] (1978) pp [illegible].
13 [illegible], A. [illegible], [illegible] Correlation analysis in Chemistry and Biology, Wiley, New York (1979).
14 [illegible], [illegible], [illegible] 17 (1977) pp [illegible]-[illegible].
15 [illegible], [illegible], [illegible] (eds) [illegible], [illegible] (1980) pp [illegible]-[illegible].
16 [illegible], [illegible], [illegible] (1980) pp [illegible].
17 [illegible], [illegible], [illegible], Biochemistry 17 (1978) pp [illegible].
18 [illegible], [illegible], [illegible], [illegible] Engineering for the European [illegible] (1980) pp [illegible].
19 [illegible], Handbook of [illegible], [illegible] Press [illegible].
20 [illegible], [illegible] 35 (1971) pp [illegible].
21 [illegible], [illegible], [illegible] (1982) pp [illegible]-[illegible].
22 [illegible], [illegible], [illegible] 144 (1984) pp [illegible].
23 [illegible], [illegible] 13 (1982) pp [illegible]-14.
24 [illegible], [illegible] 15 (1981) pp [illegible]-[illegible].
25 A. [illegible], [illegible] 32 (1985) pp [illegible].
26 [illegible], [illegible], [illegible] Encyclopedia of Chemical Technology [illegible], Wiley, New York (1978) pp [illegible].
27 [illegible] (1978) pp [illegible]-[illegible].
28 [illegible], [illegible] (1984) pp [illegible].

C. Laane, J. Tramper and M.D. Lilly (Editors), *Biocatalysis in Organic Media*,
Proceedings of an International Symposium held at Wageningen,
The Netherlands, 7–10 December 1986.

MULTIPHASE REACTORS A NEW OPPORTUNITY

M.D. LEGOY, M. BELLO, S. PULVIN and D. THOMAS

Laboratoire de Technologie Enzymatique, Université de Technologie, BP 233, 60206 Compiègne Cedex, (France)

SUMMARY

One of the most important potentialities of the use of enzymes (immobilized or not) is synthesis or modification of high added value molecules . These applications include the use of multiphasic media and of enzymes needing cofactors .

To display the feasibility of enzymic reactors needing cofactor and organic solvent, alcohol dehydrogenase has been studied for the production of long chain aldehydes . Three NAD regeneration methods have been tested : chemical, coupled enzymes and coupled substrates . Alcoholic insoluble substrate and immiscible solvent constitute the organic phase while the enzyme, the cofactor and the regenerating system were located in the aqueous one . With horse liver ADH and a thermostable ADH the possibility of using solid gas reactor has been shown .

On the other hand, hydrolysis, interesterification and ester synthesis catalyzed by specific (1-3 positions) Rhizopus arrhizus and non specific Candida cylindracea lipases have been tested in microemulsions . The microemulsions used were made of triglycerides or fatty acids, enzyme dissolved in very low water quantity, Brij 35 as surfactant and an alcoholic cosurfactant .

The interesterification rates obtained between triolein and triglycerides having chains from C_4 to C_9 were different with Rhizopus arrhizus lipase and with Candida cylindracea lipase . The Candida lipase reacted faster with C_9 triglyceride than with C_4 whereas Rhizopus lipase had the same rate for all the triglycerides . With the example of interesterification reaction the importance of water activity instead of water content has been shown . Synthesis reactions have also been tested . For example heptyl-oleate synthesis catalyzed by Candida lipase was obtained in microemulsions at a 98% yield .

INTRODUCTION

Enzymes possess unique properties . They are extremelly active catalysts, more active than most of the classic chemical ones and very specific . They act in aqueous media under mild conditions of pH, pressure and temperature, and their activity can be easily regulated .

But, in spite of these exceptional properties, most of chemical and food industries continue to use common chemical catalysts.

If more than 2000 enzymes are known, 200 are available in small quantities, only less than 20 commercialized in industrial amounts and with the exception of glucose isomerase, used for hydrolysis.

The major drawbacks of the use of enzymes as common catalysts are as follows :

- they are supposed to work in aqueous media but numerous interesting substrates are insoluble in water and to obtain synthesis or group transferring reactions the presence of water causes problem.

- they are very specific but if it is good for analytical or medical applications, it is not an advantage for the other industrial applications.

- their functional stability is not sufficient.

- one third of the known enzymes needs a cofactor. These cofactors are expensive and they are transformed during the reaction. For an industrial application of these kind of enzymes could be considered, it is necessary to regenerate the cofactor.

For solving these problems, different possibilities have been envisaged.

First of all, to improve the functional stability of the enzymes, a great number of immobilization methods have been developed (1). For an enzymatic process to be attractive, it will have to be competitive enough or better than the conventional one.

In this aim the use of different multiphasic reactors using two types of enzymes (a dehydrogenase and a hydrolase) is proposed .

BIPHASIC REACTORS

Solid-liquid reactors

It is proposed to obtain long chain or aromatic aldehydes by a continuous process using alcohol dehydrogenase. The proposed studies include the use of an enzyme needing cofactor (nicotinamide adenine dinucleotide).

The development of efficient enzyme catalyzed processes that require the participation of freely dissociable coenzymes necessitates the development of techniques not only for the immobilization of the enzyme but also for the retention of the coenzyme in the immediate vicinity of the coenzyme requiring system in order to maximize the use of expensive coenzyme and for the regeneration of the coenzyme after oxido-reduction (2-7).

As a model system the oxidation of geraniol to geranial by Horse Liver Alcohol Dehydrogenase (HLADH) has been chosen and different NAD regeneration methods tested.

Biphasic systems have been used. Enzymes and NAD (when it was possible) have been coimmobilized into albumin-glutaraldehyde foam structures. Water insoluble geraniol and hexane constituted the organic phase, while the enzyme, cofactor and regenerating system were located in the aqueous phase. This aqueous phase represented by the support was a solid phase and the organic phase was the liquid phase.

Three cofactor regeneration methods were tested in batch reactors with 80 % organic phase and 20 % aqueous phase.

(i) A chemical regeneration with 1 m-Methoxy Phenazine Methosulfate (1-mPMS) as electron carrier.

$$NADH + H^+ + 2\,O_2 \xrightarrow[\text{SOD}]{\text{1-mPMS}} NAD^+ + O_2 + H_2O_2$$

SOD : superoxide dismutase.

This regenerating method is interesting from an economical point of view because it is possible to use immobilized enzyme-coenzyme systems. In this case, one hundred cycles for each NAD molecule were obtained when 10^{-4}M NAD was used.

(ii) A bienzymatic regeneration with a lactic dehydrogenase (LDH) as second enzyme.

$$\text{Pyruvate} + \text{NADH} + H^+ \xrightarrow{\text{LDH}} \text{Lactate} + \text{NAD}^+$$

This method requires the use of a second enzyme with its substrate so the cost of the process is high and there is a problem for the separation of by-products. In this process, if the two enzymes are immobilized the cofactor must be free. With this method 450 cycles were obtained for each NAD molecule.

(iii) A coupled substrate regeneration with HLADH acting for oxidation of geraniol and reduction of acetaldehyde.

$$\text{Acetaldehyde} + \text{NADH} + H^+ \xrightarrow{\text{HLADH}} \text{Ethanol} + \text{NAD}^+$$

This very simple method permits the coimmobilization of HLADH and NAD^+. 1500 cycles were obtained with this system.

Due to the best results given by the coupled substrate regenerating method this process was tested with a fixed bed reactor. The results are presented on figure 1 for a column reactor of 10 cm^3 porous particles containing 3,6 10^{-7} moles HLADH and 5.7 10^{-6} moles NAD coimmobilized,0.1M geraniol and 0.1M acetaldehyde solubilized in hexane flowed through the support at a flow rate of 1.5 ml h^{-1}.

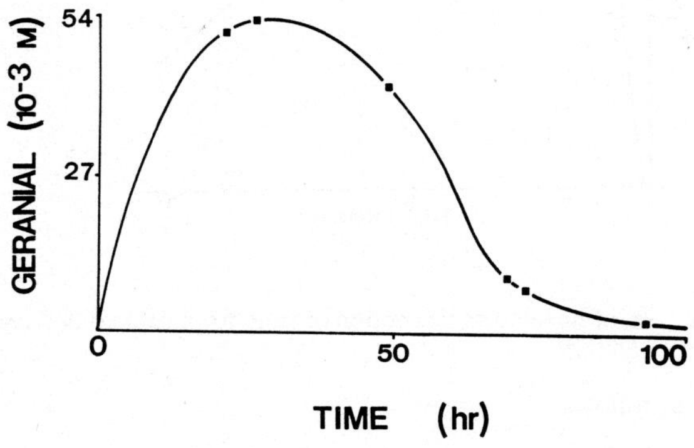

Figure 1 : Geranial production as a function of time for a 10 cm^3 column reactor.

In order to optimize the steady state duration, other reactors have been realized, showing a dependance of the steady state with the flow rate, the geraniol concentration and the NAD immobilized.

In the reaction studied, production of geranial to be used as flavor was tested as a model system, other aldehyde productions (citronellal, hexanal and so on) have also succeeded. In spite of the possibility of using this type of reactors, synthesis capacities of HLADH present important limitations specially due to the operational instability of most of the enzyme and products in water. So another type of reactor was developed.

Solid-gas reactors

To overcome the problems posed by the use of organic solvents, systems using immobilized enzymes with substrates and products in gaseous form were studied (8). For these experiments HLADH and NAD (or NADH) were coimmobilized into porous particles. In the reactors, gaseous substrates flowed through the solid support and reacted with the enzyme, and gaseous products leaved the reactor. The feasibility of the system was tested in batches for the reduction of different aldehydes and, to test the yield of regenerated cofactor, fixed bed reactors catalyzing the oxidation of butanol, pentanol or hexanol and the reduction of acetaldehyde were studied.

The results obtained for the oxidation of hexanol are presented on figure 2.

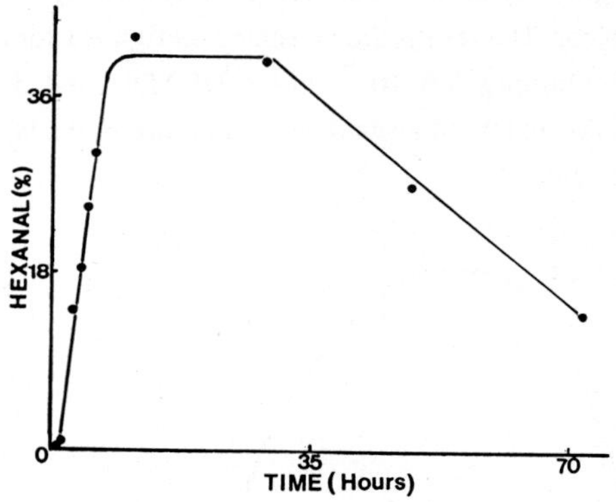

Figure 2 : % of hexanal as a function of time for a packed bed reactor.

The reactions were as follows

$$\text{Hexanol} + NAD^+ \xrightarrow{\text{HLADH}} \text{Hexanal} + NADH + H^+$$

$$\text{Acetaldehyde} + \text{NADH} + H^+ \xrightarrow{\text{HLADH}} \text{Ethanol} + \text{NAD}^+$$

In this case a 50 cm^3 column reactor filled with porous particles containing 3.6 10^{-7} moles HLADH and 5.7 10^{-6} moles NAD was used. The vapours of hexanol and acetaldehyde flowed through the coimmobilized enzyme-coenzyme support by a carrier gas saturated with water vapours. The system was maintained at 60°C. Steady state was obtained within 7 hours and lasted for about 23 hours.

The same results were obtained with immobilized NAD or NADH.

In this study the limiting step was the instability of horse liver alcohol dehydrogenase. This problem can be solved by using a thermostable enzyme for example alcohol dehydrogenase of Sulfolobus solfataricus stable at 80°C. This enzyme accepts a wide range of substrates and permits to generalize the use of the solid gas reactor to many other substrates with high boiling points.

MICROEMULSIONS

Triglycerides are major components of oils and natural fats. The development of agrofood industries have made natural fats, potential source of raw materials. Different processes have been developed for modifying the physicochemical properties of fats. Among them, transesterification and interesterification seem particularly interesting for modulating the fatty acid position and composition in triglycerides (9-17). Hydrolysis, synthesis, transesterification and interesterification are possible by a chemical way but the enzymatic approach presents many advantages as mild conditions of catalysis and specificity. Until now, different substrates have been studied but the major drawback is that they are insoluble in water.

Their solubilization in organic solvent set up the problem of enzyme denaturation so other most appropriate media have been tested. Different authors (18-19) have studied the use of reverse micelles made of isooctane, AOT (as surfactant), water and substrates.

In these cases, the enzyme is confined in the hydrophilic heart of microdroplets limited by a film of surfactant molecules.

It has been chosen to study interesterification and synthetis reactions of fats catalyzed by lipases in microemulsions. Microemulsions are monophasic, transparent, isotropic and stable from the kinetic and thermodynamic point of view. They are composed of four constituents : triglycerides or fatty acids constitute the organic phase, the enzyme in water the aqueous one, the surfactant is Brij 35 and the cosurfactant a primary or tertiary alcohol. In microemulsions, interface is very large and exchanges very rapid. When water quantity is very low, surfactant and water are dispersed in oil in form of reverse microemulsions (20-22). These microemulsions permit non only the solubilization of substrates and products but also non-conventional reactions of interesterification and synthesis. The reactions studied were as follows :

- for interesterification

$$R_1X + R_2Y \text{------------>} R_1Y + R_2X$$

- for synthesis

alcohol + fatty acid(s) ----------------> fatty acid(s) ester

One of the fundamental parameters influencing formation of new fatty acid esters is the water activity. According to the mass action law, a high yield of water brings on the hydrolysis of esters, whereas a low yield of water favors synthesis or group transfers.

Interesterifications between triolein and triglycerides having chain length fatty acids varying from 4 to 9 carbons were tested with two microbial lipases. Microemulsions composed of 33 % triolein, 1 % water containing the lipase, 30 % 2 methyl - 2 hexanol as cosurfactant, 0.1 % Brij 35 and 37 % tributyrin, tricaproin, triheptanoin or trinonanoin (w/w) were made. The reaction progresses were followed by phase gas chromatography analyses. The results obtained as a function of time are presented in figure 3 for the non specific. Candida cylindracea lipase and of figure 4 for the Rhizopus arrhizus lipase specific of 1-3- positions.

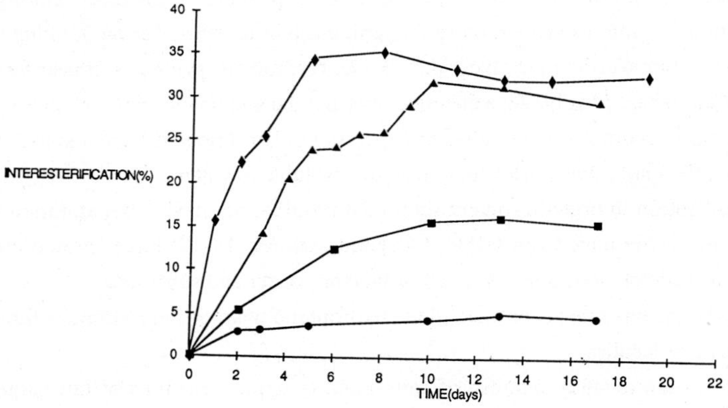

Figure 3 : Interesterification as a function of time for Candida cylindracea lipase.

for triolein-tributyrin (—●—)

for triolein-tricaproin (—■—)

for triolein-triheptanoin (—▲—)

for triolein-trinonanoin (—◆—)

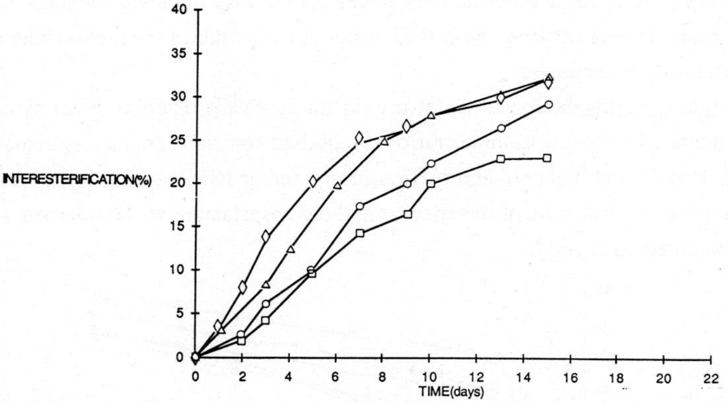

Figure 4 : Interesterification as a function of time with Rhizopus arrhizus lipase for triolein-tributyrin (— ○ —), for triolein-tricaproin (—□—), for triolein-triheptanoin (—△—), for triolein-trinonanoin (—◇—)

With Rhizopus arrhizus lipase about 40 % interesterification was obtained after 20 days reaction in all the systems tested. The non specific Candida cylindracea gave similar results for triolein-triheptanoin and triolein-trinonanoin systems but only 17 % and 6 % interesterification were seen with triolein-tricaproin and triolein-tributyrin. It seemed that this non-specific lipase under particular conditions such as reverse microemulsions, realized less interesterification with short chain lenght fatty acids.

A fundamental parameter influencing this kind of reaction is the water activity. So the triolein- triheptanoin interesterification was tested with 1.5 % water quantity but with water activities ranging from 0.25 to 0.8 by varying the ratio between T (surfactant / cosurfactant = 5 10^{-3}) and lipidic phase.

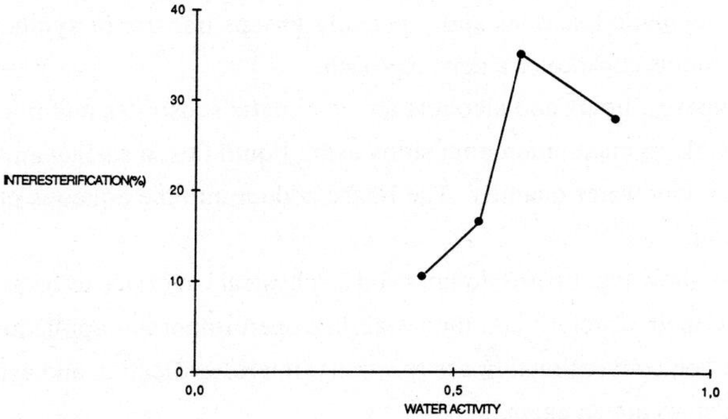

Figure 5 : Interesterification of triolein-triheptanoin as a function of water activity.

As seen on figure 5, for a water activity lower than 0.4 no interesterification occured, an optimum of reaction was obtained with 0.63 water activity, then the reaction rate decreased when the water activity increased.

The Candida cylindracea lipase was also used for synthesis reaction. Ester synthesis as a function of heptanol / oleic acid molar ratio was studied for microemulsions containing 1 % water, 0.1 % Brij 35 and heptanol and oleic acid in a molar ratio varying from 1 to 5. In this case, heptanol was the substrate of the enzyme and the cosurfactant of the microemulsion. The results are presented on figure 6.

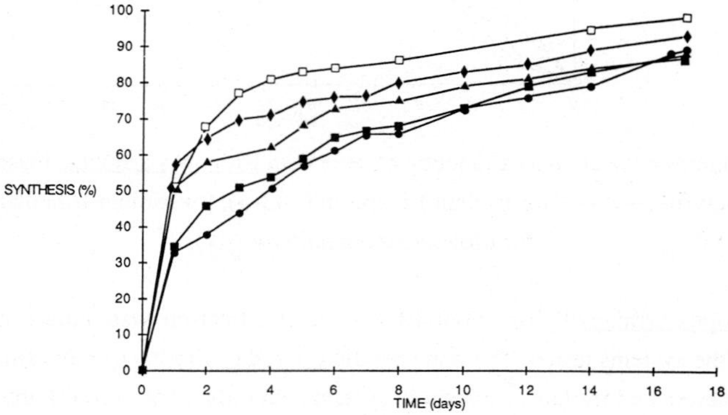

Figure 6 : Heptyl oleate synthesis as a function of time for heptanol/oleic acid ratio of 1(-●-), 2 (-■-), 3 (-▲-), 4 (-◆-), 5 (-□-).

The reaction was considerably more rapid and the ester synthesis more important when the heptanol / oleic acid ratio was high. For a molar ratio of 5, the synthesis yield was 80 % after 4 days and around 100 % after 17 days.

The obtention of enzymatic reactions and especially groups transfer or synthesis reactions with microemulsion systems constitutes a new approach.

In the case of the lipases, lipids and alcohols are enzymatic substrates and microemulsions components. It is possible to make microemulsions using liquid fats, a surfactant, an alcoholic cosurfactant and a very low water quantity. The lipase adduct into the aqueous phase doesnot break the microemulsion.

The obtained results show that lipases do not need a "physical" interface to be active.

Microemulsions,by their structure and their stability open important application fields for enzyme catalysis. The non conventional lipase reactions (interesterification and ester synthesis) observed in microemulsions are an example.

CONCLUSION

The use of biphasic reactors such as solid-liquid and solid-gas reactor and of microemulsions appears to be :

- simple , efficient and readily amenable to scale-up.

-applicable to a wide range of water insoluble substrates.

and for solid-liquid sytems and microemulsions

-usable in the presence of very low water concentration which allows the obtention of unusual reaction and greatly simplifies product recovery.

REFERENCES

1 I. Chibata, Immobilized Enzyme, I. Chibata (Ed.), Halsted Press, 1978.

2 Y. Yamazaki and H. Maeda, The co-immobilization of NAD and dehydrogenases and its application to bioreactors for synthesis and analysis, Agr. Biol. Chem. Tokyo, 46 (6), (1982), 1571-1581.

3 R. Wichmann, C. Wandrey, A.F. Buckmann and M.R. Kula, Continuous enzymatic transformation in an enzyme membrane reactor with simultaneouss NAD(H) regeneration, Biotechnol.Bioeng.,23, (1981), 2789-2802.

4 J. Grunwald and T.M.S. Chang, Immobilization of ADH, malic dehydrogenase and dextran-NAD^+ within nylon-polyethyleneimine microcapsules : preparation and cofactor recycling J. Molecular Catalysis, 11, (1981), 83-90.

5 M.O. Mansson, P.O. Larsson and K. Mosbach, Recycling by a second enzyme of NAD covalently bound to alcohol dehydrogenase, FEBS Lett., 98, (2), (1979), 309-313.

6 M.D. Legoy, J.M. Le Moullec and D. Thomas, Chemical grafting of functional NAD in the active site of a dehydrogenase. Regeneration in situ, FEBS Lett., 94, (2), (1978), 335-338.

7 G. Carrea, F. Colombi, G. Mazzola, P. Cremonesi and E. Antonini, Immobilized hydroxysteroid dehydrogenases for the transformation of steroids in water organic solvent systems, Biotechnol. Bioeng. , 21, (1979), 39-48.

8 S. Pulvin, M.D. Legoy, R. Lortie, M. Pensa and D. Thomas, Enzyme Technology and gas phase catalysis : alcohol dehydrogenase example. Biotechnol. Lett., 8, (11), (1986),783-784.

9 M.M. Hoq, T. Yamane, S. Shimizu, T. Funada and S. Ishida, Continuous hydrolysis of olive oil by lipase in microporous hydrophobic membrane reactor, J. Am. Oil Chem. Soc.,62, (6), (1985), 1016-1021.

10 A.R.Macrae, Lipase-catalyzed interesterification of oils and fats, J. Am. Oil Chem. Soc., 60, (2), (1983), 291-294.

11 C Marlot, G. Langrand, C. Triantaphylides and J. Baratti, Ester synthesis in organic solvent catalyzed by lipases immobilized on hydrophilic supports, Biotechnol. Lett., 7, (9), (1985), 647-650.

12 S. Morita, H. Narita, T. Matoba and M. Kito, Synthesis of triacylglycerol by lipase in phosphatidylcholine reverse micellar system, J. Am. Oil Chem. Soc., 61, (10), (1984), 1571-1574.

13 B. Sreenivasan, Interesterification of fats, J. Am. Oil Chem. Soc., 55, (1978), 796-805.

14 K. Takahashi, Y. Kodera, T. Yoshimoto, A. Ajima, A. Matsushima and Y. Inada, Ester-exchange catalyzed by lipase modified with polyethylene glycol, Biochem. Biophys. Res. Commun., 131, (2), (1985), 532-536.

15 T. Tanaka, E. Ono, M. Ishihara, S. Yamanaka and K. Takinami, Enzymatic acyl exchange of triglyceride in *n*-hexane, Agric. Biol. Chem., 45, (10), (1981), 2387-2389.

16 K. Yokozeki, S. Ymanaka, K. Takinami, Y. Hirose, A. Tanaka, K. Sonomoto and S. Fukui, Application of immobilized lipase to regio-specific interesterification of triglyceride in organic solvent, Eur. J. Appl. Microbiol. Biotechnol., 14, (1982), 1-5.

17 A. Zaks and A.M. Klibanov, Enzyme-catalyzed processes in organic solvents, Proc. Natl. Acad. Sci. USA, 82, (1985), 3192-3196.

18 P.L. Luisi, Enzyme hosted in reverse micelles in hydrocarbon solution, Angew. Chem. Int. Ed. Engl., 24, (1985), 439-450.

19 K. Martinek, A.V. Levashov, N.L. Klyachko, V.I. Pantin and I.V. Berezin, The principles of enzyme stabilization. IV Catalysis by water-soluble enzymes entrapped into reversed micelles of surfactants in organic solvents, Biochim. Biophys. Acta, 657, (1981), 277-294.

20 K.L. Mittal and E.J. Fendler, Solution behavior of surfactants, vol1, Plenum Press, New york, 1982.

21 P.D.I. Fletcher, G.D. Rees, B.H. Robinson and R.B. Freedman, Kinetic properties of α-chymotrypsin in water-in-oil microemulsions : studies with a variety of substrates and microemulsions systems, Biochim. Biophys. Acta, 832, (1985), 204-214.

22 M. Clausse, A. Zradba and L. Nicolas-Morgantini, Water / ionic surfactant / alkanol / hydrocarbon systems. Realism of existence and transport properties of microemulsions type media, Coll. and Polym. Sci., 263, (9), (1985), 767-770

C. Laane, J. Tramper and M.D. Lilly (Editors), *Biocatalysis in Organic Media*,
Proceedings of an International Symposium held at Wageningen,
The Netherlands, 7–10 December 1986.

CO-IMMOBILIZATION - AN ALTERNATIVE TO BIOCATALYSIS IN ORGANIC MEDIA

R. KAUL, P. ADLERCREUTZ and B. MATTIASSON
Department of Biotechnology, Chemical Center, University of Lund, P.O.Box 124, S-221 00 Lund (Sweden)

SUMMARY
Co-immobilization of biocatalyst and substrate has been suggested as a method to increase the conversion rate in systems with substrates having low aqueous solubility, especially when the inclusion of organic solvent in the reaction medium is detrimental to the overall process. The system studied was the transformation of hydrocortisone to prednisolone by *Arthrobacter simplex* when co-immobilized in polymer beads. The gaseous co-substrate was supplied directly to the bed of such a preparation. The cells were repeatedly used for the conversion after the extraction of the product from the matrix.

INTRODUCTION
Biocatalysis in organic media has been shown to be feasible and at the same time, superior to reactions in aqueous solutions, for a multitude of systems involving apolar compounds. To make the enzymes function in low water milieu, they have been mostly adsorbed to solid matrices (ref. 1), or used in the form of microemulsions (ref. 2). When cells have to be involved for bioconversions, the choice of organic solvents becomes rather critical. Brink and Tramper (ref. 3) have studied the effect of various organic solvents on the performance of *Mycobacterium* spps. A correlation was found between certain characteristics of the solvents and their inactivating ability. The studies on microbial cells in microemulsions showed that the individual enzymatic steps were catalytically active even when the cells were dead after having been exposed to such conditions (ref. 4). However, in situations when the cells are required to be in a viable state, the long term performance of the biocatalyst may be negatively affected owing to the membrane modification and other inhibitory effects caused by most organic solvents.

We examined the feasibility of two different approaches, so as to improve the substrate availability to cells in an aqueous medium, and also to maintain their viability. In both the cases, transformation of hydrocortisone to prednisolone by *Arthrobacter simplex* was studied as a model reaction system.

One of the methods which will not be presented here, dealt with running the bioconversion in an aqueous two-phase system. In a well mixed system, the reaction proceeds inside the tiny droplets of the bottom phase, surrounded by a relatively hydrophobic top phase, which serves as the steroid reservoir (ref. 5). Since the diffusional distances are very short, mass transfer in such systems is greatly facilitated.

The second approach involves the co-immobilization of the substrate with the catalyst. This arrangement exploits to advantage the proximity effects between different entities in a reaction system, observed earlier in immobilized multienzyme sequences.

METHODS

Arthrobacter simplex ATCC 6946 cells were cultivated and induced for the enzyme, steroid Δ'-dehydrogenase as described previously (ref. 6). The induced cells were harvested, washed with 0.05 M Tris · HCl, pH 7.0 and frozen as pellets until required.

The co-immobilization of hydrocortisone and *A. simplex* cells was performed in alginate (PROTAN, Norway), agar (Difco Labs, USA)and agarose (Sigma, USA) respectively.

To 10 ml of sodium alginate solution (2%, in 0.05 M Tris · HCl, pH 7.0) were added the cells (0.6 mg dry weight/ml) and the substrate (0.5 mg/ml), and mixed well. The mixture was extruded into 0.1M $CaCl_2$ solution (pH 7.0) for the preparation of the beads; which were immediately transferred to a column and drained of the buffer solution. Water saturated air was passed upward through the column at room temperature.

The reaction was followed by taking a couple of beads from the column, dissolving them in 0.1M phosphate or citrate buffer, pH 7.0, and estimating the amount of hydrocortisone and prednisolone by HPLC, after extraction into chloroform as described earlier (ref. 6).

After the completion of the reaction, all the beads were dissolved in the citrate buffer, and the cells isolated by centrifugation. After washing a few times with 0.05M Tris-HCl, pH 7.0, the cells were reimmobilized with more hydrocortisone. When required, the activation of the cells was performed by including the nutrients,1% casein peptone tryptic digest (E. Merck, Germany), 0.5% yeast extract (Fould-Springer, France), and 0.5% glucose in the co-immobilization mixture. Prednisolone could be recovered from the alginate by adsorption on Amberlite XAD-4 (BDH, England), and the alginate recycled into the system.

Both agar and agarose were used at a concentration of 4%, for the co-entrapment of bacterial cells and hydrocortisone. The beads were prepared in cold Tris-HCl solution, pH 7.0.

As a control reaction, the transformation of hydrocortisone was carried out in Tris-HCl solution, pH 7.0, by the freely suspended cells.

RESULTS AND DISCUSSION

Most of the studies on steroid biotransformation have been performed with immobilized cells (ref. 7). A perceived advantage of immobilized cell technology is the capability to confine large amounts of the biocatalyst within a small volume of the reactor. But when a majority of the biocatalyst is not functioning because the substrate (steroid) and the cosubstrate (oxygen) are in short supply owing to their poor aqueous solubility, low efficiency of the reactor is inevitable. The use of hydrophobic supports in conjunction with organic solvents for solubilizing the substrate has been suggested (ref. 7).

As observed by Lilly and Woodley (ref. 8), the mass transfer of components across the various interfaces in a multiphase system, composed of water, organic solvent and an immobilized biocatalyst, is of extreme importance in determining the performance of these reactions. Hence, the transfer would be most efficient in systems not containing a discrete aqueous phase; surface area of the biocatalyst being the determining factor under such circumstances. But then, one has to keep in mind the unwanted interactions between the solvent and the support material used for immobilization , and also the denaturing forces on the microbial cells.

By forcing the cells and the substrate together in the co-immobilized system that we propose, the rate limiting transfer of the substrate into the beads is avoided. Also the transport of oxygen to the site of catalysis could be accelerated by flushing air directly over the beads, without the presence of a discrete liquid phase. Saturation of the air with water prevented the drying of the biocatalyst and hence its inactivation.

As this kind of configuration required the product to be extracted from the insoluble particles, the support materials used were alginate, agar and agarose, because of their property of being reversibly polymerized. The mechanical stability of the beads is not a prerequisite in such a system, as they do not encounter any mechanical agitation. The reaction rate was observed to be the fastest in agarose as compared to that in the other matrices (see Fig. 1). Perhaps, the increased ionic strength in the alginate beads could have partly led to 'salting out' of oxygen (ref. 9), thus giving lower conversion rates.

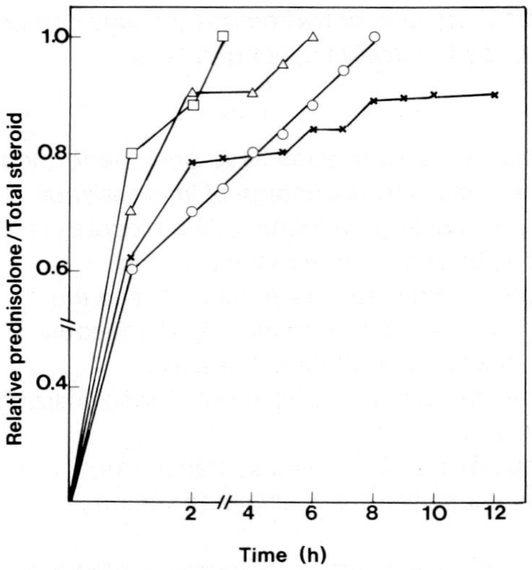

Fig. 1. Hydrocortisone transformation by A. *simplex*, co-immobilized in different matrices: (□) agarose, (△) agar, (○) alginate,(**x**) control. The steroid and the cells were used at concentrations of 0.5 and 0.6 mg dry weight/ml, respectively, in a total volume of 10 ml. Reproduced with permission from ref. 6, Copyright I986, John Wiley & Sons, Inc., New York.

Since the steroid is present in the form of a suspension, the amount of substrate that would be available for bioconversion at one particular time is rather limited. Therefore, under such conditions, the effect of substrate inhibition on the catalyst if any, is not apparent. As the reaction is allowed to proceed, the substrate continuously solubilizes, creating new substrate in the vicinity of the biocatalyst, a condition favourable for the overall process.

The studies on the co-immobilization of varying amounts of the substrate with the cells in alginate showed that a steroid concentration above 2.0 mg/ml could not be totally transformed to prednisolone (see Fig. 2). Incomplete conversions have earlier been observed when rather high concentrations of the substrate were employed, resulting in the mixed crystals of hydrocortisone and prednisolone (ref. 10). The unreacted hydrocortisone gets masked by the excess of prednisolone crystals. We observed that if the mixed crystals were milled with the glass beads and then exposed to the cells, either in a co-immobilized or a free form, almost total (98%) conversion took place.

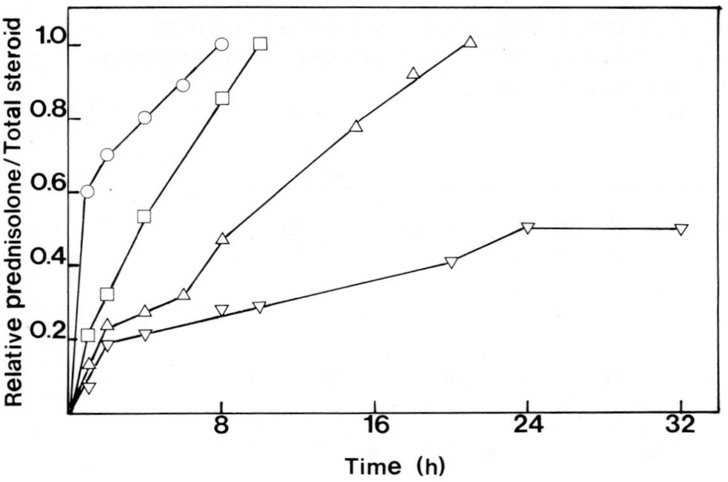

Fig. 2. Transformation of different concentrations of co-immobilized hydrocortisone. (○) 0,5 g/L, (□) 1.0 g/L, (△) 2.0 g/L, (▽) 3.0 g/L. The steroid-to-cell ratio was kept constant in all cases. Reproduced with permission from ref. 6, Copyright l986, John Wiley &Sons, Inc., New York.

The possibility of reusing the biocatalyst is one of the important attractions of the immoblization technique. For the repeated use of the cells in the co-immobilized state, the dissolution of the matrix was required in order to remove the product, and then to introduce new substrate. These studies were performed with alginate as the matrix, mainly because of the simplicity and mild conditions, both for gel formation and solubilization. It was observed that the reaction could be completed in a much shorter time during the subsequent runs (see Fig. 3). This could be due to the further induction of the enzyme in the cells by the co-entrapped hydrocortisone. After a number of runs, however, there was a decrease in the catalytic activity of the cells. Activation of the immobilized cells by incubation with the nutrient medium has been carried out before (ref. 11). The inclusion of the nutrients in the immobilization mixture during the biotransformation resulted in the revival of the enzyme activity. The wet weight of the cells thus obtained was increased about two fold; which were then immobilized in twice the volume of alginate with double the amount of hydrocortisone. Such activation was performed each time the activity was seen to decline.

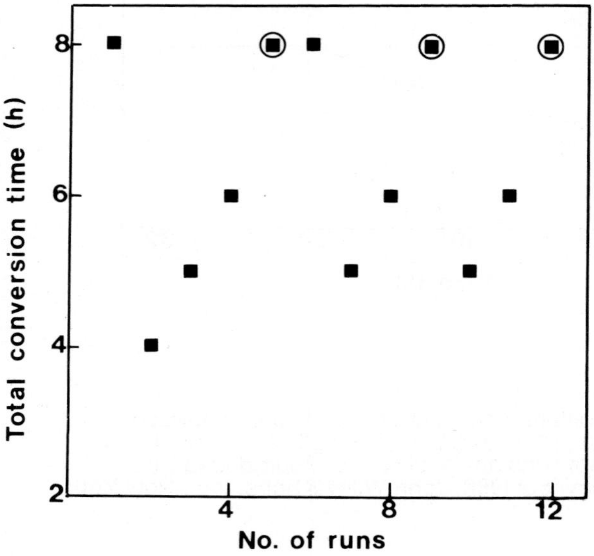

Fig. 3. Reuse of A. *simplex* cells for the conversion of hydrocortisone to prednisolone in a co-immobilized system. The concentrations of hydrocortisone and the cells used were 0.5 and 0.6 mg dry weight/ml alginate, respectively. The circled squares indicate the addition of nutrients during the particular run.

The long term operation of this process configuration would require the system to be handled in sterile conditions because of the risk of contamination during the activation steps.

In the studies reported here, prednisolone was recovered by extraction into chloroform, and repeated co-immobilization was performed using fresh alginate during each run. To make the reuse of the polymer possible, a milder system of extraction was required. We have earlier reported on the possibility of adsorbing the prednisolone on to Amberlite XAD-4, a hydrophobic polymeric resin (ref. 5). Thus, passing the alginate-prednisolone mixture over a column of the resin would lead to a simple recovery of the alginate as the eluate which can be recycled to the immobilization mixture. The prednisolone could be eluted from the column by the passage of methanol over it. Subsequent washing of the column with buffer makes it ready for the next use.

CONCLUSION

In cases where the catalyst represents a precious commodity and must be reused due to process economics, it may be difficult to perform bioconversions in organic solvents . The approach to co-immobilize the catalyst and the substrate may, at a first glance, seem quite tedious and expensive, but if only a technology to integrate the whole scheme of immobilization, solubilization and extraction of the product in a continuous manner is developed, it may open new possiblilities for efficient and cost-effective processes. In processes, like the one studied here, where a gaseous reactant is also involved it may be better to operate on the mode of a gas-solid bioreactor (ref. 12). The presence of a liquid phase may reduce the rate of diffusion of the gas through the system. However, the supply of oxygen could also be improved by using emulsions of perfluorochemicals (ref. 13). In such a case, also the shrinking of the dimensions of the catalyst-substrate co-polymerizate may be advantageous.

There are several examples of compounds that are poorly soluble in water and thus, are not being utilized efficiently by the microorganisms. One can foresee, still other configurations of a co-immobilized system. For instance, immobilizing the catalyst together with an adsorbent for binding the substrate, may create a microenvironment with a locally improved substrate concentration making it possible to carry out the bioconversions with a higher efficiency.

Co-immobilization of the substrate and the biocatalyst may not necessarily be restricted to a bead shaped conformation, but could also be carried out in the shape of fibers, membranes etc. Even the reactor configurations may be altered depending on the kind of system being studied.

ACKNOWLEDGEMENT

The financial support of the National Swedish Board of Technical Development is gratefully acknowledged.

REFERENCES

1. A.R. Macrae, Interesterification of fats and oils, in: J. Tramper, H.C. van der Plas and P. Linko (Eds.), Biocatalysts in Organic Syntheses, Elsevier, Amsterdam, l985, pp. 195-208.
2. P.L. Luisi and C. Laane, Solubilization of enzymes in apolar solvents via reverse micelles, Trends Biotechnol. 4(6) (l986) 153-161.
3. L.E.S. Brink and J. Tramper, Optimization of organic solvent in multiphase biocatalysis, Biotechnol. Bioeng., 27 (1985) 1258- 1269.
4. G. Häring, P.L. Luisi and F. Meussdoerffer, Solubilization of bacterial cells in organic solvents via reverse micelles, Biochem. Biophys. Res. Commun.,

127 (1985) 911- 915.
5. R. Kaul and B. Mattiasson, Extractive bioconversion in aqueous two-phase systems. Production of prednisolone from hydrocortisone using *Arthrobacter simplex* as catalyst, Appl. Microbiol. Biotechnol. 24 (1986) 259- 265.
6. R. Kaul, P. Adlercreutz and B. Mattiasson, Coimmobilization of substrate and biocatalyst : a method for bioconversion of poorly soluble substances in water milieu, Biotechnol. Bioeng., 28 (1986) 1432- 1437.
7. F.B. Kolot, Microbial catalysts for steroid transformations- Part 1, Process Biochem., 17(6) (1982) 12-18.
8. M.D. Lilly and J.M. Woodley, Biocatalytic reactions involving water-insoluble organic compounds, in: J. Tramper, H.C. van der Plas and P. Linko (Eds.) Biocatalysts in Organic Syntheses, Elsevier, Amsterdam, 1985, pp. 179- 192.
9. A.M. Klibanov, Enzyme stabilization by immobilization, Anal. Biochem. 93 (1979) 1- 25.
10. A. Constantinides, Steroid transformation at high substrate concentrations using immobilized *Corynebacterium simplex* cells, Biotechnol. Bioeng. 22 (1980) 119-136.
11. S. Ohlson, P.-O. Larsson and K. Mosbach, Steroid transformation by living cells immobilized in calcium alginate, Eur. J. Appl. Microbiol. Biotechnol. 7 (1979) 103- 110.
12. J.A.M. de Bont, C.G. van Ginkel, J. Tramper and K. Ch. A.M Luyben, Ethylene oxide production by immobilized *Mycobacterium Py.1* in a gas-solid bioreactor, Enzyme Microb. Technol. 5 (l983) 55-59.
13. P. Adlercreutz and B. Mattiasson, Oxygen supply by hemoglobin or emulsions of perfluorochemicals, Eur. J. Appl. Microbiol. Biotechnol. 16 (l982) 165-170.

C. Laane, J. Tramper and M.D. Lilly (Editors), *Biocatalysis in Organic Media*,
Proceedings of an International Symposium held at Wageningen,
The Netherlands, 7–10 December 1986.

ENZYMATIC PRODUCTION OF CHEMICALS IN ORGANIC SOLVENTS

ALEXANDER M. KLIBANOV

Department of Applied Biological Sciences, Massachusetts Institute of Technology, Cambridge, Massachusetts 02139, U.S.A.

Our strategy for biocatalysis in organic media is the use of solid enzymes in non-aqueous solvents (1). We believe that this methodology is superior to conventional approaches such as enzymes in biphasic systems or reverse micelles, for it is simpler (1) and affords greatly enhanced stability of enzymes (2) and strikingly different substrate specificity (3). We have successfully employed enzyme catalysis in organic solvents for a number of interesting processes (difficult or impossible in water) including:

- lipase-catalyzed regioselective acylations of glycols (4) and sugars (5);
- lipase-catalyzed asymmetric transesterifications and esterifications for the production of optically active alcohols, acids, and esters (6);
- polyphenol oxidase-catalyzed regiospecific oxidation of phenols (7);
- alcohol dehydrogenase-catalyzed stereoselective oxidoreductions for the synthesis of optically active alcohols and ketones (8); and
- peroxidase-catalyzed lignin degradation (9) and polymerization of phenols (10).

In addition to these studies, our laboratory's current research in the area of non-aqueous enzymology also involves:

- biosensors based on enzymes functioning in non-aqueous milieu (e.g., see refs. 11 and 12);
- enzymatic modification of sugars for the production of new biosurfactants;
- enzymatic synthesis of peptides; and
- enzyme-catalyzed gas phase reactions.

1. A.M. Klibanov, "Enzymes that work in organic solvents", CHEMTECH **16**, 354-359 (1986).

2. A. Zaks and A.M. Klibanov, "Enzymatic catalysis in organic media at 100°C", Science **224**, 1249-1251 (1984).

3. A. Zaks and A.M. Klibanov, "Substrate specificity of enzymes in organic solvents vs. water is reversed", J. Amer. Chem. Soc. **108**, 2767-2768 (1986).

4. P. Cesti, A. Zaks and A.M. Klibanov, "Preparative regioselective acylation of glycols by enzymatic transesterification in organic solvents", Appl. Biochem. Biotechnol. **11**, 401-407 (1985).

5. M. Therisod and A.M. Klibanov, "Facile enzymatic preparation of monoacylated sugars in pyridine", J. Amer. Chem. Soc. **108**, 5638-5640 (1986).

6. G. Kirchner, M.P. Scollar and A.M. Klibanov, "Resolution of racemic mixtures via lipase catalysis in organic solvents", J. Amer. Chem. Soc. **107**, 7072-7076 (1985).

7. R.Z. Kazandjian and A.M. Klibanov, "Regioselective oxidation of phenols catalyzed by polyphenol oxidase in chloroform", J. Amer. Chem. Soc. **107**, 5448-5450 (1985).

8. J. Grunwald, B. Wirz, M.P. Scollar and A.M. Klibanov, "Asymmetric oxidoreductions catalyzed by alcohol dehydrogenase in organic solvents", J. Amer. Chem. Soc. **108**, 6732-6734 (1986).

9. J.S. Dordick, M.A. Marletta and A.M. Klibanov, "Peroxidases depolymerize lignin in organic media but not in water", Proc. Natl. Acad. Sci. USA **83**, 6255-6257 (1986).

10. J.S. Dordick, M.A. Marletta and A.M. Klibanov, "Polymerization of phenols catalyed by peroxidase in non-aqueous media", Biotechnol. Bioeng. **29**, in press (1987).

11. R.Z. Kazandjian, J.S. Dordick and A.M. Klibanov, "Enzymatic analyses in organic solvents", Biotechnol. Bioeng. **28**, 417-421 (1986).

12. C.G. Boeriu, J.S. Dordick and A.M. Klibanov, "Enzymatic reactions in liquid and solid paraffins: application for enzyme-based temperature abuse sensors", Bio/Technology **4**, 997-999 (1986).

SESSION IV

ENGINEERING ASPECTS

Chairman : P.B. Poulsen

Co-chairman : J. Kloosterman

C. Laane, J. Tramper and M.D. Lilly (Editors), *Biocatalysis in Organic Media*, Proceedings of an International Symposium held at Wageningen, The Netherlands, 7-10 December 1986.

ENZYME ACTION AND ENZYME REVERSAL IN WATER-IN-OIL MICROEMULSIONS

C. Oldfield[1], G. D. Rees[2], B. H. Robinson[2], and R. B. Freedman[1]

[1]Biological Laboratory, University of Kent, Canterbury CT2 7NJ, UK

[2]Chemical Laboratory, University of Kent, Canterbury CT2 7NH, UK

INTRODUCTION

The exploitation of enzymes for biocatalysis in systems of low water content has been limited by lack of fundamental knowledge of enzyme properties in such systems. The majority of such systems are not amenable to study by methods which provide fundamental information on the structural and kinetic properties of enzymes. Water-in-oil (w/o) microemulsions are an exception. They provide a defined physical system ideally suited to the characterisation of enzyme properties by the methods used for aqueous solutions (refs. 1-2). The physical properties of w/o microemulsions are described elsewhere in this volume. Here we emphasise the usefulness of w/o microemulsions for determining fundamental kinetic properties of enzymes in predominantly organic media.

Microemulsions have the following unique advantages; i) they are reproducible, thermodynamically-stable systems whose properties are fully determined by their composition; ii) they are optically-transparent, facilitating physical and kinetic studies; iii) they have a low and controllable overall water content, usually in the range 0.1 - 10% (v/v); iv) the water activity can be reduced independently of the water content by addition of polar cosolvents; v) enzyme molecules solubilized in microemulsions are located within the dispersed droplets in essentially an aqueous micro-environment; vi) the large interfacial area between the continuous (oil) and dispersed (aqueous) phases permits rapid substrate and product exchange without stirring.

In previous work, we studied the properties of α-chymotrypsin in w/o micro-emulsions. This was a useful model system in that both the enzyme and the substrates used were essentially confined to the water droplets and hence kinetic studies gave information on this environment and on enzyme properties within it. Early work indicated that careful purification of surfactants (or selection of pure surfactants), good buffering and thorough controls for enzyme inactivation were essential if reliable kinetic data were to be obtained (ref.3). Taking these precautions, and using a broad approach drawing on several substrates differing in charge and polarity, and employing several microemulsion systems (involving respectively anionic, non-ionic and cationic surfactants) we have studied chymotrypsin activity in microemulsions as a function of R, temperature, pH, water:oil ratio etc. (ref. 4). The results indicate that the fundamental kinetic properties of the enzyme (k_{cat}, ΔH^*, pH-dependence) are essentially unchanged from those in bulk water (refs. 4, 5).

LIPASE ACTION IN MICROEMULSIONS; THE SIGNIFICANCE OF INTERFACIALLY-BOUND SUBSTRATE

Bacterial lipase *(C.viscosum)* solubilised in w/o microemulsions was active towards p-nitrophenylalkanoate substrates from C_2 to C_{18} in chain length. The activity was stable and there was a small, R-dependent shift in pH optimum from that in bulk water. The substrates partition into the oil phase, so that the effective substrate concentrations at the enzyme are << K_m and hence kinetic data are best expressed as second order rate constants k_2 (=k_{cat}/K_m). Values of k_2 were determined at the pH optima in bulk water and in micro-emulsions for the various chain length substrates (Figure 1). It was found that in microemulsions, k_2 varied little with chain length. For short and medium chain substrates (C_2-C_8), this is because the effects of partitioning

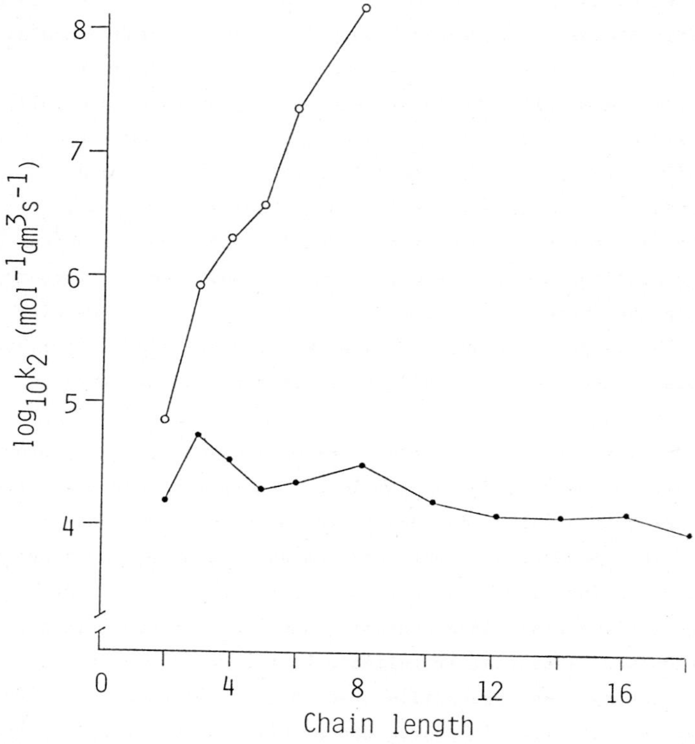

Fig. 1. The variation with chain length of the rate constant for lipase-catalysed hydrolysis of p-nitrophenyl alkanoates (O) in bulk aqueous solution, (●) in R=12.5, 0.1M AOT microemulsions, at optima pH in each case.

and intrinsic enzyme specificity cancel - the shorter-chain esters are poorer substrates but achieve higher concentrations in the droplets because of their more favourable partition coefficients. We have previously shown that the observed k_2 values for these substrates are entirely consistent with the assumption that the enzyme kinetic properties in the microemulsion are unchanged from those in bulk water, and that the effective substrate concentration at the enzyme is wholly determined by the measured substrate partition coefficient between bulk heptane and water phases (ref. 6).

For longer-chain substrates (C_{10}-C_{18}) the partition coefficients are so unfavourable that the concentrations free in the aqueous interior of the droplets is negligible; nevertheless the observed k_2 values are comparable to those for shorter chain esters. In these cases the enzyme is acting predominantly on substrate molecules bound at the interface. The presence of such molecules can be demonstrated directly by ultracentrifugation, and their accessibility to hydrolysis from within the droplet can be shown by monitoring glycylglycine-catalysed aminolysis in microemulsions containing such substrates.

The significance of interfacial binding of substrates can also be seen in studies of lipase-catalysed ester synthesis, using systems in which both overall water content and water activity are low. Microemulsions containing 5% (v/v) ethanol overall were used to study esterification of simple carboxylic acids (0.1 mol.dm^{-3} overall). Microemulsions stabilised by the cationic surfactant CTAB (0.2 mol.dm^{-3}, R = 10) were used for ease of product recovery by ether extraction. For acids of chain length C_6-C_{16}, recoveries of ethyl esters were in the range 50-70%; lower values were observed for esters of shorter chain length acids.

The system was also used to study esterification and lactonization with acids and alcohols where both parties are expected to be concentrated at the droplet interface. Octanol and octanoic acid in this system were esterified in 71% yield, and there was also a good yield (43%) of ester formed between hexanol and the carboxylic acid group of 16-hydroxy-hexadecanoic acid. However this long-chain hydroxy acid did not lactonize in detectable yield in these conditions, nor was any condensation observed between hexanoic acid and the hydroxy group of the hydroxy acid. The evidence suggests that only the carboxylic acid group of the C_{16} hydroxy-acid is accessible to the lipase in microemulsions; presumably the substrate is located in the surfactant layer oriented such that the -COOH group is hydrated, but that the $-CH_2OH$ is remote from the aqueous droplet.

We have previously demonstrated lipase-catalysed glyceride synthesis in microemulsions containing oleic acid and a dispersed phase consisting of water + glycerol (ref. 7). Table 1 gives product analyses at equilibrium in such a system before and after addition of water to change the water content of the polar phase from 10% to 50% (v/v); this emphasises that the enzyme is stable in this system and that the equilibrium product composition is determined by the water activity. In summary, all these studies indicate that *C.viscosum* lipase is stable in w/o microemulsions and is a versatile catalyst of hydrolyses and condensations in this environment through its ability to act on substrates which partition into the polar droplets and to the interface.

TABLE 1

Glyceride compositions at equilibrium in lipase-containing microemulsions, as a function of (H_2O)

Original microemulsion composition: (AOT) = 0.42 mol.dm^{-3}, (glycerol) = 0.82 mol.dm^{-3} , (oleic acid) = 0.32 mol.dm^{-3}, (H_2O) = 0.56 mol.dm^{-3}, (lipase) = 0.09 g.dm^{-3}. Incubation temperature = 40°C; equilibrium attained typically with $t_{1/2}$ = 5 hr.

	(H_2O)= 0.56 mol.dm^{-3}	(H_2O) = 2.8 mol.dm^{-3}
Oleic acid	5%	50%
Monoolein	32%	20%
Diolein	60%	29%
Triolein	2%	< 1%

STUDIES ON BILIRUBIN OXIDASE

In addition to their suitability for carrying out 'enzyme-reversal', microemulsions are useful for other bio-transformations of non-polar and hence water-insoluble substrates, where the prime difficulty is to co-solubilise substrate and catalyst. Previous work on microemulsions has concentrated on hydrolases. To test the more general usefulness of microemulsions with other classes of enzymes and in other applications, we have studied bilirubin oxidase in microemulsions. This enzyme is difficult to study in aqueous systems, since bilirubin is highly insoluble. However it dissolved readily in the oil phase (50:50 chloroform:heptane) of water-in-oil microemulsions stabilised by CTAB. Bilirubin oxidase dissolved in the water droplets, and oxidised the substrate to biliverdin, giving a readily measurable absorbance change at 460nm, with an isobestic at ∿380 nm. Reactions were linear for several minutes and rates were proportional to both enzyme and substrate concentrations. The second-order rate constant increased as R decreased (Figure 2) implying that the enzyme acted on interfacially-bound substrate rather than that dissolved in the aqueous domain.

GENERAL CONCLUSIONS

Microemulsions are versatile and permit exploitation of a range of enzyme-catalysed processes. Enzymes are stable and active in microemulsions, and retain essentially bulk aqueous kinetic properties, consistent with the location of enzymes within the droplets. Aqueous solubility of substrates is not a pre-requisite for conversion; in some cases interfacially-bound molecules are effective substrates.

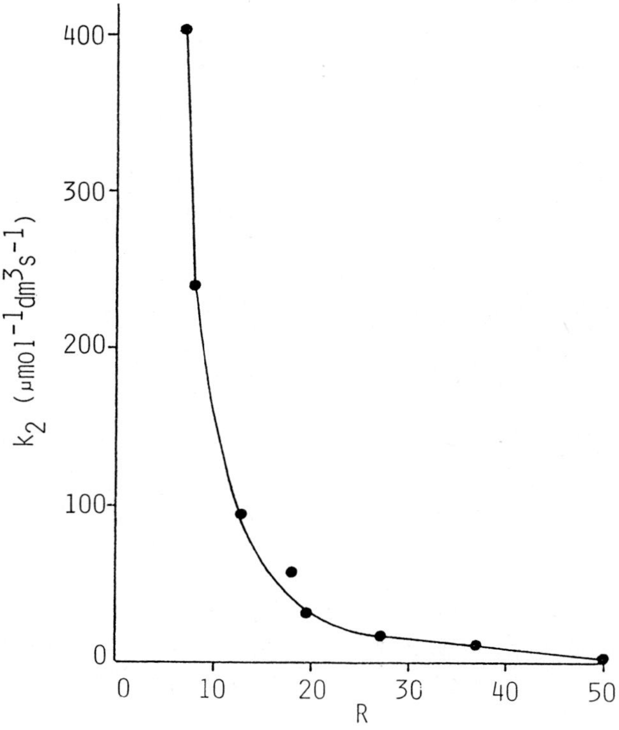

Fig. 2. The variation with R of the second-order rate constant for bilirubin oxidase activity in CTAB microemulsions.

REFERENCES

1 P. L. Luisi, Angew. Chem., 24 (1985) 439-450
2 P. L. Luisi and C. Laane, Trends Biotech, 4 (1986) 153-161
3 P. D. I. Fletcher, R. B. Freedman, J. Mead and B. H. Robinson, Colloids and Surfaces, 10 (1984) 193-203
4 P. D. I. Fletcher, G. D. Rees, B. H. Robinson and R. B. Freedman, Biochim. Biophys. Acta., 832 (1985) 204-214
5 P. D. I. Fletcher, R. B. Freedman and B. H. Robinson, submitted for publication
6 P. D. I. Fletcher, R. B. Freedman, C. Oldfield and B. H. Robinson, J. Chem. Soc., (Faraday), 81 (1985) 2667-2679
7 P. D. I. Fletcher, G. D. Rees, R. Schomacher, R. B. Freedman and B. H. Robinson, Biochim, Biophys. Acta., (1987) in the press.

C. Laane, J. Tramper and M.D. Lilly (Editors), *Biocatalysis in Organic Media*, Proceedings of an International Symposium held at Wageningen, The Netherlands, 7–10 December 1986.

WATER ACTIVITY IN BIPHASIC REACTION SYSTEMS

P.J. HALLING

Department of Bioscience and Biotechnology, University of Strathclyde, 204 George Street, Glasgow G1 1XW, UK.

SUMMARY

Mass action effects of water can contribute to shifts in the equilibria catalysed by hydrolytic enzymes acting in biphasic reaction mixtures. A convenient treatment uses the thermodynamic activity of water (a_w), which may be measured via relative humidities. To improve synthetic yields a_w must be reduced significantly below 1, and this is not necessarily achieved simply by low water contents.

Enzyme activity at measured low a_w in biphasic systems has been demonstrated for lipase catalysed esterification and interesterification, and peptide synthesis with proteases. A number of other reaction mixtures reported in the literature probably have a_w significantly less than 1.

INTRODUCTION

There is substantial industrial interest in the use of readily available hydrolytic enzymes to catalyse synthetic reactions. For example, lipases can catalyse transesterification or ester synthesis, while proteases can form peptide bonds. With hydrolases in dilute aqueous solution the hydrolysis reactions are usually dominant, but the Gibbs energy changes are not large, and practicable changes in conditions can shift the equilibrium position sufficiently to give useful yields of synthetic products.

One way of altering the equilibrium position is to introduce a second liquid phase into the reaction mixture, containing non-polar water-immiscible organic solvents and/or reactants. This approach was first demonstrated by Klibanov et al. (ref. 1), who synthesised amino acid esters with chymotrypsin in a water-chloroform biphasic system. The equilibrium is shifted in favour of the synthetic product because it is relatively better solvated in the organic phase than the starting materials. The change in equilibrium position may be analysed in terms of partition coefficients, as done in detail by Martinek et al. (ref. 2.)

Water is a product of the reverse hydrolysis reaction, and is often present only at very low levels in biphasic systems. As a result it is tempting to think that water mass action also helps to shift the equilibrium towards synthesis. However, low water contents are not a sufficient condition for water mass action to give an increased yield of synthetic product. Even in systems containing less than 1% of water, its mass action effect can often be negligibly changed from that in dilute aqueous solution. It is, however, possible to construct biphasic reaction mixtures in which water mass action does make a valuable contribution to increased yield; and enzymes can remain active under these conditions. A convenient treatment of water mass action in these systems uses the thermodynamic water activity (a_w), rather than concentrations (ref. 3).

Use of water activity

The use of a_w offers 3 advantages compared with the more conventional concentrations:-

1. By definition, it is a_w that directly determines the effects of water on the equilibrium position.
2. Unusually, in biphasic systems, it is often easier to measure a_w rather than water concentrations; a_w may be determined via the relative humidity or water vapour pressure of an equilibrated gas phase, while low water concentrations in organic liquids can be difficult to measure in our water-rich world.
3. At equilibrium a_w is by definition equal in all phases, while concentrations of water may be widely different.

Rather than an absolute activity, it is easiest to use a relative activity conventionally compared with pure water at the same temperature, which has a_w of 1.

For a reaction:

$$A + H_2O \rightleftharpoons B + C$$

we can write :

$$K\ (\text{organic phase}) = \frac{[B]_o\ [C]_o}{a_w\ [A]_o} \qquad (1)$$

using concentrations of A, B and C in the organic phase.

For the aqueous phase we have a similar equation:

$$K_c = \frac{[B]_a\ [C]_a}{a_w\ [A]_a}$$

When this phase is a dilute solution a_w will be close to 1, so it may be seen that K_c is the conventional equilibrium constant for aqueous solution. Furthermore, under these conditions K_c and K (organic phase) will be related by a set of concentration ratios between the phases for A, B and C. These will essentially equal the partition coefficients, so:

$$K\ (\text{organic phase}) = \frac{P_a\ P_b}{P_c} \cdot K_c$$

Equation 1 shows that water mass action can only contribute to an increase in yield of A if a_w is reduced significantly below 1. The conditions required to achieve this reduction in a_w are more stringent than might appear. It is not sufficient for the water concentration in the organic phase to be low, but rather it must be reduced below saturation. (Any water-saturated organic liquid has an a_w of 1 even if the water concentration is 100 ppm or less). Equally a low volume fraction of the aqueous phase does not necessarily mean a low a_w. In a biphasic system the ratio of phase volumes may be varied at will without affecting a_w, provided the composition of each phase is unchanged. (Similarly, varying the quantity of one phase, while maintaining constant composition, will not affect the properties and composition of the other). As a consequence of the above two effects, it can be seen that the total water content of the system can be very low without significantly reducing a_w below 1.

As noted at the start of this section, at equilibrium a_w is equal in both phases, and it is often easier to predict its value by considering the nature of the aqueous phase. If a_w is to be significantly less than 1, then the aqueous phase must become a very concentrated solution or dispersion of hydrophilic species; for example, saturated (5M) NaCl and 75% glucose syrup both have an a_w of about 0.8, while a mixture of equal weights of protein and water will give a value of about 0.9.

Enzyme activity at measured low a_w in biphasic systems

Despite the stringent conditions required, it is possible to construct biphasic enzyme reaction mixtures in which a_w is well below 1; furthermore, enzymes can remain active in them and an increase in synthetic yield results. Halling & Macrae (ref. 4) describe a reaction mixture for lipase-catalysed fat interesterification that contained less than 0.3% water, but which had an a_w of 0.84 to 0.89. If, however, more water was removed from the system by recirculating and drying the reactor headspace gases then a_w could be reduced progressively to 0.07, while enzyme activity continued, at least at first. Table 1 shows the effect of this reduction in a_w on the position of the lipase-catalysed hydrolysis equilibrium: an increased yield of triacylglycerol, and a major reduction in levels of the hydrolytic by-product diacylglycerol, which is particularly undesirable in many fats. The same patent reported an enzymic esterification in which a_w was reduced by addition of dry silica gel to the reaction mixture (Table 2), giving an increased yield of oleyl ricinoleate. Macrae (ref. 5) describes measurements of a_w at various positions within a packed bed reactor for enzymic interesterification. Most of the bed was at an a_w of about 0.5, and the lipase catalyst was as active as in the initial section of the bed where a_w was higher. (Water consumption in the hydrolytic side-reaction leads to a fall in a_w).

TABLE 1

Effect of water removal by headspace gas recirculation during enzymic interestification

		Water Removal	Control
PRODUCT FAT	Triacylglycerol	68.2	51.7
COMPOSITION	Diacylglycerol	1.2	8.7
(%)	Free fatty acid	30.6	40.0
FINAL a_w		0.07	0.84-0.89

REACTION MIXTURE : 288 ml PETROLEUM ETHER, 120 g FAT, 12.5 g CELITE-LIPASE, 1 ml WATER.

TABLE 2

Effect of adding dry silica gel on enzymic esterification

	Silica Added	Control
Yield of oleyl ricinoleate (%)	76.2	41.2
Water activity	0.6	0.95

REACTION MIXTURE : 5 ml SOLUTION OF ACID AND ALCOHOL IN HEXANE, 250 mg CELITE-LIPASE, 25 μl WATER.

We have demonstrated similar effects during peptide synthesis catalysed by the protease thermolysin (Cassells & Halling, unpublished). Our standard reaction mixture consists of 21 ml of a solution of 10 mM of each of CBZ-phenylalanine and phenylalanine methyl ester in ethyl acetate. Samples of the organic phase are analysed periodically for dipeptide formed and for residual starting materials. The equilibrium position can be summarised using the ratio of the concentration of dipeptide to the product of those of the two protected amino acids. The catalyst consists of 6 mg of thermolysin, dried onto the surface of 6 g of silica gel particles. The reaction is carried out in a sealed reactor with a gas headspace (Fig. 1): the headspace can be recirculated via a drying column to remove water from the reaction system. A humidity sensor (Endress & Hauser DY40E) in the headspace can be used to measure a_w when the external pump is off and the gas space equilibrates with the reaction mixture.

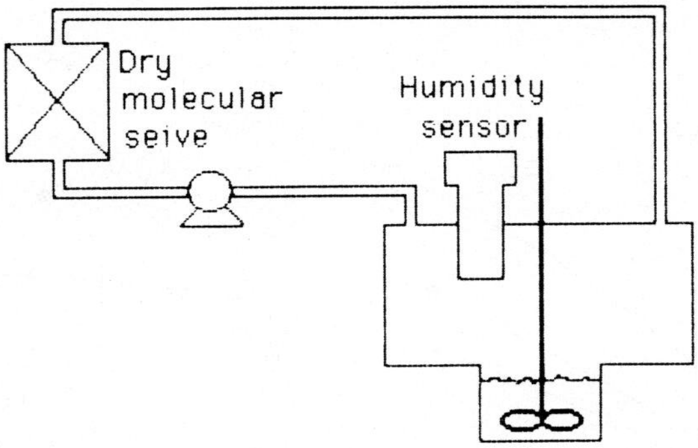

Fig. 1. Scheme of apparatus used for reactions at reduced water activity.

In a control reaction a total of 1.8 g of water is present, but this is adsorbed on the silica and dissolved in the organic solvent, so that there is no visible water phase. Nevertheless, the humidity sensor indicates that a_w is close to 1. A series of 3 reactions gave equilibrium concentration ratios of 1.9, 2.9 and 2.5 mM^{-1}. As expected for a_w close to 1, the equilibrium position is similar to that in a biphasic system where a dilute aqueous solution makes up 50% of the volume.

After equilibrium has been reached, if water is removed by headspace gas circulation, then a further reaction occurs giving an increase in dipeptide concentration and a fall in starting materials. In reactions where a_w was reduced to 0.3 and 0.7, the concentration ratios rose to 7.1 and 3.3 mM^{-1} respectively.

A reduced a_w can also be obtained by adding less water to the system initially (the ethyl acetate was dried and the water used to pre-hydrate the catalyst). Fig. 2 shows that the enzyme is still active (though with reduced rates) at a_w at least down to 0.5. As expected the equilibrium positions are shifted in favour of dipeptide (Table 3).

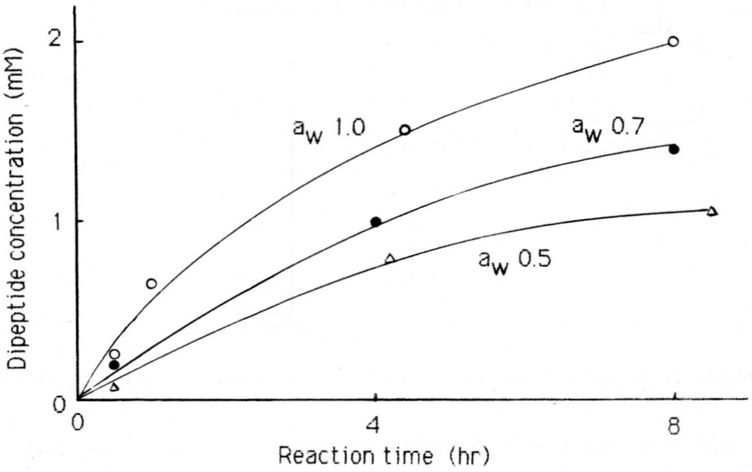

Fig. 2. Effect of water activity on rates of peptide synthesis. Varying amounts of water with 21 ml organic phase and 6 g silica-thermolysin.

TABLE 3

Effect of water on equilibrium of peptide synthesis.

Varying amounts of water were added to 21 ml organic phase and 6 g silica-thermolysin.

Amount of Water (g)	Water Activity	Equilibrium Concentration Ratio (mM^{-1})
1.8	1.0	1.9
0.9	0.7	2.7
0.6	0.5	5.0

Comments on other literature systems

Even though measurements have not been reported, it is possible to estimate whether a_w values are significantly below 1 in most other biphasic reaction mixtures described in the literature. Many systems have been prepared by combining a dilute aqueous solution with a water-saturated organic phase;

both of these have a_w close to 1, so the resulting system will have a similar value, even though the organic phase volume fraction may be 99% or more. If the organic phase is initially dry, it could take up water from the aqueous phase, concentrating this and reducing a_w, but this is not usual with non-polar organic liquids that dissolve little water. Some reduction in a_w may occur when polar species such as EtOH are included in the initial organic phase, and these subsequently partition to reach significant concentrations in the aqueous phase.

A number of the biphasic reaction systems described in the literature undoubtedly have a_w substantially below 1. Some examples of those in which continued enzyme activity has been demonstrated will be mentioned here. Patterson et al. (ref. 6) described ester synthesis using a fungal cell bound lipase in an organic solution of the acid and alcohol; the catalyst was freeze dried mycelium, the organic phase was dried with molecular seive before use, and activity was observed even if no water was added back. Gatfield (ref. 7) reported that a freeze dried lipase powder suspended in a liquid mixture of dry acid and alcohol catalysed ester synthesis. In a study of enzymic transesterification in an organic solution of tributyrin and heptanol, Zaks & Klibanov (ref. 8) found that the acetone dried lipase powders used had water contents between 0.5% and 6.1%; the reaction took place even if no other water was added.

If low a_w reaction mixtures are to be widely exploited for increasing the yield of syntheses catalysed by hydrolytic enzymes, then clearly it is essential that they retain activity under these conditions. The retention of activity may often not be greatly influenced by the organic liquid, but rather depend on the hydrphilic species in the immediate environment of the enzyme. Thus, it may be relevant that in the absence of an organic phase, esterase activity has been demonstrated in food products at measured a_w as low as 0.1. From a theoretical viewpoint, though, a few water molecules are probably essential for the activity of an enzyme; they are tightly bound to the protein and will remain in place down to low a_w. Much of the water probably plays a general solvent/environmental role, and may be replaced by other hydrophilic species such as polyols, sugars or other proteins. Thus, the retention of activity in low water biphasic systems probably depends on the nature of the solutes or dispersants in the aqueous phase, rather than a_w per se.

Acknowledgements

I am grateful to the SERC for financial support.

REFERENCES

1 A.M. Klibanov, G.P. Samokhin, K. Martinek and I.V. Berezin, A new approach to preparative enzymatic synthesis, Biotech. Bioeng. 19 (1977), 1351-1361.

2 K. Martinek, A.N. Semenov and I.V. Berezin, Enzymatic synthesis in biphasic aqueous organic systems. I. Chemical equilibrium shift, Biochim. Biophys. Acta, 658 (1981), 76-89.

3 P.J. Halling, Effects of water on equilibria catalysed by hydrolytic enzymes in biphasic reaction systems, Enzyme Microb. Tech., 6 (1984), 513-516.

4 P.J. Halling and A.R. Macrae, Fat processing, Eur. Pat. Appl. 64 855 (1982).

5 A.R. Macrae, Interesterification of fats and oils, in: J. Tramper, H.C. van der Plas and P. Linko (Eds.), Biocatalysts in Organic Synthesis, Elsevier, Amsterdam (1985), 195-208.

6 J.D.E. Patterson, J.A. Blain, C.E.L. Shaw, R. Todd and G. Bell, Synthesis of glycerides and esters by fungal cell-bound enzymes in continuous reactor systems, Biotech. Lett., 5 (1979), 211-216.

7 I.L. Gatfield, The enzymatic synthesis of esters in nonaqueous systems, Ann. NY Acad. Sci., 434 (1984), 569-572.

8 A. Zaks and A.M. Klibanov, Enzyme catalysed processes in organic solvents, Proc. Nat. Acad. Sci. US, 82 (1985), 3192-3196.

C. Laane, J. Tramper and M.D. Lilly (Editors), *Biocatalysis in Organic Media*, Proceedings of an International Symposium held at Wageningen, The Netherlands, 7–10 December 1986.

DESIGN OF AN ORGANIC-LIQUID-PHASE/IMMOBILIZED-CELL REACTOR FOR THE MICROBIAL EPOXIDATION OF PROPENE

L.E.S. BRINK and J. TRAMPER

Department of Food Science, Food and Bioengineering Group, Agricultural University Wageningen, De Dreyen 12, 6703 BC Wageningen (The Netherlands)

SUMMARY

The production of stereospecific propene oxide by non-growing *Mycobacterium* cells entrapped in calcium alginate has been used as a model to study the application of water-immiscible organic solvents in biotechnological processes. The far-reaching consequences of the solvent choice are discussed. High activity retentions of the immobilized cells relate to low polarities and high molecular weights of the used solvents. Simple models for the intrinsic epoxidation kinetics and for the effects of external and internal-diffusion limitations are integrated in a macrokinetic model to describe the behaviour of a packed-bed immobilized-cell reactor. Depletion of the limiting substrate, oxygen, along the length of the bioreactor can be prevented by using an organic solvent, n-hexadecane, as the transport medium. Model predictions of the oxygen conversion in the bioreactor at various degrees of external and internal-diffusion limitation, at various liquid space times and with water or n-hexadecane as the continuous phase are in good agreement with experimentally obtained values.

INTRODUCTION

A promising, current trend in the field of biocatalysis is the replacement of part of the aqueous reaction medium by a water-immiscible organic solvent (refs. 1-4). There may be many reasons to justify such a dramatic change of the reaction environment. High concentrations of poorly water-soluble substrates and/or products are possible in organic solvents. Reaction equilibria may be shifted favourably, and substrate and/or product hydrolysis can be largely prevented. Furthermore, substrate or product inhibition may be reduced as a consequence of a lower inhibitor concentration in the aqueous environment of the biocatalyst. Finally, recovery of product and biocatalyst may be facilitated. An obvious, major drawback of introducing an organic solvent is denaturation of the enzymes responsible for the desired bioconversion. Immobilization of the biocatalyst may then become interesting in view of a possible, protective effect by the support material.

Though many examples have already been presented of the use of an organic phase in biocatalytic conversions (e.g. refs. 5-9) design rules for bioreactors operating with an organic solvent as a main constituent of the reaction medium are still scarce. The introduction of an additional liquid phase could in theory complicate the design of these reactors. An elementary classification of two-liquid-phase reaction systems and two-liquid-phase bioreactors has recently been given by Lilly and Woodley (ref. 4). The classification is based on the form of the theoretical concentration profile and on the type of aqueous phase configuration.

In this paper some fundamental aspects of organic-liquid-phase biocatalytic conversions are discussed using the production of chiral propene oxide from propene and oxygen by non-growing *Mycobacterium* cells entrapped in calcium alginate as a model (refs. 10-11). Several of the above-mentioned advantages of the use of an organic solvent play a role in this model system: high solubility of oxygen and, especially, propene in a suitable solvent, decreased product inhibition by *in situ* extraction of the toxic propene oxide into the organic phase and, finally, straightforward recovery of the formed epoxide. Furthermore, quantitative gas analysis (ref. 12) and the prediction of thermodynamic and physical properties are facilitated as a result of the relatively simple molecules involved in the epoxidation (propene, oxygen and propene oxide).

Various features of the multiphase system have been investigated by measuring consumption rates of oxygen and propene and accumulation rates of epoxide under various reaction conditions in a recirculation batch reactor (bubble column or packed bed) with continuous feed of the two substrates. First, factors influencing the choice of an organic solvent have been studied (ref. 9). The effects of external and internal diffusion on the intrinsic epoxidation kinetics of the immobilized cells are theoretically predicted and, subsequently, experimentally validated (refs. 13-14). The reactor design equations of the organic-liquid-phase/immobilized-cell packed-bed reactor are formulated by combining the derived micro-kinetic model with a model for the hydrodynamics of the packed bed (ref. 15). The predictions of this reactor model compared favourably with experimentally obtained oxygen conversions.

OPTIMIZATION OF THE ORGANIC SOLVENT

The type of organic solvent will have significant effects on the reaction kinetics and stability of the biocatalyst, the chemical and mechanical stability of the support material (calcium alginate) and the partition of propene, oxygen and epoxide between the different phases. A large number (34) of water-immiscible solvents were considered for use in the organic-liquid-phase/immobilized-cell system (ref. 9). Many solvents cause rapid inactivation of the free, propene-epoxidizing cells. This appears also to be the case if the cells are immobilized in calcium alginate. However, the support material prevents direct cell-organic solvent contact and the associated aggregation and clotting of cells, mostly accompanied with loss of activity. A stable structure of the calcium alginate gel was found in all the organic solvents tested, provided that these solvents were saturated with water. Examples of solvents which deactivate the immobilized cells very rapidly are: n-hexane, toluene, butyl acetate and chloroform. Only a few solvents did not result in a serious loss of immobilized--cell activity: n-hexadecane, di-iso-pentyl ether, di-phenyl ether, the larger di-n-alkyl phthalates and two perfluoro chemicals. To avoid time-consuming and expensive testing of large numbers of solvents, a relation was sought between the measured activity retentions of immobilized cells and readily available solvent properties. The combined use of two characteristics of organic solvents made it possible to explain the experimentally found activities (Figure 1). The polarity, as expressed here by the Hildebrand solubility parameter δ, has already been mentioned in a qualitative manner (refs. 1, 16). The second characteristic, dimension of the solvent molecule (as expressed by the molecular weight or the molar volume), may relate to the degree of hindrance a solvent molecule could experience on its way from the organic bulk phase to the catalyzing enzymes in the immobilized cells. Most of the low activities can be found in an area of high polarity ($\delta > 8$) and small molecular weight ($M < 200$). Outside this area the higher activity retentions are found. The experiments with the larger di-n-alkyl phthalates and perfluoro chemicals provide good examples of this finding.

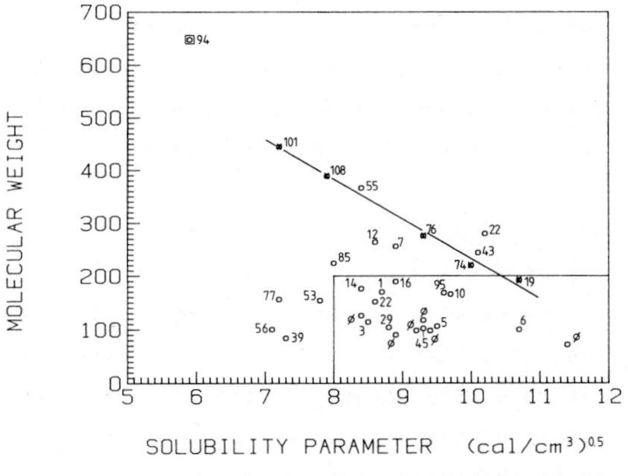

Fig. 1. Measured activity retentions (%) of immobilized cells exposed to organic solvents as function of molecular weight and solubility parameter of solvents; line: di-n-alkyl phthalates.

Recently, Laane, Boeren and Vos (ref. 17) and Laane et al. (ref. 18) suggested that the logarithm of the partition coefficient of a given compound in the standard octanol/water two-phase system is a more meaningful parameter for describing the solvent polarity. It was concluded that organic solvents having a log P < 2 (high polarity) are in most cases not suited for use in two-liquid--phase bioreactions, whereas solvents having a log P > 4 (low polarity) are generally suitable for any biocatalytic system.

The solubilities of oxygen and propene oxide in organic solvents could also be related to the solubility parameters of these solvents. The capacity for propene is less well described by the Hildebrand solubility parameter, but is also less relevant, as the capacity of the solvents for propene is always about two orders of magnitude higher than that of water, and thus limitation of the rate by an insufficient supply of propene is less likely to occur. A low solvent polarity (low δ) corresponds with a high solubility of oxygen but also with a low extraction capacity for the epoxide (ref. 9). Therefore, optimization of the solvent polarity is essential, taking into account the polarities of substrates and product and the influence of solvent polarity on the activity retention of the immobilized cells. The final choice of the solvent must also depend on solvent properties which will affect the epoxide recovery from the organic phase (boiling point, water capacity) and on some general properties like viscosity, toxicity, flammability and price.

DESIGN OF AN ORGANIC-LIQUID-PHASE/IMMOBILIZED-CELL REACTOR

The apparent kinetic activity of the gel-entrapped cells may be limited due to diffusion of substrates and product to, in and from the biocatalyst particles. Modelling of the biocatalyst performance must therefore be based on a combination of intrinsic epoxidation kinetics and mass transfer rates. Ultimately, this biocatalyst model (the microkinetics) should be conjoined with a reactor flow model (the macrokinetics) to take into account concentration profiles at the scale of the bioreactor.

Intrinsic epoxidation kinetics

The used *Mycobacterium* (strain E3) possesses a mono-oxygenase, which can produce epoxides from gaseous alkenes and oxygen (ref. 10). It was found experimentally, that about one mole of oxygen is needed for the epoxidation of one mole of propene, and that the reaction is nearly irreversible. As a result of the high solubility of propene in organic solvents, as compared to that of oxygen, oxygen-limited reaction conditions will be emphasized below.

The following kinetic model was found to give a satisfactory description of observed propene-epoxidation rates (v_1), oxygen--consumption rates (v_2) and propene oxide-production rates (v_P) (ref. 14):

$$v_1 = -\frac{dS_1}{dt} = \frac{V\,S_2^2}{K_A\,(1 + P/K_P') + S_2^2\,(1 + P/K_P'')} \qquad (1)$$

$$v_2 = -\frac{dS_2}{dt} = -\frac{dS_1}{dt} + \frac{\alpha\,V\,S_2}{K_R + S_2} \qquad (2)$$

$$v_P = \frac{dP}{dt} = -\frac{dS_1}{dt} - v_C\,(P,\,S_2) \qquad (3)$$

with S_1 and S_2 the propene and oxygen concentration, respectively, P the propene oxide concentration, V the maximum propene-epoxidation rate, α and K_A parameters, K_R the Michaelis-Menten constant of the cell respiration, K_P' and K_P'' the epoxide inhibition parameters and v_C the epoxide-consumption rate.

Parameter estimates were obtained by fitting (non-linear regression) this kinetic model to different types of experimental time-course data ($S_1(t)$, $S_2(t)$, $P(t)$) (ref. 14). The experimental reaction conditions were adjusted carefully to eliminate influences of diffusion

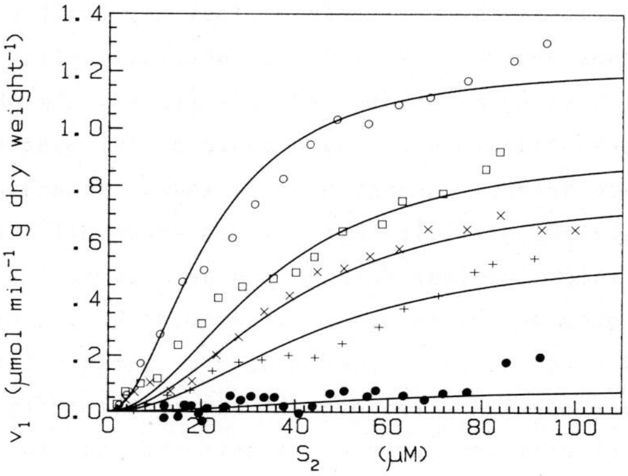

Fig. 2. Propene consumption rate, v_1, versus the oxygen concentration, S_2, at various propene oxide levels; P (mM): 0 (o), 3.0 (□), 5.3 (x), 10.6 (+), 116 (●); model fit: equation (1) with $K_A = 4.6 \times 10^2$ µM^2, $K_P' = 1.3$ mM, $K_P'' = 9.6$ mM and $V = 1.2$ µmol min^{-1} (g dry weight)$^{-1}$.

limitation, cell deactivation and, sometimes, product inhibition.
An example of the obtained model fit provided by equation 1 is given in Figure 2. The predictive capacity of the kinetic model (excess propene, equations 1-3) was tested by comparing independently determined propene, oxygen and epoxide time-course data to the corresponding calculated values (Runge-Kutta integration). See Figure 3. A good agreement is found between experimental concentrations and model predictions.

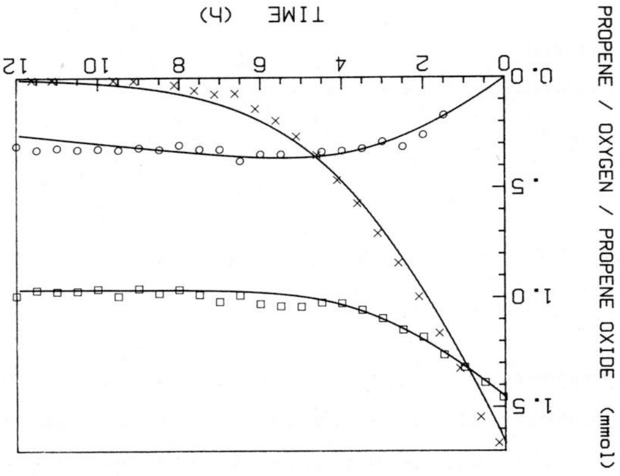

Fig. 3. Amount of unconverted propene (□) and oxygen (x) and accumulated propene oxide (o) in the experimental system versus time (1 mmol propene corresponds to S_1=192 µM, 1 mmol oxygen corresponds to S_2=48.1 µM, 1 mmol propene oxide corresponds to P=4.0 mM); kinetic model predictions: equations (1-3) with V=2.5 µmol min^{-1} (g dry weight)$^{-1}$, K_A=5.1 x 10^2 µM^2, K_P'=1.3 mM, K_P''=9.6 mM, α=1.9, K_R=27 µM and v_c=0.3 µmol min^{-1} (g dry weight)$^{-1}$.

Mass-transfer effects

Several transport steps in the multiphase system may limit the above described, intrinsic epoxidation rate of the immobilized cells. First, the gaseous substrates have to be transfered to a liquid bulk phase. In Figure 4 it is assumed that this is the organic phase. Subsequently, the substrate molecules must diffuse through organic and/or aqueous stagnant films around the biocatalyst particle. Finally, to reach all the entrapped cells, the substrates should diffuse through the gel matrix. In addition, though not shown in Figure 4, the formed propene oxide has to diffuse out of the biocatalyst and into the bulk liquid. Some of these external and internal diffusion steps may limit the overall epoxidation rate, thereby lowering the biocatalyst efficiency. In our system the most serious limitation was found to be the intraparticle diffusion of the two substrates in the calcium alginate gel (ref. 13). This will be quantified below in case of a solvent and epoxide-free, ideally-mixed reaction system (bubble column). The effect of the organic solvent on the oxygen transfer in a packed-bed bioreactor will be dealt with in the following section. The combined influences of diffusion limitation and product inhibition will not be dealt with here, but have been treated elsewhere (ref. 14).

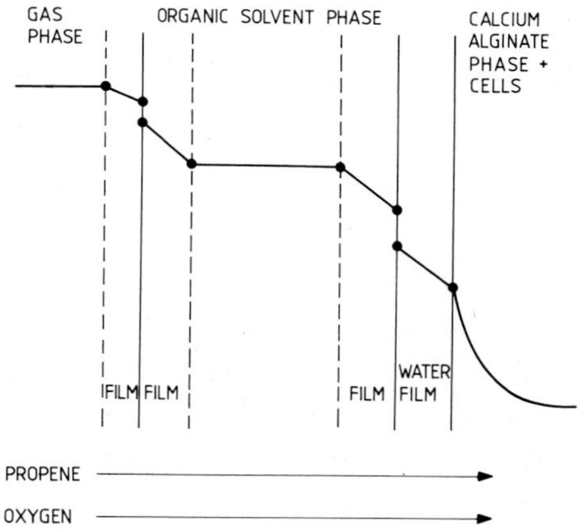

Fig. 4. Mass-transfer steps in the multiphase system.

The theoretical analysis of coupled reaction and internal pore diffusion in permeable biocatalysts, if required in series with external film diffusion, is well-established. Reviews of this area have recently been given by Kasche (ref. 19), Radovich (ref. 20) and Karel, Libicki and Robertson (ref. 21). Theoretical results are often presented as relationships between the internal effectiveness factor of the immobilized enzymes or cells and a Thiele-type modulus, which is determined by system characteristics as the particle diameter (d_P), effective diffusion coefficient in the particle (D_E), reaction kinetics (usually Michaelis-Menten: V_M and K_M) and, sometimes, the bulk substrate concentration. The effects of external diffusion steps are generally described by an external effectiveness factor, which is often related to the Sherwood number. The overall effectiveness factor, η, of the biocatalyst (the ratio of observed reaction rate to the diffusion-free rate) is defined as the product of external and internal effectiveness factors, and can be derived from the solution of the following, steady-state differential equation, describing the coupled one-substrate diffusion and reaction in a porous, spherical medium:

$$D_E \left(\frac{d^2S}{dr^2} + \frac{2}{r}\frac{dS}{dr} \right) = X \; v(S) \qquad (4)$$

with the boundary conditions:

$$D_E \, dS/dr = \beta (S^* - S_S) \quad \text{at } r = 0.5\, d_P \qquad (5)$$

$$dS/dr = 0 \quad \text{at } r = 0 \qquad (6)$$

and the overall (i.e.: gas/liquid and liquid/solid) external mass-transfer coefficient β:

$$\beta = (a_S(1/(k_L a) + 1/(k_S a_S)))^{-1} \qquad (7)$$

where r is the radial distance from the centre of the particle, X the cell density, S_S and S^* the substrate concentrations at the gel surface and at the gas/liquid interface, respectively, a and a_S the gas/liquid and liquid/solid specific surface areas, respectively, and k_L and k_S the mass-transfer coefficients in the stagnant liquid films adjacent to the gas/liquid and liquid/solid interfaces, respectively.

The mass-transfer effects of propene and oxygen were investigated using the above described effectiveness factor concept, in combination with a pseudo-one-substrate model. The much more complex case of two limiting substrates will not be considered. The high solvent capacity for propene, as compared to that of oxygen, makes it likely that only oxygen will limit the reaction rate in an organic-liquid-phase/immobilized-cell reactor. Under pseudo-one-substrate conditions both the propene and oxygen-consumption rate of the calcium alginate-entrapped cells (v(S)) could be described by Michaelis-Menten kinetics (ref. 13).

An analytical solution of equation 4 (combined with Michaelis-Menten kinetics) is unavailable. However, an elegant, approximative approach to this boundary-value problem is the collocation method (refs. 22-24). This method can be used to calculate the effectiveness factors (internal, external, overall) of the biocatalyst bead by solving a set of algebraic equations, if values for V_M, K_M, X, d_P, D_E, β and S^* are available. The overall reaction rate, v_R, can then be calculated using the following relation:

$$v_R = \eta \, (V_M S^* / (K_M + S^*)) \qquad (8)$$

The effective diffusion coefficients of propene and oxygen in the calcium alginate beads were set equal to the corresponding diffusion coefficients in water at 30°C (1.5×10^{-9} and 2.4×10^{-9} m^2s^{-1}, respectively) in view of the high water content of the support material and the small, molecular dimensions of the two substrates (ref. 13).

This assumption is supported by three recent, experimental studies (refs. 25-27). The volumetric gas/liquid mass-transfer coefficients, k_La, were determined experimentally using a dynamic oxygen electrode method. The liquid/solid mass-transfer coefficients, k_S, were estimated using Sherwood relations (ref. 13).

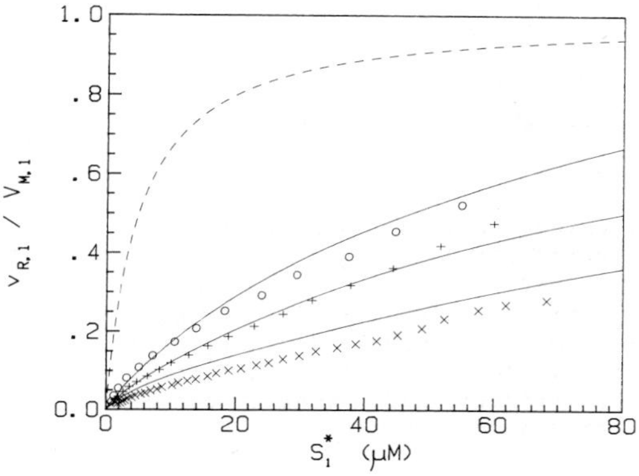

Fig. 5. Dimensionless propene consumption rate, $v_{R,1}/V_{M,1}$, versus the propene concentration at the gas/liquid interface, S_1^* ($V_{M,1}$=3.7 μmol min^{-1} (g dry weight)$^{-1}$, d_P=2.0 mm); experimental reaction rates, X(x10^4 g dry weight(m^3 particle)$^{-1}$): 3.2 (o), 6.5 (+), 13.0 (x); diffusion model predictions (——); negligible diffusion resestance (---).

The predicted consumption rates of propene and oxygen were correlated to observed rates obtained from two types of time-course data. At a constant, high oxygen concentration the decrease in propene level was measured at regular time intervals, whereas at a constant, high propene concentration the oxygen level in the experimental system was monitored. The pronounced effect of the cell load of the biocatalyst particles on experimental and predicted propene-consumption rates is shown in Figure 5. In Figure 6 the combined effects of particle diameter and cell load on observed and calculated oxygen-consumption rates are demonstrated. A good agreement is seen between experiment and theory at all levels of propene and oxygen. The absolute reaction rates as well as the effects of variations in cell load and/or particle diameter are satisfactorily predicted. The several assumptions made in the diffusion model (like the effective diffusion coefficients, the collocation method and the pseudo-one-substrate conditions) do not seem to give rise to unacceptably large errors. The predicted, external effectiveness factors are usually higher than 0.85 ($S^* > 20$ μM). Therefore, the sometimes low overall effectivities (observed and calculated) are mainly caused by severe pore-diffusion limitation.

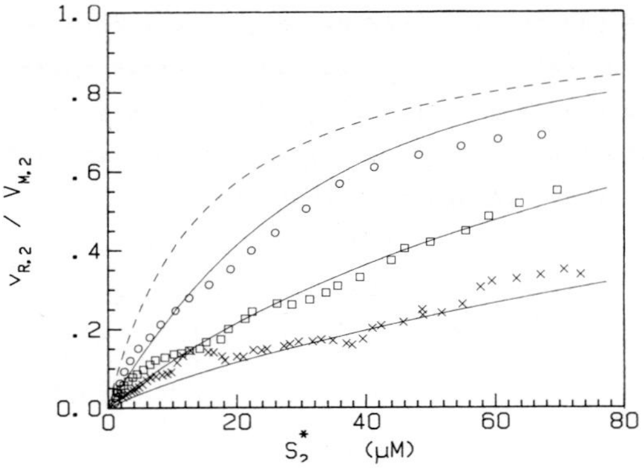

Fig. 6. Dimensionless oxygen consumption rate, $v_{R,2}/V_{M,2}$, versus the oxygen concentration at the gas/liquid interface, S_2^* ($V_{M,2}$=12.6 μmol min^{-1} (g dry weight)$^{-1}$); experimental reaction rates, (o) X=1.2 x10^4 g dry weight (m^3 particle)$^{-1}$, d_P=0.8 mm, (□) X=1.2 x10^4 g dry weight (m^3 particle)$^{-1}$, d_P=2.3 mm, (x) X=3.6 x10^4 g dry weight (m^3 particle)$^{-1}$, d_P=2.5 mm; diffusion model predictions (——); negligible diffusion resistance (---).

Reactortype and integral bioreactor model

A large interfacial area between alginate gel and solvent is one of the main requirements for high rates of substrate transport from the organic bulk to the immobilized cells. This will be easier achieved in a liquid/ solid, fixed-bed reactor than in a slurry reactor, as a well-defined suspension of the hydrophilic particles in the hydrophobic organic phase cannot be obtained in the latter reactor type. In a packed bed, however, the solvent can be forced to flow through a stationary bed of biocatalyst particles, thereby continuously supplying the substrates and extracting the product (refs. 28-30). Other factors favouring the selection of a packed-bed bioreactor for the present system of investigation are: high reactor productivity, decreased influence of product inhibition, no critical pH or temperature control required, hardly any clogging problems and, finally, absence of gaseous products.

A serious drawback of employing a packed-bed reactor, rather than a gas/liquid/solid fixed bed (or trickle bed), could be depletion of the two gaseous substrates at the end of the column. However, the use of an organic solvent with a high capacity for gases as substrate reservoir and transport medium could eliminate the need for a separate gas phase in the bioreactor. This would simplify the design of the immobilized-cell reactor markedly.

A reactor model of the organic-liquid-phase/immobilized-cell packed bed should be based on a combination of microkinetics and macrokinetics. The in the previous sections described, microkinetic model of the immobilized biocatalyst can be used for this purpose, though the modelling of the external diffusion steps (see equations 5 and 7) must be adopted to the situation in the packed-bed reactor (e.g. Sherwood relation) (ref. 15). An obvious model for the hydrodynamics of a packed-bed bioreactor is plug flow (ref. 15). Assuming steady-state conditions, negligible product inhibition and only one limiting substrate (oxygen), the reactor design equation can then be written as:

$$U dS_B/dz + \varepsilon_a . \eta_c . \eta(S_B) . X . V_M . (S_B/K)/(K_{M,2}+S_B/K) = 0 \qquad (9)$$

with the boundary condition:

$$S_B = S_I \quad \text{at } z=0 \qquad (10)$$

and with U the superficial fluid velocity, S_B and S_I the oxygen concentrations in the bulk liquid and in the inlet of the reactor, respectively, z the axial distance from the bottom of the biocatalyst bed, ε_a the volume fraction of calcium alginate beads, η_c the liquid/solid contacting efficiency and K the partition coefficient of oxygen between the organic solvent and water. The biocatalyst bed was diluted with small (~ 1 mm) glass beads (ε_g ~ 0.6, ε_a ~ 0.05) to reduce axial dispersion and to force the organic solvent to flow around each individual gel particle. Hence, the liquid/solid contacting efficiency, η_c, can be assumed to approach unity. The biocatalyst effectiveness factor, η, can be predicted at any given bulk substrate concentration, S_B, using the above described microkinetic model. This makes it possible to integrate numerically the oxygen mass balance (equation 9) and, subsequently, to obtain the oxygen concentration profile in the bulk liquid, $S_B(z)$, and the oxygen conversion, ξ, for different values of the oxygen inlet concentration, S_I, and reactor space time, τ (= ε L/U).

The epoxidation experiments were carried out in an organic-liquid-phase/immobilized-cell tubular reactor with recycle of an organic solvent or, for comparison, recycle of an aqueous 0.05 M calcium chloride solution (ref. 15). N-hexadecane was used as the organic phase as this solvent satisfies many requirements: little loss of immobilized-cell activity (ref. 9), no effect on the epoxidation kinetics ($K_{M,2}$ ~ 15 µM), high normal boiling point (287°C), limited flammability, low toxicity and high capacity for propene. The oxygen capacity of this solvent is rather small (K ~ 5). Propene oxide levels were kept negligible by hydrolysis in a sulphuric--acid solution (removal from the gas phase), such that no inhibition occurred. The propene concentration was controlled at a high, constant level, thus pseudo-

-one-substrate conditions were again applicable. The oxygen level was allowed to drop with reaction time to examine the effects of various oxygen reactor-inlet concentrations, S_I, on the degree of oxygen conversion, ξ^{EXP}:

$$\xi^{EXP} = v^{EXP} \tau / (\varepsilon S_I) \tag{11}$$

The volumetric oxygen-consumption rate, v^{EXP}, was extracted from the measured time-couse data (ref. 15).

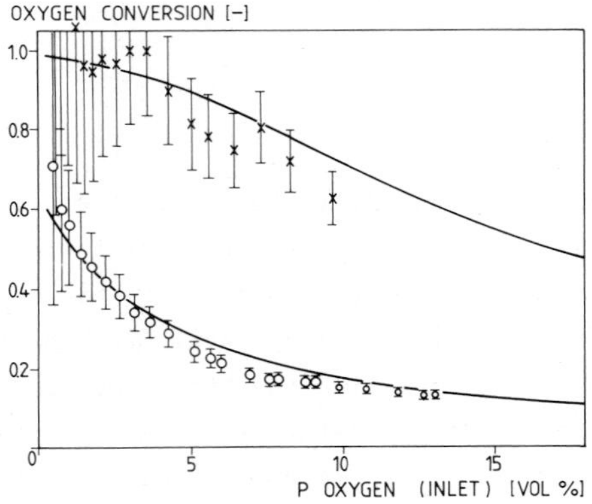

Fig. 7. Comparison between predicted and observed oxygen conversions at various oxygen partial pressures of the reactor-inlet stream; transport medium: water (x), n-hexadecane (o); vertical lines: estimated errors in experimental oxygen conversions.

Predictions of the performance of the packed-bed bioreactor, based on a combination of microkinetics and macrokinetics, were verified using experimental oxygen conversions (Figure 7). A satisfactory agreement was found between the model calculations and observed conversions in case of an aqueous reaction medium as well as an organic medium. The theoretical values are mostly within the experimental errors of the observed oxygen conversions. Axial oxygen depletion of the aqueous transport medium along the length of the packed bed is severely limiting the epoxidation rate even at the higher oxygen inlet concentrations (or oxygen partial pressures). However, oxygen depletion is prevented by employing n-hexadecane as the substrate reservoir at all oxygen levels of the inlet stream. The experimental (and predicted) degrees of oxygen conversion are all significantly lower than the corresponding oxygen conversions obtained with water as the continuous phase, though the absolute amount of oxygen (and propene) consumed is higher. The advantages of using an organic substrate reservoir would be even more pronounced when an immobilized-cell-compatible solvent with a higher oxygen capacity would be applied. Perfluoro chemicals, with their very inert properties and their extreme high oxygen capacities, would provide an (expensive) alternative.

Acknowledgements. The authors wish to thank Prof. Ir. K.Ch.A.M. Luyben and Prof. Dr. Ir. K. Van 't Riet for helpful discussions. These investigations were supported (in part) by the Netherlands Technology Foundation (STW).

REFERENCES

1 E. Antonini, G. Carrea and P. Cremonesi, Enzyme catalysed reactions in water-organic solvent two-phase systems, Enzyme Microb. Technol., 3 (1981) 291-296.
2 G. Carrea, Biocatalysis in water-organic solvent two-phase systems, Trends in Biotechnol., 2 (1984) 102-106.
3 M.D. Lilly, Two-liquid-phase biocatalytic reactions, J. Chem. Tech. Biotechnol., 32 (1982) 162-169.
4 M.D. Lilly and J.M. Woodley, Biocatalytic reactions involving water-insoluble organic compounds, in: J. Tramper, H.C. van der Plas and P. Linko (Eds), Biocatalysts in Organic Syntheses, Elsevier, Amsterdam, 1985, pp. 179-192.
5 B.C. Buckland, P. Dunnill and M.D. Lilly, The enzymatic transformation of water-insoluble reactants in nonaqueous solvents. Conversion of cholesterol to cholest-4-ene-3-one by a *Nocardia* sp., Biotechnol. Bioeng., 17 (1975) 815-826.
6 J.M.C. Duarte and M.D. Lilly, The use of free and immobilized cells in the presence of organic solvents: the oxidation of cholesterol by *Nocardia rhodochrous*, in: H.H. Weetall and G.P. Royer (Eds), Enzyme Engineering, Vol. 5, Plenum Press, New York, 1980, pp. 363-367.
7 B. Cambou and A.M. Klibanov, Preparative production of optically active esters and alcohols using esterase-catalyzed stereospecific trans-esterification in organic media, J. Am. Chem. Soc., 106 (1984) 2687-2692.
8 S. Fukui and A. Tanaka, Application of biocatalysts immobilized by prepolymer methods, Adv. Biochem. Eng./Biotechnol., 29 (1984) 1-33.
9 L.E.S. Brink and J. Tramper, Optimization of organic solvent in multi-phase biocatalysis, Biotechnol. Bioeng., 27 (1985) 1258-1269.
10 A.Q.H. Habets-Crützen, L.E.S. Brink, C.G. van Ginkel, J.A.M. de Bont and J. Tramper, Production of epoxides from gaseous alkenes by resting-cell suspensions and immobilized cells of alkene-utilizing bacteria, Appl. Microbiol. Biotechnol., 20 (1984) 245-250.
11 J. Tramper, L.E.S. Brink, R.S. Hamstra, J.A.M. de Bont, A.Q.H. Habets--Crutzen and C.G. van Ginkel, Production of (chiral) epoxy alkanes in second-generation bioreactors, in: Third European Congress on Biotechnology, Vol. 2, Verlag Chemie, Weinheim, 1984, pp. 269-276.
12 L.E.S. Brink, J. Tramper, K. van't Riet and K.Ch.A.M. Luyben, Automation of an experimental system for the microbial epoxidation of propene and 1-butene, Anal. Chim. Acta, 163 (1984) 207-217.
13 L.E.S. Brink and J. Tramper, Modelling the effects of mass transfer on kinetics of propene epoxidation of immobilized *Mycobacterium* cells: 1. pseudo-one-substrate conditions and negligible product inhibition, Enzyme Microb. Technol., 8 (1986a) 281-288.
14 L.E.S. Brink and J. Tramper, Modelling the effects of mass transfer on kinetics of propene epoxidation of immobilized *Mycobacterium* cells: 2. product inhibition, Enzyme Microb. Technol., 8 (1986b) 334-340.

15 L.E.S. Brink and J. Tramper, Facilitated mass transfer in a packed-bed immobilized-cell reactor by using an organic solvent as substrate reservoir, J. Chem. Tech. Biotechnol., 1987, in press.
16 T. Omata, A. Tanaka and S. Fukui, Bioconversions under hydrophobic conditions: effect of solvent polarity on steroid transformations by gel-entrapped *Nocardia rhodocrous* cells, J. Ferment. Technol., 58 (1980) 339-343.
17 C. Laane, S. Boeren and K. Vos, On optimizing organic solvents in multi-liquid-phase biocatalysis, Trends in Biotechnol., 3 (1985) 251-252.
18 C. Laane, S. Boeren, K. Vos and C. Veeger, Rules for the optimization of biocatalysis in organic media, Biotechnol. Bioeng., in press.
19 V. Kasche, Correlation of experimental and theoretical data for artificial and natural systems with immobilized biocatalysts, Enzyme Microb. Technol., 5 (1983) 2-13.
20 J.M. Radovich, Mass transfer effects in fermentations using immobilized whole cells, Enzyme Microb. Technol., 7 (1985) 2-10.
21 S.F. Karel, S.B. Libicki and C.R. Robertson, The immobilization of whole cells: engineering principles, Chem. Eng. Sci., 40 (1985) 1321-1354.
22 J.V. Villadsen and W.E. Stewart, Solution of boundary-value problems by orthogonal collocation, Chem. Eng. Sci., 22 (1967) 1483-1501.
23 P.A. Ramachandran, Solution of immobilized enzyme problems by collocation methods, Biotechnol. Bioeng., 17 (1975) 211-226.
24 V. Kasche, A. Kapune and H. Schwegler, Operational effectiveness factors of immobilized enzyme systems, Enzyme Microb. Technol., 1 (1979) 41-46.
25 H. Tanaka, M. Matsumura and I.A. Veliky, Diffusion characteristics of substrates in Ca-alginate gel beads, Biotechnol. Bioeng., 26 (1984) 53-58.
26 P. Adlercreutz, Oxygen supply to immobilized cells: 5. theoretical calculations and experimental data for the oxidation of glycerol by immobilized *Gluconobacter oxydans* cells with oxygen or *p*-benzoquinone as electron acceptor, Biotechnol. Bioeng., 28 (1986) 223-232.
27 B.J.M. Hannoun and G. Stephanopoulos, Diffusion coefficients of glucose and ethanol in cell-free and cell-occupied calcium alginate membranes, Biotechnol. Bioeng., 28 (1986) 829-835.
28 G. Bell, J.R. Todd, J.A. Blain, J.D.E. Patterson and C.E.L. Shaw, Hydrolysis of triglyceride by solid phase lipolytic enzymes of *Rhizopus arrhizus* in continuous reactor systems, Biotechnol. Bioeng., 23 (1981) 1703-1719.
29 B.P. Ter Meulen and G.J. Annokkee, Process for carrying out an enzymatic reaction, Eur. Patent No. 68594 (1983).
30 A.R. Macrae, Interesterification of fats and oils, in: J. Tramper, H.C. van der Plas and P. Linko (Eds), Biocatalysts in Organic Syntheses, Elsevier, Amsterdam, 1985, pp. 195-208.

C. Laane, J. Tramper and M.D. Lilly (Editors), *Biocatalysis in Organic Media*,
Proceedings of an International Symposium held at Wageningen,
The Netherlands, 7–10 December 1986.

PHASE TOXICITY IN A WATER-SOLVENT TWO-LIQUID PHASE MICROBIAL SYSTEM

Raphael BAR

Department of Chemical Engineering
University of Virginia
Charlottesville, VA 22901

The use of organic media in biocatalytic processes seems to be continuously increasing both in a one-liquid (organic) phase[1] and in a water-organic solvent two-liquid phase[2,3] systems. In the later systems, a solvent is employed for carrying out either an extractive fermentation or a microbial transformation. Non-toxicity of this solvent to microorganisms is a <u>sine qua non</u> condition for a successful microbial process. However, biocompatibility of the solvent may not guaranty the operational success of the process. Indeed, the basic requirements from solvents employed in extractive fermentations and microbial transformations would be different for purposes of product separation. So far, successful extractive fermentation included low-boiling metabolites such as ethanol[4,5] acetone and butanol[6] and high-boiling ones such as organic acids.[7] In all these cases, product recovery can be done by either distillation of the former metabolites or back-extraction of the later ones. Therefore, since the products are separated from the solvent, the proper organic solvent employed in extractive fermentations should be high-boiling and polar. However, an organic substrate to be microbially transformed, is often by-itself high-boiling and therefore, an ideal solvent for microbial transformations would be one which is separated from the product, i.e., a low-boiling solvent. Since a low-boiling temperature can be associated with a low molecular weight, it appears that the ideal solvent requirements of polarity and low-boiling temperature - for extractive fermentations and microbial transformations respectively - contradict the biocompatibility requirements of low polarity in combination with a high molecular weight as suggested by Brink and Tramper[8].

In this paper, a distinction is made between solvent toxicity exerted on the microorganisms at the molecular level and at the phase level. A solvent non-toxic at the molecular level could exhibit a negative effect when used in excess. Methods to protect microorganisms from the negative effects exclusively due to the presence of a separate organic phase will be discussed.

Physical Characterization of a Two-Liquid Phase System

The characterization of a two-liquid phase system is centered around the aqueous solubility of the organic liquid (Scheme 1). If the organic liquid (either a solvent or a substrate to be microbially transformed) is insoluble in water, agitation is required for contacting the two phases and a dispersion system is thus formed.

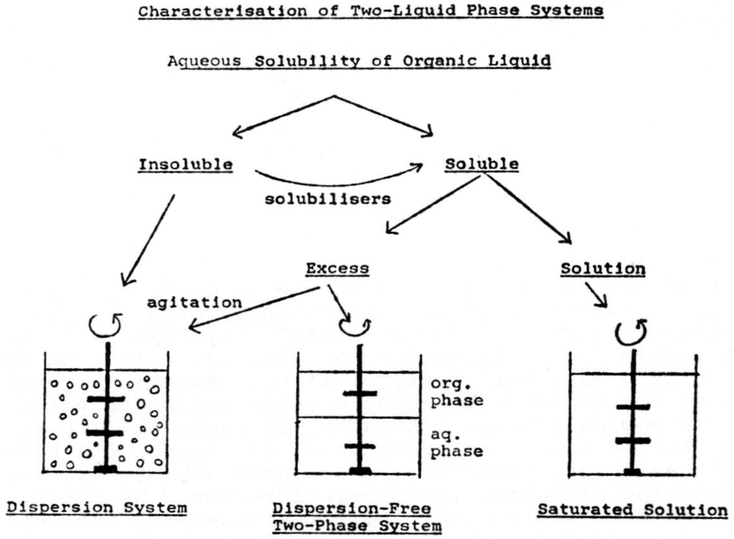

Scheme 1

When the organic liquid is somewhat soluble in water, either a sufficiently small of it is used to form an aqueous solution or an excess of it - to form a two-liquid phase system. In the latter case, the organic phase can be separately stirred to form a dispersion-free two-liquid phase system or intermixed with the aqueous phase to form again a dispersion system.

It will be shown that microorganisms suspended in the aqueous phase can be subject to different solvent effects in each of the aforementioned three systems even though the aqueous phase is initially solvent-saturated.

Sometimes, it may be possible to switch from the "insoluble" to the "soluble" two-liquid phase system by adding a small amount of a water-miscible solvent called "solubilizer" (Scheme 1). Thus, an insoluble organic substrate to be microbially transformed may slightly dissolve in the aqueous phase to which a co-solvent was added. These co-solvents must, of course, be non-toxic and Table 1 shows that many such co-solvents are harmless to yeast cells, when employed at low concentrations.

Table 1. Effect of co-solvents ("solubilizers") on growth of Baker's yeast*

Co-solvent	Biomass formed g/L	Ethanol formed g/L
control	5.4	22.3
acetone	5.0	21.0
1-propanol	4.1	22.5
methanol	5.4	21.5
dioxane	5.5	21.2
acetonitrile	5.0	21.7
ethylacetate	4.0	21.5
1,2 dimethoxyethane	5.8	21.0
dimethylsulfoxide	4.8	21.3
dimethylformamide	4.9	21.2
methylethylketone	4.6	22.2
methylisobutylketone	0.8	5.8

* Complex medium with 50 g/L glucose and 1.5% v/v of co-solvent. Temp.: 30°C.

Solvent Toxicity

As Scheme 2 shows, molecular toxicity is caused by the dissolved solvent molecules in the aqueous phase. This type of toxicity prevails in the solvent-saturated cell suspension and in the dispersion-free two-liquid phase system (Scheme 1). It could be due to some enzyme inhibition[9], or protein denaturation or some membrane modification[10]. Attempts to correlate this toxicity with various solvent parameters were previously done[11,12] and recently, a correlation with log P (P = distribution coefficient of the organic liquid in octanol/water system) is re-suggested in a graphical form[13].

It is, however, important to note that cells suspended in a solvent-saturated aqueous phase may absorb some of the dissolved solvent especially into their lipid membranes which consequently, may expand[10]. Since cells interact and equilibrate with the dissolved organic molecules, the resulting concentration of the dissolved solvent may not be toxic. Indeed, cells of <u>Saccharomyces cerevisiae</u> do not grow in a water-octane dispersion system[14], but do grow when placed in an octane-saturated broth[15]. In the dispersion system, the agitation continuously replenishes the amount of dissolved solvent absorbed by the biomass and saturates the aqueous phases up to a concentration which toxic to the cells. If, however, the dissolution rate of the organic solvent is too slow, microbial growth can take place even in a dispersion-free two-liquid phase system.

SOLVENT TOXICITY

Solvent

Dissolved	Excess
Molecular Toxicity	Phase Toxicity
* Enzyme inhibition * Protein denaturation * Membrane modification: - membrane expansion - structure disorder - permeability change - etc.	* Extraction of nutrients * Disruption of cell wall: - extraction of outer cellular components * Limited access to nutrients: - cell attraction to interface - emulsion formation - cell coating

Scheme 2

In addition to the previously mentioned toxic effects, other negative effects can be exclusively due to the presence of a separate organic phase. These negative effects, referred to as phase toxicity, will partly or totally inhibit the biocatalyst activity. Phase toxicity is experimentally observed in systems in which molecular toxicity is absent.

Extraction of nutrients from the aqueous environment into the organic layer is likely to slow down the growth of the microorganisms. Tributylphosphate, for example, is a good extractant of metals[4] such as calcium. A solvent is also likely to extract an organic precursor required by the microorganisms.

Microbial cells can also experience a limited access to nutrients from within the aqueous phase. The attraction of cells to interfaces[16] and the formation of emulsions may impede a free uptake of nutrients. It has been reported that cells can even be coated with an organic layer[4], which could lead to a starvation due to blockage of nutrient diffusion.

A serious damage of phase toxicity is a result of the organic phase functioning as an extractant of the biomass. In this case, some outer components of the cell wall would be first extracted, thus, bringing about a cell wall disruption and alternately, cell death. The attraction of cells to the interface is likely to initiate and enhance this process. This is further enhanced by a vigorous agitation in a dispersion system.

Methods of Protection from Phase Toxicity

The following means can be employed for reducing phase toxicity:

1. Use of concentrated nutrients. This is done when a nutrient is partly extracted into an organic phase. This method was used by Cho and Shuler[4] who grew yeast in a two-liquid phase system. Adding excess nutrients may be needed both in a dispersion and dispersion-free systems.

2. Weak agitation and a small volume of organic phase. By doing so, contact between the two phases is reduced and so, the negative effects of phase toxicity are also reduced in a dispersion system.

   In a dispersion-free system, interphase mixing is totally absent and thus, phase toxicity in the form of cell killing, is minimized. Such a system was used to carry out a microbial transformation in which benzaldehyde was reduced to benzyl alcohol. When 300 ml of dodecanol or a mixture (50%V) of tributylphosphate and hexadecane which contained 10 g of aldehyde were brought in contact with 1 liter of yeast broth, both microbial growth and aldehyde reduction (67% yield when the initial glucose concentration was 50 g/L) took place. However, when the stirring speed was increased to form a dispersion system, neither growth nor reduction occurred.

3. Rich inoculum. Since cells can be killed by the agitation as previously mentioned, application of a rich inoculum may compensate for the killed cells and sustain further growth. In this case, two competing processes may simultaneously take place: killing and growth process. The later one is favored when increased amounts of healthy, and exponentially growing cells are employed especially in a dispersion system.

4. Cell immobilization. Entrapment of cells inside gel beads clearly prevents direct contact with the eddies of the agitated organic phase. It is important to note that this method is useful only when the solvent does not exhibit molecular toxicity. Otherwise, the dissolved solvent molecules which are capable of diffusing into the gel bead, will exert their toxic effect.

Indeed, as previously reported[16], cells of Lactobacillus delbrueckii which are inhibited by carbon tetrachloride and tributylphosphate in a solvent-saturated broth, were not protected by immobilization in carrageenan gel beds. However, the cells which are inhibited in dodecanol-broth dispersion system but not in dodecanol-saturated broth, were protected by entrapment in gel beads. Anaerobic microorganisms may be better protected due the fact they can grow deeply into the gel beads. In this case, however, limitation of nutrient diffusion is to expected.

Immobilization of cells in the form of biofilm onto an inert support offers too a protection from phase toxicity. However, since the external surface of the biofilm is exposed to the eddies of the solvent, some damage does take place even though the overall growth is sustained. This is probably the case depicted by Minier and Goma[17] who used immobilized yeast cells in an agitated two-liquid phase.

5. Use of Membranes. A hydrophobic membrane can divide between an aqueous cell suspension and an organic phase, thus forming a dispersion-free two liquid phase system. Membranes in the form of a flat sheet[4] or hollow fibers[5] were previously used for extractive ethanol fermentation. Similarly, a hydrophilic membrane was found to protect yeast cell from phase toxicity in a dispersion system[15].

Conclusions

A clear distinction is to be made between solvent toxicity exerted on the microorganisms at the molecular and phase levels. Phase toxicity can be detected by comparison of the biocatalyst activities observed in a solvent-saturated aqueous phase, in a dispersion-free two liquid-phase system and in dispersion system. Measures to reduce phase toxicity can subsequently be employed as previously discussed.

Literature Cited

1. Klibanov, A. M., CHEMTECH, 16, 354 (1986).

2. Carrea, G., Trends in Biotechnol., 2, 102 (1984).

3. Roffler, S. R., Blanch, H. W. and Wilke, C. R., Trends in Biotechnol., 2, 129 (1984).

4. Cho, T. and Shuler, M. L., Biotechnol. Progress, 2, 53 (1986).

5. Frank, G. T. and Sirkar, K. K., Biotechnol. Bioeng. Symp., 15, in press.

6. Ishii, S., Taya, M. and Kobayashi, T., J. Chem. Eng. Japan, 18, 125 (1985).

7. Levy, P. F., Sanderson, J. E. and Wise, D. L., Biotechnol. Bioeng. Symp., 11, 239 (1981).

8. Brink, L. E. S. and Tramper, J., Biotechnol. Bioeng., 27, 1258 (1985).

9. Nagamoto, H., Yasuda, T. and Inoue, H., Biotechnol. Bioeng., 28, 1172 (1986).

10. Seeman, P., Pharmacol. Rev., 24, 583 (1972).

11. McGowan, J. C., Nature, 200, 1217 (1963).

12. Hanch, C., Muir, R. M., Fujita, T., Maloney, P. P., Geiger, F., and Streich, M., J. Amer. Chem. Soc., 85, 2817 (1963).

13. Laane, C., Boeren, S. and Vos, K., Trends in Biotechnol., 3, 251 (1985).

14. Finn, R. K., J. Ferm. Technol., 44, 305 (1966).

15. Bar, R., to be published.

16. Bar, R. and Gainer, J. L,. Biotechnol. Progress, 1987, in press.

17. Minier, M. and Goma, G., Biotechnol. Bioeng., 24, 1565 (1982).

SESSION V

CONVERSIONS II

Chairman : E. Vandamme

Co-chairman : J. Tramper

C. Laane, J. Tramper and M.D. Lilly (Editors), *Biocatalysis in Organic Media*,
Proceedings of an International Symposium held at Wageningen,
The Netherlands, 7–10 December 1986.

BIOCATALYSIS IN WATER-ORGANIC SOLVENT TWO-PHASE SYSTEMS

G. CARREA

Istituto di Chimica degli Ormoni, Via Mario Bianco 9, 20131 Milano, Italy.

SUMMARY

Enzymes can be employed in media containing high proportions of water-immiscible organic solvents for the transformation or synthesis of compounds poorly soluble in water. A variety of reactions such as oxidoreductions, epoxidations, isomerizations, interesterifications, and syntheses of peptide and ester bonds have been carried out with good efficiency in such media. The paper describes the regio- and stereospecific oxidoreduction of the hydroxyl-keto groups of steroids catalyzed by hydroxysteroid dehydrogenases. The effect of organic solvents on the activity and stability of the free or immobilized enzymes is discussed together with the problems connected with the solubility, partition and transfer of substrates and products between phases. The enzymatic systems used for the simultaneous regeneration of NAD(P)(H), which are essential cofactors of the hydroxysteroid dehydrogenases, are illustrated.

INTRODUCTION

The potential of enzyme catalyzed reactions in the field of organic synthesis is now remarkable. This is due, among other reasons, to the increased number of available enzymes, to their use in the immobilized form, and to the existence of efficient systems of coenzyme regeneration. However, substrates poorly soluble in water such as steroids offer special difficulties in this area since the enzyme-catalyzed transformations must be carried out using large reaction volumes and consequently large amounts of biological catalysts and cofactors, whenever the latter is required. Also, the use of large reaction volumes makes more difficult the recovery of products. Attempts to increase steroid solubility by adding water-miscible organic solvents to the reaction medium have given unsatisfactory results since high solvent concentrations progressively give rise to inhibition, decreased specificity, and denaturation of the enzymes (ref. 1).

In contrast, when organic solvents practically immiscible or poorly miscible with water are used, the situation becomes substantially different. In this case, a two-phase system consisting of water and an organic solvent is established. The water phase contains enzymes and hydrophilic cofactors, and the organic solvent high concentrations of hydrophobic substrates. On shaking or stirring, substrates diffuse into the aqueous phase where they undergo the enzyme catalyzed transformation and then the formed products return to the organic phase. In the system, the concentration of solvents in water is low and not dependent on the ratio of the two phases, even if the partial volume of the organic phase is much greater than that of the aqueous phase, and the inhibitory and denaturing effects are much lower than those induced by comparable concentrations of miscible solvents (refs. 2-10).

In the present article, regio- and stereospecific oxidoreductions of the hydroxyl-keto groups of steroids (Fig. 1), catalyzed by NAD(P) dependent hydroxysteroid dehydrogenases (HSDH), are described. The reactions are carried out in two-phase systems and are coupled to the simultaneous enzymatic regeneration of the coenzymes. The effect of organic solvents on the activity and stability of the free or immobilized enzymes is discussed together with the problems connected with the solubility, partition and transfer of substrates and products between phases.

ENZYME KINETICS

The effect of various water-immiscible organic solvents on the activity of hydroxysteroid dehydrogenases has been examined (refs. 2,7). In particular, the effect of ethyl acetate on the K_m and V_{max} values of the enzymes has been studied. Table 1 shows that the solvent behaves as a competitive inhibitor for steroid substrates in the case of β-HSDH, 20β-HSDH and 12α-HSDH whereas in the case of 3α-HSDH there is a mixed inhibition with an increase of K_m values and a decrease, particularly high with androsterone, of V_{max} values. The decrease of V_{max} values is, from a practical point of view, a serious drawback since even at very high concentrations of substrate the enzyme is not fully active. Ethyl acetate did not substantially influence the K_m values of the hydroxysteroid dehydrogenases for NAD(P)(H) and also did not affect the activity of the enzymes used for the regeneration of the nicotinamide cofactors (lactate dehydrogenase, glutamate dehydrogenase, formate dehydrogenase and glucose dehydrogenase). Butyl acetate and ethyl ether were found to act similarly to ethyl acetate with 12α-HSDH (ref. 9) and steroid isomerase (ref. 11).

ENZYME STABILITY

As exemplified in Table 2, the stability of the enzymes in two-phase systems

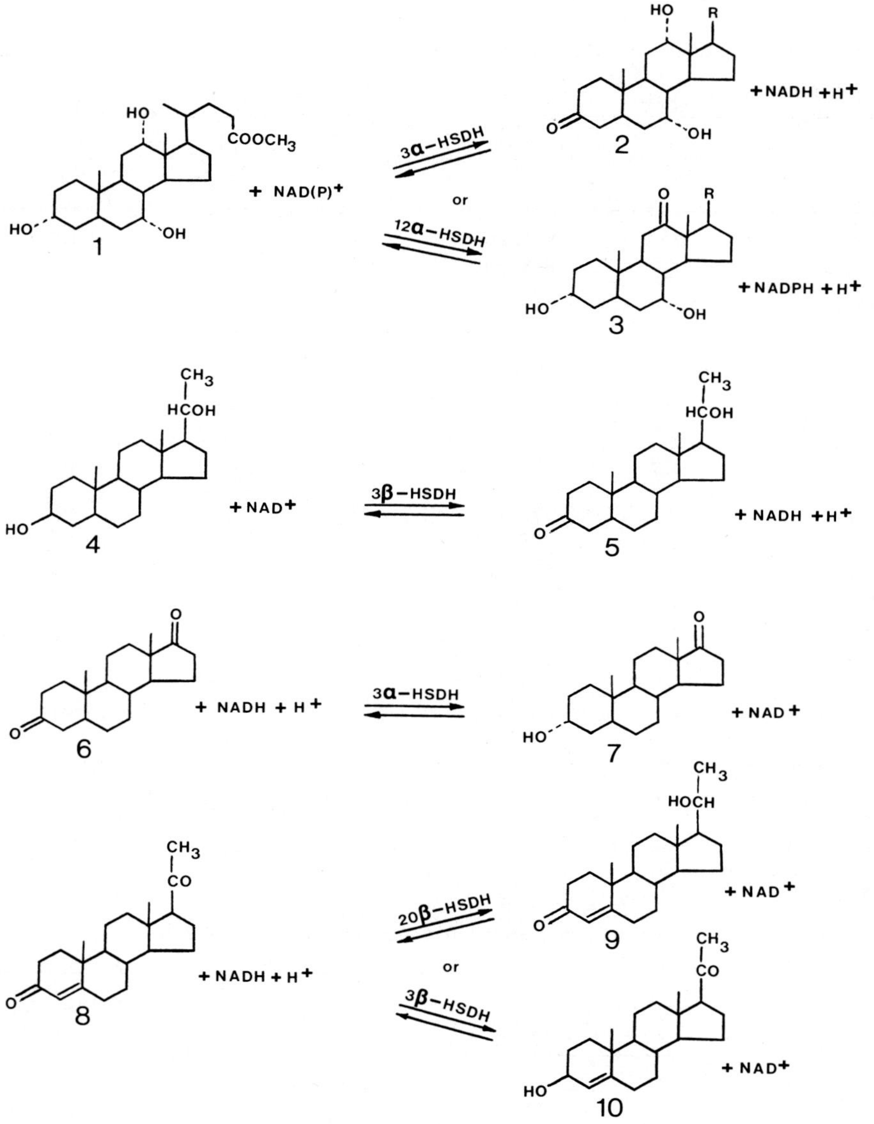

Fig. 1. Examples of regiospecific oxidations and regio- and stereospecific reductions of steroids. 3α-HSDH, 3α-hydroxysteroid dehydrogenase; 3β-HSDH, 3β-hydroxysteroid dehydrogenase; 12α-HSDH, 12α-hydroxysteroid dehydrogenase; 20β-HSDH, 20β-hydroxysteroid dehydrogenase. 1, cholic acid methyl ester; 2, 7α,12α-dihydroxy-3-oxo-5β-cholanic acid methyl ester; 3, 12-ketochenodeoxycholic acid methyl ester; 4, pregnan-3β,20α-diol; 5, pregnan-20α-ol-3-one; 6, androstandione; 7, androstan-3α-ol-17-one; 8, progesterone; 9, 4-pregnen-20β--ol-3-one; 10, 4-pregnen-3β-ol-20-one.

TABLE 1

Effect of ethyl acetate on the kinetic parameters of β-HSDH, 20β-HSDH, 12α-HSDH and 3α-HSDH

Enzyme	Solvent conc. (M)	Substrate	K_M (μM)	V_{max}, rel.
β-HSDH	0	Estradiol	2.1	100
	0.6	Estradiol	9.2	100
20β-HSDH	0	Cortisone	29.0	100
	0.7	Cortisone	95.0	95
	0	Progesterone	3.5	100
	0.7	Progesterone	9.5	140
12α-HSDH	0	Cholic methyl ester	59.2	100
	0.7	Cholic methyl ester	390.0	100
	0	Cholic acid	64.0	100
	0.7	Cholic acid	420.0	125
3α-HSDH	0	Androsterone	2.8	100
	0.7	Androsterone	3.4	21
	0	Cholic methyl ester	3.0	100
	0.7	Cholic methyl ester	6.8	37
	0	Cholic acid	4.3	100
	0.7	Cholic acid	31.0	60

TABLE 2

Influence of various organic solvents on the stability of β-HSDH

Solvent	Relative stability	Solvent	Relative stability
Water only	100	Butyl acetate	60
n-Hexane	96	Ethyl ether	54
Isooctane	95	Ethyl acetate	53
Chlorobenzene	90	Chloroform	26
Trichloroethylene	88	Benzene	19
Carbon tetrachloride	76		

depends on the nature of the organic solvent. Generally, solvents of low polarity and low water-solubility do not strongly affect enzyme stability. Shaking (and also the rate of shaking, ref. 3) influences the enzyme stability in biphasic systems, as illustrated in Table 3 where it is shown that 3α-HSDH is less stable when it is shaken.

TABLE 3

Effect of shaking on the stability of 3α-HSDH

Time (days)	Activity (rel.)			
	Buffer		Buffer:Ethyl acetate (1:1)	
	Unshaken	Shaken [a)]	Unshaken	Shaken [a)]
0	100	100	100	100
1	87	84	71	59
2	79	79	53	38
3	73	76	44	28

[a]150 strokes per min.

The stability can be improved by adding albumin, coenzymes or substrates (4,6). Immobilization was found to markedly increase the stability of several hydroxysteroid dehydrogenases (refs. 7,12). For instance, β-HSDH and 12α-HSDH showed half-life of about two months, under working conditions, in biphasic systems (7,12). The improvement of stability induced by immobilization can be due, among other reasons, to the prevention of protein aggregation that takes place with free enzymes in organic solvents. The activity recoveries of immobilized hydroxysteroid dehydrogenases were found to be in the 30-60% range (refs. 7,12,13).

SOLUBILITY AND PARTITION OF REAGENTS AND PRODUCTS BETWEEN PHASES

Fundamental requirements for an organic solvent suitable for preparative purposes are a high capacity to solubilize reagents and products and reduced inhibiting and denaturing effects on enzymes. For steroid transformations with hydroxysteroid dehydrogenases, butyl and ethyl acetate gave the best results (refs. 2,6,10).

The partition coefficients (α) of substrates and products between the organic and the aqueous phase depend on the nature of the solvent and of the steroid

(Table 4). When there is high enzyme inhibition by substrate or product, high α values are favourable, provided that the concentration of the substrate in the aqueous phase is still high enough to assure good enzymatic activity. This is the case, for instance, of testosterone with β-HSDH and of progesterone with 20β-HSDH which have very low K_m values also in the presence of saturating concentrations of organic solvents (ref. 3, Table 1). By contrast, high α values represent a drawback when also the K_m values of the substrates are high, as in the case of cholic acid methyl ester with 12α-HSDH (ref. 9).

TABLE 4

Partition coefficients (α) of steroids in two-phase systems

Organic solvent	α		
	Testosterone	Progesterone	Cholic acid methyl ester
Butyl acetate	563	4400	147
Ethyl acetate		1210	

Noncompetitive inhibition by steroid products is reduced by the organic solvent not only because it reduces the concentration of the product in the aqueous phase, but also because it increases its K_i value i.e., it reduces the affinity of the product for the enzyme. Table 5 shows the inhibition pattern of 20β-HSDH by product in the presence or absence of ethyl acetate. The K_i value of the product is doubled in the presence of the organic solvents, which means that the concentration of product that will lower to 50% the apparent concentration

TABLE 5

4-Pregnen-20β-ol-3-one as product inhibitor of 20β-HSDH using progesterone as substrate

Inhibitor conc. (μM)	Buffer			Buffer containing 0.7 M ethyl acetate		
	K_m (μM)	V_{max},rel	K_i (μM)	K_M (μM)	V_{max},rel [a]	K_i (μM)
0	3.5	100		9.5	100	
42	3.5	45	35	9.7	63	72
84	3.6	30	33	9.3	46	71

[a] In 0.7 M ethyl acetate, V_{max} is 140% of that in buffer.

of the enzyme will be, in the presence of the organic solvent, two times higher than in buffer alone.

TRANSFER BETWEEN PHASES

The transfer of reagents between the aqueous and organic phase can be achieved by shaking or stirring. When free enzymes are employed, the rate of shaking or stirring that gives the maximum transformation is generally dependent on the nature of the solvent and substrate and on the amount of enzyme utilized (refs. 3,6). Variations of the aqueous/organic solvent ratios scarcerly influence the reaction rate (ref. 10). When immobilized enzymes are used, the situation is more complex and the penetration of reagents into matrices may be rate limiting and dependent on the nature of the matrix and the aqueous/organic solvent ratios (ref. 13). The reaction rates are usually higher in ethyl acetate than in butyl acetate (ref. 13).

With enzymes immobilized onto porous beads, fixed bed reactors can also be employed (ref. 13). Enzymes, cofactors and hydrophilic compounds remain inside the hydrophilic matrix beads filled with the aqueous buffer (saturated with the organic solvent), while the hydrophobic substrates, which are dissolved in the organic solvent (saturated with the aqueous buffer) pumped through the reactor, diffuse to the aqueous phase where they are transformed (Fig. 2). Eupergit C, Sepharose CL-4B, Hyflo Super Cel and Glycophase G are all suitable to the end. Enzymes can be simply confined inside the beads or covalently linked to the matrix. The effect on the reactor performance of parameters such as enzyme load, chemical nature of the matrix, bead and pore size, and flow rate is currently being examined.

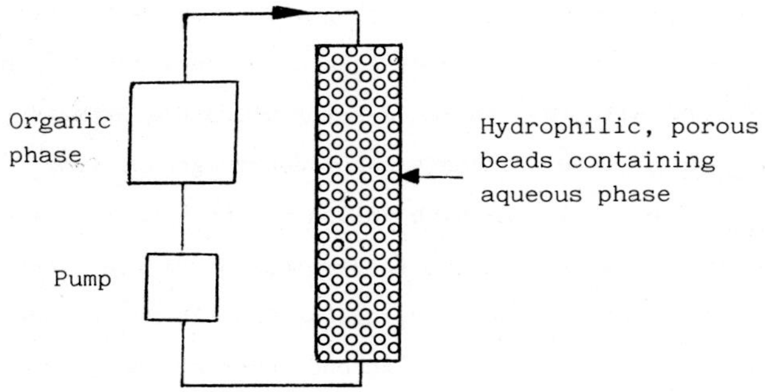

Fig. 2 Scheme of fixed bed reactor.

Table 6 shows the rates of androsterone oxidation, catalyzed by 3α-HSDH, with free or immobilized enzymes, in shaken vessels or fixed bed reactors (ref. 13). The highest reaction rate (initial) was obtained with the free enzyme in a shaken vessel and the lowest with the covalently immobilized enzyme used in a shaken vessel. However, it should be emphasized that the free enzyme can not be reused and that its stability is lower.

TABLE 6

Reaction rates of androsterone oxidation catalized by 3α-HSDH

Conditions	Reaction rate (rel.)
Free enzyme, shaken vessel	100
Enzyme confined into Eupergit C [a], fixed bed reactor	65
Enzyme covalently linked to Eupergit C, fixed bed reactor	45
Enzyme covalently linked to Eupergit C, shaken vessel	30

[a]Eupergit C was treated with ethanolamine before use.

ENZYMATIC REGENERATION OF NICOTINAMIDE COFACTORS

The in situ regeneration of the costly nicotinamide cofactors has been achieved by enzymatic means (refs. 4-7,12,13). For instance, the $NAD(P)^+$ consumed in a specific oxidation catalyzed by a hydroxysteroid dehydrogenase can be regenerated at the expense of a much cheaper compound such as α-ketoglutarate, in the presence of glutamate dehydrogenase from bovine liver (refs. 9,12).

$$\alpha\text{-Ketoglutarate} + NAD(P)H + H^+ + NH_3 \rightleftharpoons \text{L-glutamate} + NAD(P)^+ + H_2O$$

NAD^+ can be successfully regenerated also using the pyruvate/lactate dehydrogenase system (refs. 4,13) or the acetaldehyde/alcohol dehydrogenase system (ref. 13), and $NADP^+$ with the system α-ketoglutarate/glutamate dehydrogenase from Proteus or with acetone/alcohol dehydrogenase from Thermoanaerobium brockij (ref. 13). The high values of the equilibrium constants of the $NAD(P)^+$ regenerating reactions drive to completion the reactions of interest catalyzed by hydroxysteroid dehydrogenases. The $NAD(P)^+$ regenerating systems using acetaldehyde or acetone as the second substrate are particularly suitable for packed bed reactors since these compounds are soluble in organic solvents and therefore can be feeded together with the steroids. However, it should also be emphasized that acetaldehyde tends to inactivate the enzymes.

Similarly, the NAD(P)H consumed in a specific reduction catalyzed by a hydroxysteroid dehydrogenase can be regenerated at the expence of glucose in the presence of glucose dehydrogenase, with complete reduction of the substrate of interest (ref. 13).

$$\text{Glucose} + \text{NAD(P)}^{+} \rightleftharpoons \text{gluconolactone} + \text{NADH} + \text{H}^{+}$$

Very effective for the regeneration of NADH is also to formate/formate dehydrogenase system (refs. 12,13). The turnover numbers (moles of product generated per mole of coenzyme) obtained in NAD(P)(H) regeneration in biphasic systems were up to 500 (ref. 12).

The use of immobilized enzymes can affect the recycling efficiency of cofactors because very often immobilization causes an increase of the K_m values of NAD(P)(H), as exemplified in Table 7 with 12α-HSDH and glutamate dehydrogenase. However, this limitation can be overcome by using coimmobilized enzymes (an hydroxysteroid dehydrogenase and an enzyme suitable for cofactor regeneration) instead of separately immobilized enzymes (refs. 14,15).

TABLE 7

Michaelis constants of free and Sepharose CL-4B immobilized 12α-HSDH and glutamate dehydrogenase (GlDH) from Proteus

Enzyme	Substrate or coenzyme	Free enzyme K_m (μM)	Immobilized enzyme $K_{m,app}$ (μM)
12α-HSDH	Cholic acid	110	120
	NADP	15	45
GlDH	α-Ketoglutarate	340	310
	NADPH	14	150

EXAMPLES

The experimental conditions and the results obtained in the transformation of some steroids are summarized here

Reduction of progesterone. The reduction of progesterone is based on the following reactions:

$$\text{Progesterone(pregn-4-ene-3,20-dione)} + \text{NADH} + \text{H}^+ \xrightleftharpoons{20\beta\text{-HSDH}} \text{pregn-4-ene-20}\beta\text{-hydroxy-3-one} + \text{NAD}^+$$

$$\text{HCOOH} + \text{NAD}^+ \xrightleftharpoons{\text{FDH}} \text{CO}_2 + \text{NADH} + \text{H}^+$$

The overall equilibrium constant for the coupled reactions is about 10^{15}. The reaction system was made up as follows. Aqueous phase: 15 ml of 0.1 M potassium phosphate buffer, pH 7, containing 0.2 M formate, 1 mM DTT, 15 mg serum albumin, 3 µmol NAD, 21 U 20β-HSDH and 24 U of formate dehydrogenase (FDH). organic phase: 15 ml of butyl acetate containing 1.2 mmol (375 mg) progesterone. After 30 h of shaking, progesterone was completely and specifically converted to the 20β-hydroxyderivative. The turnover number (moles of product generated per mole of coenzyme) was 400 (ref. 12). At the end of the run, the residual activity of 20β-HSDH and FDH was 60% and 65% respectively.

Oxidation of testosterone. The process is based on the following reactions:

$$\text{Testosterone} + \text{NAD}^+ \xrightleftharpoons{17\beta\text{-HSDH}} \text{androstenedione} + \text{NADH} + \text{H}^+$$

$$\text{Pyruvate} + \text{NADH} + \text{H}^+ \xrightleftharpoons{\text{LDH}} \text{lactate} + \text{NAD}^+$$

The overall equilibrium constant for the coupled reactions is 9.4×10^3. The reaction system was made up as follows. Aqueous phase: 10 ml of 0.1 M potassium phosphate buffer, pH 8.5, containing 0.2 M sodium pyruvate, 1 mM DTT, 15 mg serum albumin, 3 µmol NAD, 20 U 17β-HSDH and 100 U lactate dehydrogenase (LDH). Organic phase: 20 ml of butyl acetate containing 1.04 mmol (300 mg) testosterone. After about 25 h of shaking, testosterone was completely converted to androstendione. The turnover number for NAD was 350. At the end of the run, the residual activity of 17β-HSDH and LDH was 54% and 70% respectively (ref. 12).

Oxidation of cholic acid methyl ester to 12-ketochenodeoxycholic acid methyl ester. The oxidation was carried out with coimmobilized enzymes in a shaken vessel or in a fixed bed reactor. The process carried out in the vessel is based on the following reactions:

$$\text{Cholic acid methyl ester} + \text{NADP}^+ \xrightleftharpoons{12\alpha\text{-HSDH}} \text{12-ketochenodeoxycholic acid methyl ester} + \text{NADPH} + \text{H}^+$$

$$\alpha\text{-Ketoglutarate} + \text{NADPH} + \text{NH}_3 + \text{H}^+ \xrightleftharpoons{\text{GlDH}} \text{L-glutamate} + \text{NADP}^+ + \text{H}_2\text{O}$$

The overall equilibrium constant for the coupled reactions is about 3.5×10^6. The reaction system was made up as follows. Aqueous phase: 10 ml of 0.1 M potassium phosphate buffer, pH 8, containing 0.2 M ammonium acetate, 0.1 M α-ketoglutarate, 0.75 mM ADP, 1 mM DTT, 1 μmol NADP and 5.5 U 12α-HSDH coimmobilized onto Sepharose CL-4B (5 ml) with 28 U of glutamate dehydrogenase (GlDH). Organic phase: 10 ml of butyl acetate containing 0.47 mmol (200 mg) of cholic acid methyl ester. A complete and specific (12α-OH group) oxidation of cholic acid methyl ester to 12-ketochenodeoxycholic acid methyl ester was achieved in about 4 days. The turnover number for NADP was 470. After three months of continuous use, the residual activity of 12α-HSDH and GlDH was 48% and 60% respectively (ref. 12).

In the process carried out in a fixed bed reactor, NADP was regenerated with the following reaction catalyzed by alcohol dehydrogenase from T. brockij (ADHTB):

$$\text{Acetone} + \text{NADPH} + \text{H}^+ \xrightleftharpoons{\text{ADHTB}} \text{isopropanol} + \text{NADP}^+$$

The reactor was made up as follows. 12α-HSDH (20 U) and ADHTB (30 U), coimmobilized onto Eupergit C (4 ml), were packed in a small column and equilibrated with 0.05 M potassium phosphate buffer (previously saturated with ethyl acetate), pH 7.8, containing 2 mM DTT and 0.6 mM NADP. Then 32 ml ethyl acetate (saturated with buffer) containing 0.38 mmol (160 mg) steroid and 0.4 M acetone were percolated and, after removal of excess aqueous phase from the matrix (V_o = 0.3 ml per ml of matrix), recirculated through the column at a flow rate of 90 ml per h. The volume of aqueous phase retained by Eupergit C was 0.55 ml per ml matrix. A complete oxidation of cholic acid methyl ester was achieved in about 4 days. The organic phase was recovered and the column replenished with a fresh solution of steroid and acetone. The column was used for three consecutive runs and the turnover number for NADP was 860 (ref. 13).

Oxidation of 5α-androstane-3α,17β-diol. The oxidation was carried out with coimmobilized enzymes in a fixed bed reactor. The process is based on the following reactions:

$$5\alpha\text{-Androstane-}3\alpha,17\beta\text{-diol} + \text{NAD}^+ \xrightleftharpoons{3\alpha\text{-HSDH}} 5\alpha\text{-androstan-}17\beta\text{-ol-3-one} + \text{NADH} + \text{H}^+$$

$$\text{Pyruvate} + \text{NADH} + \text{H}^+ \xrightleftharpoons{\text{LDH}} \text{lactate} + \text{NAD}^+$$

The overall equilibrium constant for the coupled reactions is 2×10^3. The reactor was made up as follows: 3α-HSDH (100 U) and LDH (200 U), coimmobilized onto Eupergit C (6 ml), were packed in a column and equilibrated with 0.1 M potassium phosphate buffer (previously saturated with ethyl acetate), pH 8.5, containing 0.2 M sodium pyruvate, 2 mM DTT and 0.3 mM NAD. Then, 16 ml ethyl acetate (saturated with buffer) containing 0.55 mmol (160 mg) steroid were percolated and, after removal of excess aqueous phase from the matrix, recirculated through the column at a flow rate of 90 ml per h. A complete and specific (3α-OH group) oxidation of the substrate was achieved in about 12 h. The turnover number was 550. Other runs were carried out after substitution of both aqueous and organic phase. After 16 days of continuous use the residual activity of 3α-HSDH and LDH was 70% and 85% respectively. Similar results were obtained with Sepharose CL-4B immobilized enzymes (ref. 13).

The use of the acetaldehyde/alcohol dehydrogenase system for NAD^+ regeneration, with acetaldehyde dissolved in the organic phase, made possible to carry out consecutive runs without substitution of the aqueous phase. However, acetaldehyde affected the stability of the enzymes.

CONCLUSIONS

The described steroid transformations exemplify the feasibility of carrying out enzymatic oxidoreductions of steroids in two-phase systems. Many other steroids can be converted, under similar conditions, using the examined enzymes or other enzymes. In all cases, the use of two-phase systems drastically reduces reaction volumes and allows an easy separation of the products, that are predominantly present in the organic phase, from the enzymes and cofactors, that are present in the aqueous phase. In some cases, also the negative effect of product inhibition is reduced.

The regeneration of the costly coenzymes markedly reduces the cost of the process and should make it possible to fully exploit the great specificity of hydroxysteroid dehydrogenases for preparative-scale transformations of steroids in two-phase systems. Also, the use of coupled enzymatic reactions highly improves transformation yields when overall equilibrium constant values are favorable. The immobilization of the enzymes makes it possible to use them repeatedly, both in shaken vessels and fixed bed reactors, and increases their stability. It should also be emphasized that, contrary to what happens with separately immobilized enzymes, coimmobilized enzymes regenerate the coenzymes very effectively. It is likely that the methodology is advantageously usable also with other NAD(P) dependent dehydrogenases acting on hydrophobic substrates.

AKNOWLEDGEMENTS

The author gratefully aknowledges the discussions and help of R.Bovara, P. Pasta and S. Riva in preparation of this manuscript.

REFERENCES

1 L.G. Butler, Enzymes in non-acqueous solvents, Enzyme Microb. Technol., 1 (1979) 253-259.

2 P. Cremonesi, G. Carrea, G. Sportoletti and E. Antonini, Enzymatic dehydrogenation of steroids by β-hydroxysteroid dehydrogenase in a two-phase system, Arch. Biochem. Biophys., 159 (1973) 7-10.

3 G. Lugaro, G. Carrea, P. Cremonesi, M.M. Casellato and E. Antonini, The oxidation of steroid hormones by fungal laccase in emulsion of water and organic solvents, Arch. Biochem. Biophys., 159 (1973) 1-6.

4 P. Cremonesi, G. Carrea, L. Ferrara and E. Antonini, Enzymatic dehydrogenation of testosterone coupled to pyruvate reduction in a two-phase system, Eur. J. Biochem., 44 (1974) 401-405.

5 G. Carrea, P. Cremonesi, M.M. Casellato and E. Antonini, Enzymatic reactions in heterogeneous phase. Preparation of 5-androstene-3,17-dione, Steroids Lipids Res., 5 (1974) 162-166.

6 P. Cremonesi, G. Carrea, L. Ferrara and E. Antonini, Enzymatic preparation of 20β-hydroxysteroids in a two-phase system, Biotechnol. Bioeng., 17 (1975) 1101-1108.

7 G. Carrea, S. Colombi, G. Mazzola, P. Cremonesi and E. Antonini, Immobilized hydroxysteroid dehydrogenases for the transformation of steroids in water-organic solvent systems, Biotechnol. Bioeng., 21 (1979) 39-48.

8 E. Antonini, G. Carrea and P. Cremonesi, Enzyme catalyzed reactions in water-organic solvent two-phase systems, Enzyme Microb. Technol., 3 (1981) 291-296.

9 G. Carrea, R. Bovara, P. Cremonesi and R. Lodi, Enzymatic preparation of 12-ketochenodeoxycholic acid with NADP regeneration, Biotechnol. Bioeng., 26 (1984) 560-563.

10 G. Carrea, Biocatalysis in water-organic solvent two-phase systems, Trends Biotechnol., 2 (1984) 102-106.

11 P. Cremonesi, G. Mazzola and L. Cremonesi, Enzyme catalyzed reactions in water-organic solvent heterogeneous system, Annali Chimica, 67 (1977) 415-422.

12 G. Carrea and P. Cremonesi, Enzyme-catalyzed steroid transformations in water-organic solvent two-phase systems, Methods Enzymol. (K.Mosbach, ed.), 136 (1987) in press.

13 G. Carrea, S. Riva, R. Bovara and P. Pasta, in preparation.

14 G. Carrea, R. Bovara, R. Longhi and S. Riva, Preparation of 12-ketochenodeoxycholic acid from cholic acid using coimmobilized 12α-hydroxysteroid dehydrogenase and glutamate dehydrogenase with NADP cycling at high efficiency, Enzyme Microb. Technol., 7 (1985) 597-600.

15 J.R. Wykes, P. Dunnill and M.D. Lilly, Cofactor recycling in an enzyme reactor. A comparison using free and immobilized dehydrogenases with free and immobilized NAD, Biotechnol. Bioeng., 17 (1975) 51-68.

SESSION VI

SPECIFIC/FUTURE APPLICATIONS

Chairman : A. Ballesteros

Co-chairman : E.M. Meijer

C. Laane, J. Tramper and M.D. Lilly (Editors), *Biocatalysis in Organic Media*,
Proceedings of an International Symposium held at Wageningen,
The Netherlands, 7–10 December 1986.

Commercial Aspects of Biocatalysis in Low-Water Systems.

P Critchley
Applied Enzymology Section, Unilever Research Laboratory,
Colworth House, Sharnbrook, Bedford, MK44 1LQ, UK

Summary
This paper addresses the way in which industrialists use information and the wide framework of reference in which most industrial decisions are made. The technical fact that something can be done is often the first step on a long ladder of progress. Similarly, whilst the use of lipases for acyl exchange reactions has gathered momentum and produced many references over the past 12 years it is only recently that clear signs of commercial success have appeared.

The science of low water systems is in its infancy and will depend heavily on biochemists and physical chemists communicating and finding solutions to problems together. Critical areas that require attention are:
a) The conformation changes in protein (enzymes) that occur in concentrated aqueous, non-aqueous and interfacial systems, and
b) The various physical chemical effects of a matrix including charge, partition, Nernst diffusion layer effects and pore diffusion.

Introduction

It is not always possible for industrialists to know the detailed mechanisms controlling all parts of a process when they first exploit the science. Perhaps the best examples are the 'industrial' application of biocatalysts for bread, cheese and wine making that preceded the science biochemistry by more than two thousand years. More recently, in the latter part of the 19th century, rennet was first isolated on a commercial scale from the stomach of calves to standardise milk coagulation for cheese making, and takadiastase from *Aspergillus oryzae* was available for saccharification of starch, although the usage was small compared with the use of acid hydrolysis for a further 45 years. The fermentation industry developed rapidly during World War II to produce the very large volume of antibiotics required and this in turn had a beneficial effect on enzyme production both for the saccharification of starch and for proteolytic enzymes used in detergents.

Project selection is vital in good industrial research and a number of points need to be considered. The factors involved in making commercial decisions are very different from those made in an academic research laboratory and must involve engineers, medical/safety experts and marketing executives as well as researchers. Questions such as the cost of competitive technologies must be considered and may include a consideration of the flexibility of the plant to make a range of related products depending on the intended capacity and marketable tonnage per annum. A major question to be asked must be, is the technology to be added onto existing downstream extraction and purification plant or will it require a 'green-field' site which could increase the cost of production and many of the overheads very significantly. The price of the product, if it exists today, is a very important consideration and how much this fluctuates on the World market. The same questions must be asked of the cost of starting materials that will make up the feedstream. If the product does not exist in todays market a more complicated equation must be constructed to put a value on the market niche identified so that projected cash flows can be analysed.

A further set of questions revolve around the safety of the process and the legislative requirements that must be satisfied both during manufacture and for the final product. Where a product is prepared as a food ingredient or pharmaceutical a considerable mass of data is required to ensure that good manufacturing practices are satisfied and that a high quality, safe product is produced reproducibly. It is important for industry to protect investments of time and money spent on development and this is often done by patenting. However, this will polarize the research development in a certain way to generate the examples and once a patent is granted the cost to maintain it internationally are significant. In some instances it is better not to publish anything and to rely on trade secrets and lead-time in a field. This kind of thinking means that very few of the details of any process are published for many years.

Final considerations are the political and socio-economic factors that are important considerations for any company thinking of investing millions of pounds and several years of effort. An important question in this area must be is the market likely to be responsive to the changes caused by the introduction of a new or substitute material? Recent experiences of some European companies producing high fructose syrup from starch (Ref 1) and early responses in Europe to alcohol production for fuel (Ref 2) suggest that at least in some instances the technological problems are the simple part of the commercialisation of a process.

Many of the early processes relied on adding crude enzyme to the substrate and mixing for a fixed time before heating to deactivate the enzyme and where relevant drying. Whilst some of these early applications contained a significant amount of water others did not. The lipolysis of butter oil to produce cheese flavours (Ref 3) involved the use of emulsions that contained 25-95% fat. The water content as a percentage of the total weight of reaction mixture was low but the reactions of interest were largely hydrolytic ones producing a range of free fatty acids from triglycerides.

Another example is the use of mixed carbohydrates, particularly cellulases and pectinases, to degrade plant cell walls and release more oil from oil seeds during processing. The use of 500-1000g of enzyme per tonne of seeds at temperatures of 30-50°C can produce 2-6% increase in oil yield (Ref 4). The author makes the point that the process is most useful for oil extractions from palm and olive that are more fleshy (wet) fruits and the process will tolerate the addition of minimal amounts of extra water.

This paper concentrates on the use of enzymes rather than cells as biocatalysts. The level of information available for commercial exploitation of immobilised living cells is small, particularly in low water systems.

Two widely different enzyme reactions are used to illustrate the advantage to industry of using low water systems. In the first example a free enzyme, α-galactosidase is required to modify a galactomannan. This hydrolytic reaction was too costly in dilute solution and requires purified enzyme.

The second example is very different and involves the use of a crude immobilised lipase preparation to upgrade triacyl glycerols. Whilst a very small amount of water seems to be necessary to maintain enzyme activity and promote acidolysis (acyl exchange), the amount of water must be controlled.

Enzymic Modification of Galactomannan

For several years the availability of locust bean gum has been variable and hence the price has climbed significantly. Locust bean gum is a galactomannan having a backbone of ß1→4 linked mannose units to which single galactosyl residues are attached at intervals through ß1→6 linkages. A number of galactomannans occur in nature but the two available in thousands of tonne quantities are guar gum which is cheap and comes from the annual *Cyamopsis tetragonolobus* grown in India and USA and locust bean gum from perennial trees *Ceratonia siliqua* grown around the Mediterranean and in the Middle East. In both species the polysaccharides are deposited in the cell walls of the seed as a major food reserve for germination. The main differences between the two polymers are the degree of substitution of the mannan backbone by galactose residues and the molecular weight. In guar 38 to 40% galactose is normal whereas locust bean gum has far fewer galactose residues (normally 22-24%). The molecular weight of locust bean gum, calculated from viscosity measurements, averages about two thirds the value obtained for native guar gum. The properties of interest to industry are the gelling characteristics both of the native locust bean gum and in combination with other polysaccharides such as agar, carageenan and xanthan (Ref 5). Since pure ß1→4 mannan forms insoluble fibrils similar to ß1→4 glucan (cellulose) it is important to retain some galactosyl residues along the backbone to ensure solubility. However, guar gum forms viscous solutions and pastes rather than gels and interacts with other polysaccharides rather poorly. Therefore, a commercial objective was to find a cost effective way to remove some of the galactosyl residues from the guar gum without significant cleavage of the backbone to produce a polysaccharide with properties similar to those of locust bean gum.

Substrate Concentration

No chemical method was available that would selectively remove galactose residues without breaking many of the mannan backbone linkages and hence losing most of the useful properties. The use of an endo∝-galactosidase was an obvious choice but dewatering an enzyme-modified viscous solution containing about 1% of product added too much cost to a process and so it was unacceptable. A more concentrated system was required and therefore it was important to know whether substrate or product inhibition was a problem. A plot of maximum velocity against concentration of substrate at 40°C (Fig 1) showed that a concentration of about 10% substrate was required to saturate an ∝-galactosidase II (Ref 6) purified from guar seed. No inhibition was detected within the range of concentrations tested. Partially degraded, water-soluble, fractions from guar gum were used to avoid problems of interpreting diffusion effects in viscous solutions and pastes (Ref 7).

Purification of ∝-galactosidase II was essential to remove ß-mannanase activity which would cleave the backbone of the polysaccharide and reduce its value.

It was not possible to consider immobilised enzyme systems in packed beds or fluidised bed reactors with concentrations >0.5 to 1% of guar because of its physical state. Also, the activity of the product to form gels by interactions between neighbouring chains would be a problem. Therefore, a batch process was developed using enzyme from guar seed and heat deactivation to stop the reaction when sufficient galactose had been removed. Since some heat and alcohol deactivated ∝-galactosidase is present in guar gum used in food preparations it was reasonable to assume that the addition of a little more of the same enzyme would not cause too many problems in in processing or safety.

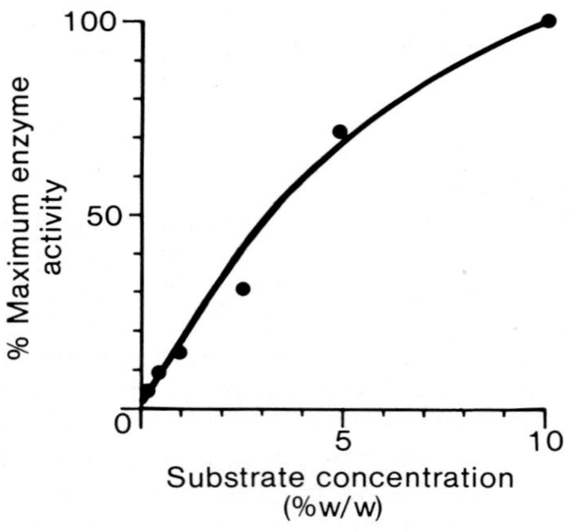

Figure 1
Diagram showing the influence of substrate concentration on the rate of hydrolysis of guar galactomannan by ∝-galactosidase II. (13 nkat on p-nitro phenyl ∝-galactoside in 0.1M sodium acetate buffer pH 4.5 for 5 min at 40°C).

An interesting property of ground guar galactomannan is that at high concentration (>20% wt. in water) the material becomes particulate and resembles bread crumbs which can be easily handled in food mixers and extruders (Ref 7). The ground galactomannan (flour) is formed by grinding guar splits (ie. seeds from which the embryo and the shell have been removed mechanically) in a colloid mill. Aleurone cells around the periphery of the cotyledons produce both ∝-galactosidase and ß-mannanase when hydrated and can cause degradation of polysaccharide to lower molecular weight fractions. Therefore, the splits or the flour had to be boiled in aqueous alcohol (80% v/v) for 10 minutes to destroy the aleurone cells, enzymes and any contaminating microorganisms.

Effect of Temperature
When ∝-galactosidase II was sprayed onto 40% solid bread crumb made by prehydrating guar flour, and mixed to a final substrate concentration of 20% a galactose to mannose ratio of 21:79 was obtained after 48 hours at 40°C (Ref 7). The long time of incubation and the low temperature were ideal conditions to allow microbial contamination and growth and therefore ways to speed-up the reaction were investigated to make it more suitable for commercial exploitation. Raising the temperature was an obvious option but the published optimum for guar crude ∝-galactosidase was 38°C (Ref 8) and 40°C for purified ∝-galactosidase II (Ref 5) and the enzyme rapidly lost activity above 45°C. When enzyme was added to a high concentration of prehydrated guar flour (50-60% solid) to give a final concentration of 40% substrate, after mixing in the enzyme, a rapid reaction was observed at temperatures up to 55°C for 24 hrs. The enzyme lost activity after a few hours at 60°C. Other ∝-galactosidase enzymes purified from lucerne and from fenugreek showed similar activity curves on 40% guar flour at 55°C (Ref 7), although not surprisingly the kinetics of the guar enzyme were faster than the other two for a given dosage of enzyme (Fig 2).

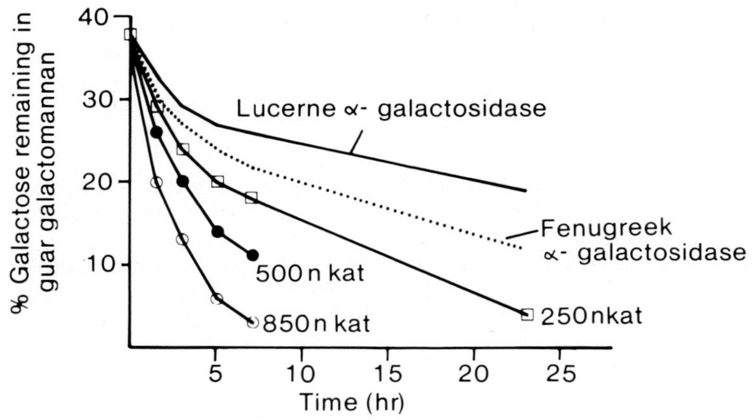

Figure 2
Diagram showing the rate of galactose removal from galactomannan (40% w/w) at 55°C by different activities of ∝-galactosidase from guar (250, 500, 850 nkat/g. guar) and by ∝-galactosidase (250 nkat/g. guar) purified from lucerne and fenugreek.

The stabilising influence of soluble polymers to combat thermal deactivation of a number of enzymes is known (Ref 9, 10) and therefore the mechanism of stabilisation of plant ∝-galactosidase in the presence of high concentrations of galactomannan may be a general one involving polymer entanglement, which restricts conformational change, rather than a specific substrate-enzyme interaction. Since the objective was to design a commercial process no experiments were performed to separate any contribution from a general versus a specific interaction.

Water activity
The influence of water activity on the hydrolytic reaction was not examined directly but substrate concentration was varied between 40 and 70% at a given temperature and enzyme level (Table I).

Table 1
The effect of substrate concentration on the rate of guar galactose removal at 55°C with an enzyme level of 850 nkat/g.

Guar concentration w/w %	% 0	Galactose remaining in galactomannan after time 1.5 hr	3 hr	5 hr	7 hr	23 hr
40	38	20	13	6	3	nd
50	38	25	21	17	14	nd
60	38	33	32	32	32	31
70	38	36	36	36	36	35

(nd = not determined)

Whilst reaction rate was considerably slower at water contents of less than 50% some reaction was detectable (Ref 7). The water activity in a 60% guar flour and enzyme reaction mixture is 0.96.

The Product

Depending on the substrate concentration, time and temperature of incubation a range of products were available with galactose: mannose ratios from 3 up to 30. Enzyme-modified galactomannan with a similar galactose: mannose ratio to locust bean gum showed very similar gelling properties (Table 2). Samples from which more galactose had been removed showed increased gel strength (Ref 7). Further tests have confirmed that whilst native guar would not form gels with xanthan or agarose the enzyme-modified guar produced firm gels at low concentration (Ref 11).

Table 2

The interaction of galactomannans (1.0% w/v) with xanthan (0.5% w/v) measured by a compression test in an Instron materials tester 24 hr after forming the gel plugs. (Cross head speed of the parallel plates was 20 mm/min).

Galactomannan	Galactose/mannose ratio	Yield stress (N)
Guar gum	40:60	No gel formed
Locust bean gum	23:77	3.3
Enzyme modified guar gum	23.5:76.5	3.0
Enzyme modified guar gum	8:82	5.3

Acidolysis and Interesterification

A number of attempts to change the acyl groups of inexpensive triacylglycerols to synthesise high value fats, such as those used in confectionery, have been published (Refs 12, 13, 14). The reactions are catalysed by 1:3 specific lipases and therefore substitute only esters on the primary hydroxyl groups of glycerol (Refs 15, 16). The experiences of these workers shows that a major problem is firstly the low solubility of substrate and products in an aqueous phase, and secondly the rapid hydrolytic reaction that produces diacylglycerols and monoacyl glycerols. Some hydrolysis is present in reactors where the water activity is less than one although the contribution of this reaction to the product can be significantly reduced under these conditions. Most of the attempts to design a commercial process have concentrated on ways to minimise hydrolysis of triacyl glycerols by reducing the quantity of water present around the enzyme. However, this can lead to loss of activity of the enzyme and ways to optimise activity without excessive diacyl glycerol formation have been a focus of much research. Reducing the amount of water can have a significant benefit in reducing the problem of poor mass transfer of substrate.

Provided that the water activity is less than one any hydrolysis uses water and rapidly swings the hydrolytic reaction towards its equilibrium point as water is consumed. This was demonstrated (Ref 17) in a packed-bed acidolysis reactor in which an almost water saturated feed stream of hexane, triacylglycerol and stearic acid was used to start the reaction in a hydrated bed of lipase precipitated on Celite. The initial level of diacylglycerol produced fell rapidly to an equilibrium value of 6% over the first 1-2 hrs as the reactor reached a steady state and water was consumed within the catalyst bed. Once a steady state was reached the feed stream was rapidly dehydrated as it passed through the reactor and the water activity changed from 0.95 in the feed stream to 0.5 in the deeper layers of the bed. Diacylglycerols change the properties of confectionery fat and must be removed during the down stream processing. The 1:2, 2:3 diacyl glycerols initially formed by partial hydrolysis of substrate can be transformed by chemical isomerisation into 1:3 diacyl glycerols. A short exposure time to the lipase reduces the deleterious side reactions including the amount of diacyl glycerols formed and therefore a packed-bed reactor was preferred rather than a continuous stirred tank of immobilised enzyme, although the latter also permits catalyst recovery and reuse (Ref 16).

Immobilisation of Lipases

Lipases require an interface for reaction and the measured activity of a lipase, assuming that precautions have been taken to avoid denaturation, has been shown frequently to be proportional to the surface excess of substrate (Ref 18). Therefore lipases only show a significant hydrolytic activity when correctly interfaced eg. on a silicone-coated glass particle (Ref 18) or in an emulsion of fine droplets (Ref 19). A wide range of materials have been examined in a search for commercially useful supports on which to immobilise lipases with varying levels of success. The material must be cheap and must not involve sophisticated cross-linking chemistry as this could prevent the food applications such as production confectionery fat. Using an acetone precipitation method from aqueous solution (Ref 12) 1:3 specific lipase from Aspergillus sp. was immobilised on four different silica materials. The lipase is not eluted because it will not dissolve in triacyl glycerol or hexane. The resulting catalysts were compared for both hydrolytic activity on a triacylglycerol, and acidolysis with myristic acid on palm mid-fraction (Ref 20). One of the supports was silane-coated, controlled pore silica (ex. Corning) and the other three materials were grades of Celite (Filtercel, Celite 560 and Hyflo Supercel, ex Johns Manville). Filtercel has a large surface area (20-30m^2 /g, about 10x that of Celite 560 and Hyflo Supercel) because it had not been subjected to a flux-calcination procedure which causes some fusing of the internal structure of the particles. The particle size of Celite 560 is significantly larger (5-10x) than that of Hyflo Supercel (which has 80% of particles $<20\mu$) but this is twice the size range of Filtercel. The lipolytic activity of the catalysts was more than ten times higher than the interesterification rate for all three samples of Celite but the controlled pore glass gave a low value for lipolytic activity and no interesterification was detected. It is likely that the combination of small pores (mean 375Å) and the silane-coating retained enzyme only at the surface of the particles. Celite 560, showed a lower interesterification activity although its lipolytic activity was comparable with that of the other Celite catalysts. The poor interesterification activity was almost certainly due to poor mass transfer of substrate within the larger particles of this support. Over a short time span (3 hrs) Hyflo Supercel proved to be the better support for interesterification with the high surface area Filtercel a close second.

In a study of the influence of the method of immobilisation on the hydrolytic activity of *Candida cylindracae* lipase (Ref 21) very few methods were suitable from a range including adsorption, covalent attachment and entrapment. Out of 18 materials investigated only Celite 535 as an adsorbent, a carbodiimide-condensed alkylamine Spherosil, and two hydrophobic entrapment polymers (made by photo-cross linking resin prepolymers) showed activities of more than 10% of the hydrolytic activity of enzyme in emulsion. However, the authors clearly showed that the rate of hydrolysis of triacyl glycerol was slower with immobilised/entrapped lipase and this is probably due to poor penetration of substrate emulsion into the particles.

In an earlier study (Ref 22), when 3 different Spherosils (one involving adsorption, one ionic and one covalent bonding) and 9 different polymer entrapment supports were compared with Celite 535 in an acidolysis ester exchange system with *Rhizopus delemar* lipase, a disappointing range of activities was demonstrated. Only ionic bonding of lipase to quaternary nitrogen groups on Spherosil and the use of hydrophobic photo-cross linked polymer gave results better than 20% compared with lipase activity on Celite alone. The method of synthesising the hydrophobic photo-cross linked polymer requires the addition of Tween 80 as an emulsifier, between the lipase in aqueous buffer and the polypropylene (PPG) ester polymer and photo sensitizer in hexane. It is well known that surface active agents can modify lipase activity by changing the nature of the interface and therefore it is difficult to interpret some of these results.

The access of substrate to lipase entrapped within a porous support is a potential problem. To date most supports used had a high affinity for water and therefore one would expect the substrate concentration in the microenvironment of the enzyme to be low and mass transfer into the aqueous phase to be the rate limiting step for reaction. The approach (Ref 23) of using a hydrophobic prepolymer to entrap catalyst should increase the hydrophobic interaction in the environment of the enzyme and thereby increase substrate concentration. Whilst hydrophobic prepolymer could be used to entrap lipase already immobilised on Celite the method proved very variable for free lipase (Ref 22). In neither case was the interesterification activity of the enzyme as good as that measured on Celite. Work in our Laboratory (Ref 24) has shown similar results for *Mucor meihei* lipase precipitated onto Celite (Hyflo Supercel) and then entrapped in a hydrophobic photo-cross linked prepolymer.

The solvent used for the interesterification reaction was important. In our experience, only trichloroethane was suitable for polyethylene glycol (PEG) and PPG polymers. This solvent led to no adverse effect on lipase and gave favourable partitions of substrate in both polymer systems and with Celite. Whilst hexane was a good solvent for reactions with Celite and Celite in PPG it did not work for PEG probably because the polymer dehydrates the solvent. (Table 3).

Table 3
The % myristic acid incorporated into olive oil over 24 hrs at 40°C using a catalyst of *Mucor meihei* lipase precipitated onto Celite (Ref 12). The lipase on Celite catalyst (0.5g) was entrapped in polymer made from PPG or PEG diacrylate ester (1.0g) following the method of Yokozeki et al (Ref 22).

Solvent	Celite	Celite in PPG polymer	Celite in PEG polymer
1:1:1-Trichloroethane	16.8	10.8	9.1
Toluene	8.7	2.9	0.2
Hexane	18.4	11.3	0

When the amount of pre-polymer used was increased above the ratio quoted in Table 3 there was a significant loss of lipase activity and reducing the size of the particles by chopping had no effect.

However, very low enzyme activities were obtained when a solution of crude lipase or a purified enzyme were entrapped in a similar range of diacrylate esters of PEG or PPG. Reducing the polymer particle size showed no increase in activity, in contrast to findings with lipase immobilised on Celite and then entrapped in polymer. When polymer material was extracted with aqueous buffer about 50% of the lipase activity (assessed by hydrolysis) was eluted from PEG or PPG polymers, compared with 50-70% recovery from Celite showing that a significant amount of potentially active enzyme was present.

Both native and SDS-polyacrylamide gel electrophoresis patterns of the lipase extracted from polymers were comparable with non-immobilised enzyme showing that no major, irreversible change had occurred. Therefore it seems likely that the lack of activity of lipase in the polymer must be a physical mechanism operating within the polymer matrix. Studies on swelling and diffusion of substrate into pre-swollen polymer showed that this was not a problem provided that the correct solvent (1:1:1 trichloroethane) was selected to match the hydrophobicity of the polymer and that the solvent did not deactivate lipase. Microscopic investigation using interference contrast and autofluorescent techniques (Ref 25) showed that complex phases were created in the lipase - prepolymer reaction mixture and that these prepolymer emulsions were very sensitive to polar surfaces. They broke down rapidly on contact with clean glass surfaces and therefore silcone-treated glass had to be used. Two main effects were observed when components of the reaction mixture were altered.

a) A change of surfactant from Tween 80 (HLB 15.0) to the much more hydrophobic Span 80 (HLB 4.3) in the same proportion, produced a simple emulsion rather than the complex type seen with Tween 80 (Fig 3).

b) Use of prepolymers with more hydrophilic groups (PEG) caused precipitation of material from the emulsion which has a yellow autofluorescence when excited with blue light, strongly suggesting protein material (Fig. 4). The physical nature of the precipitate was very shear sensitive; large lumps appeared on gentle agitation whereas a fine, filamentous precipitate, having the same autofluorescent property, was formed after vigorous stirring. It is probable that withdrawal of water by the PEG prepolymer causes phase separation of protein.

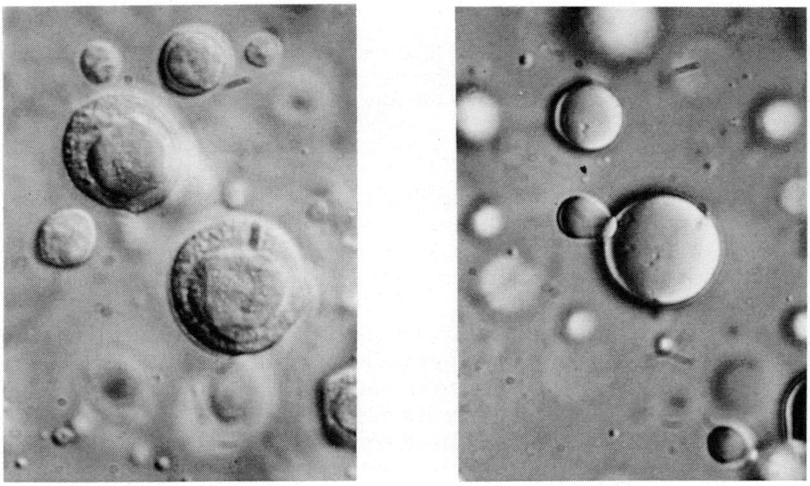

(a) (b)

Figure 3a
Interference contrast of micrograph of complex emulsion formed in prepolymer mixes when Tween 80 was used as the surfactant to form the emulsion droplets (largest droplet 18μ diameter).

Figure 3b
Interference contrast micrograph of a simple emulsion formed in prepolymer mixes when Span 80 was used as the surfactant (largest droplet 17μ diameter).

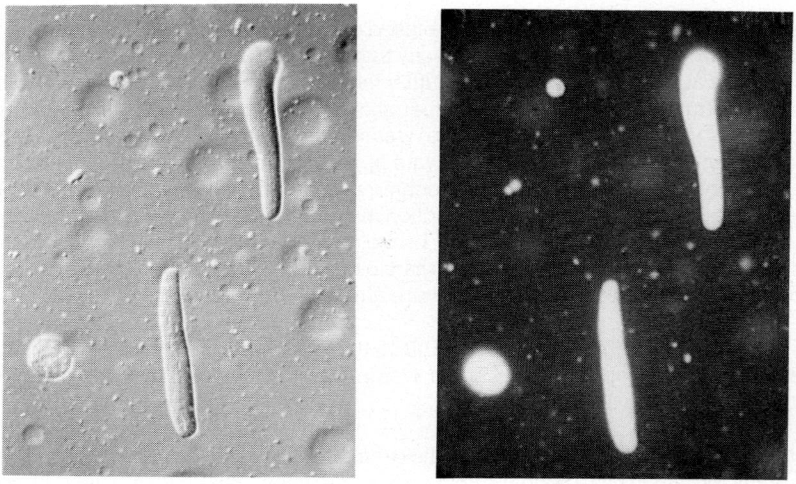

(a) (b)

Figure 4a
Interference contrast micrograph of phase separated material in prepolymer with a high PEG content at low sheer (length 140μ).

Figure 4b
Autofluorescence micrograph of the same material as seen in 4a, showing the strong yellow autofluorescence in blue light suggesting that the precipitate is protein (length 140μ).
Because of black and white reproduction the yellow regions of Figure 4b reproduce in white.

Whilst the use of photo cross-linkable prepolymers has proved successful for steroid transformations using immobilised cells (Ref 23) there is little published evidence to suggest that this method of immobilisation offers major advantages for commercial exploitation of lipases at this time other than the improved stability claimed by Yokozeki et al (Ref 22).

A new catalyst has been developed (Ref 26) that can be used for interesterification and ester synthesis reactions. The catalyst is made by binding *Mucor meihei* lipase to a macroporous, positively charged resin with a particle size of over 200 μm. The catalyst had an optimum temperature of 60-70°C. The lipase activity of a selected fraction of larger particles (300-600 μm diameter) was lower than the small particle fraction due to diffusional resistance in the porous particles. This problems has been observed by many researchers and can be analysed mathematically using numerical calculus (Ref 27). The macroporous resin showed relatively little compression and a good flow rate was maintained. This property, coupled with the high temperature optimum, enable an interesterification of olive oil with lauric acid (5:2 wt/wt) to be carried out at 60°C continuously for 3200 hrs. without solvent. The half life of the catalyst was 1600 hrs. and the productivity was 3.5 tonnes of triglyceride per kilogramme of enzyme run at a flow rate controlled to give 65% conversion (ie. 21% lauryl oleyl oleyl triglyceride). The working temperature, conversion, productivity and half-life are very comparable with figures from working reactors with other immobilised enzymes such as glucose isomerase and penicillin amylase, although the lipase reactor (14g. of catalyst) used was very small compared with those used in the sugar and antibiotic industries.

Acknowledgements
The author is indebted to Dr BV McCleary, who was on sabbatical from the Department of Agriculture, New South Wales, Australia, Dr PV Bulpin, now at Amersham International, Mr THC Windust and Mr D Cooke, who carried out the experimental work on the galactomannan transformation.

References

1) TJ Palmer, High Fructose Syrups, Proc. 1st National Conf. on Biotechnology in the Food and Drink Industries, 19-20th Sept 1983, Brintex Ltd, London pp. 117-124.
2) L von Bremen and M Schmoltzi, Economics and Politics of the Ethanol Market, Trends in Biotechnol. January (1986) 16-23.
3) JH Nelson, Enzymatically Produced Flavours for Fatty Systems, J Amer. Oil Chem. Soc, 49 (1972) 559-562.
4) T Godfrey, Edible Oils in: T Godfrey and J Riechelt (Ed), Industrial Enzymology: The Application of Enzymes in Industry, Macmillan Publishers Ltd, Surrey, 1983. pp 424-427.
5) BV McCleary, R Amado, R Waibel and H Neukom, Effect of Galactose Content on The Solution and Interaction Properties of Guar and Carob Galactomannans, Carbohyd Res. 92 (1981) 269-285.
6) BV McCleary, Enzymatic Interactions in The Hydrolysis of Galactomannan in Germinating Guar: The Role of Exo-ß-mannanase, Phytochemistry 22 (1983) 649-658.
7) BV McCleary, P Critchley and PV Bulpin, European Patent Application EP 0 121 960 A3 (1984).
8) HS Dugal and JW Swanson, Enzymatic Modification of Locust Bean and Guar Gums, Ippta 11 (1974) 29-35.
9) VV Mozhaev and K Martinek, Structure-Stability Relationships in Proteins: New Approaches to Stabilising Enzymes, Enzyme Microb. Technol, 6 (1984) 50-68.
10) G Greco jr and L Gianfreda, The Use of Synthetic Polymers for Preventing Enzyme Thermal Inactivation; some comments in the paper by Alfani and co-workers, Biotechnol. Letters, 7 (1985) 65-68.
11) BV McCleary, ICM Dea, J Windust and D Cooke, Interaction Properties of D-Galactose-Depleted Guar Galactomannan Samples, Carbohyd. Polymers, 4 (1984) 253-270.
12) MH Coleman and AR Macrae, UK Patent 1, 577, 933 (1980).
13) Fuji Oil Co, Japanese Patent 56025643 (1981).
14) T Tanaka, E Ono, M Ishihara, S Yamanaka and K Takinami, Enzymatic and Exchange of Triglyceride in N-Hexane, Agric. Biol. Chem. 45 (1981) 2387-2389.
15) RW Stevenson, FE Luddy and HL Rothbart, Enzymatic Acyl Exchange to Vary Saturationin Di- and Triglycerides, J Amer. Oil Chem. Soc, 56 (1979) 676-680.
16) AR Macrae, Lipase-Catalysed Interesterification of Oils and Fats, J Amer. Oil Chem. Soc, 60 (1983) 291-294.
17) AR Macrae, Interesterification of Fats and Oils; in: J Tramper, HC van der Plas and P Linko (Eds.), Biocatalysis in Organic Synthesis, Elsevier, Amsterdam, 1985, pp 195-208.
18) HL Brockman, JH Law and JF Kezdy, Catalysis by Adsorbed Enzymes. The Hydrolysis of Tripropionin by Pancreatic Lipase Adsorbed to Siliconised Glass Beads, J Biol. Chem. 248,(1973) 4965-4970.
19) G Benzonana and P Desnuelle, Etude Cinetique de L'action de La Lipase Pancretique Sur Des Triglycerides en Emulsion. Essai d'une Enzymologie En Milieu Heterogene, Biochem. Biophys. Acta, 105 (1965) 121-136.
20) RA Wisdom, P Dunnill, MD Lilly and AR Macrae, Enzymic Interesterification of Fats: Factors Influencing The Choice of Support for Immobilised Lipase, Enzyme Microb. Technol, 6 (1984) 443-446
21) Y Kimura, A Tanaka, K Sonomoto, T Nihura and S Fukui, Application of Immobilised Lipase To Hydrolysis of Triglyceride, Eur. J Appl. Microbiol. Biotechnol, 17 (1983) 107-112.
22) K Yokozeki, S Yamanaka, K Takinami, Y Hirose, A Tanaka, K Sonomoto and S Fukui, Application of Immobilised Lipase to Regio-Specific Interesterification of Triglyceride in Organic Solvent. Eur. J Appl. Microbiol. Biotechnol, 14 (1982) 1-5.
23) S Fukui and A Tanaka, Applications of Biocatalysts Immobilised by Prepolymer Method, Advances in Biochem. Engineering Biotechnol, 29 (1984) 1-33.
24) A Peilow, J Bosley and P Critchley, Unpublished Observations.
25) M Asquith and P Critchley, Unpublished Observations.
26) TT Hansen and P Eigtved. A New Immobilised Lipase for Interesterification and Ester Synthesis, in: AR Baldwin (Ed), Proc. World Conference on Emerging Technologies in the Fats and Oil Industry, Cannes, France, Nov 3-8, 1985, Amer. Oil Chem. Soc. pp 365-369.
27) A Bodalo, JL Gomez, E Gomez, J Bastida, JL Iborra and A Manjon, Analysis of Diffusion Effects on Immobilised Enzymes on Porous Supports with Reversible Michaelis-Menten Kinetics, Enzyme Microb. Technol, 8 (1986) 433-438.

C. Laane, J. Tramper and M.D. Lilly (Editors), *Biocatalysis in Organic Media*, Proceedings of an International Symposium held at Wageningen, The Netherlands, 7–10 December 1986.

INTEGRATION OF ENZYME CATALYSIS IN AN EXTRACTIVE FERMENTATION PROCESS

M.R. AIRES BARROS[1], A.C. OLIVEIRA[2] and J.M.S. CABRAL[2]

[1]Departamento de Energias Renováveis, LNETI, 1600 Lisboa, Portugal

[2]Laboratório de Engenharia Bioquímica, Instituto Superior Técnico, 1000 Lisboa, Portugal

SUMMARY

A novel recovery process for fermentation products, namely alkanols and organic acids, utilizing extraction and enzyme reaction is described. This enzyme-catalysed separation process involves the use of a lipase in the extractive fermentation system, to esterify the product with the extractant. *In situ* recovery of ethanol from fermentation of high glucose concentration (400 g/l) was used as a model and water-immiscible fatty acids were selected as organic solvents for ethanol. The effect of the extractants on cell viability of *Saccharomyces bayanus*, was tested in free and entrapped cell preparations. Immobilization seems to protect yeast cells against solvent toxicity, when oleic acid was used as solvent. The distribution coefficient of ethanol between the aqueous and the organic phases can be improved ten fold by exterification with a lipase of *Mucor miehei*. The enzyme exhibits a pronounced substrate specificity and only esterifies ethanol with long chain fatty acids (C_8C_{18}), the ethyl ester increasing with the chain length. The ethanol esterification with oleic acid by a lipase, at pH 4,5 and 30^oC, was studied, the enzymatic reaction being only inhibit at high concentration of both substrates. Ethyl oleate formation was favored at low water activities, in the presence of high glucose concentration. An extractive fermentation of ethanol using kcarrageenan entrapped *S. bayanus* cells, a lipase and oleic acid was carried out and the productivity of this process compares favorably with the ethanol fermentation in the absence of the enzyme

INTRODUCTION

Biotechnological processes are characterized by being performed in aqueous solutions, usually, at low temperaturas and pressures, and by giving diluted product streams.

Product inhibition is a classic situation in biochemistry, in which the metabolic activity of the organism producer is reduced by the end-product(s), solvents and organic acids being substances causing general inhibiton phenomena. In order to improve the productivity, several technological approaches have been studied, such as extractive fermentation(1), on line adsorption(2), vacuum(3) and flash fermentation(4).

In situ extraction of bioproducts is an approach used to reduce the end-product inhibiton and to integrate the fermentation stage with those of downstream processing in order to optimize the whole fermentation process(5). Liquid-liquid extraction is one unit operation which utilizes effectively the partioning of components between two immiscible solvents or the difference of the partition coefficients between components. The most important characteristics for a liquid extraction process, when establishing an integrated fermentation-separation process, are:

- High capacity for the product, described by the equilibrium distribution coefficient (k_D);
- High selectivity for the product over water, described by the separation factor (α); and
- No toxicity to the microorganism, described by the solubility parameter (δ)

The two first characteristics may be improved by extraction with reaction(6). The purpose of this work is to develop a novel effective extraction process of fermentation end-products, namely solvents (alkanols) and organic acids, utilizing liquid-liquid extraction and enzyme reaction. The biochemical process model considered is the production of ethanol, as this compound is a non-competitive inhibitor of the glucose-to-ethanol pathway.

Extractive fermentation of ethanol, by gel entrapped Saccharomyces bayanus cells, using water-immiscible organic acids as extractants is described to ellucidate the principles of this novel recovery process. Organic acids were chosen as solvents for ethanol, as they provide a high selectivity for ethanol over water(7,8) and allow the enzymatic esterification of ethanol by lipases.

MATERIAL AND METHODS

Microorganism

Saccharomyces bayanus, from L'Institut d'Oenologie de Paris, was used as a typical ethanol producing yeast.

Enzyme

A lipase (EC 3.1.1.3) of Mucor miehei, SP 225 a gift from NOVO Industri A/S, Copenhagen, was used in the esterification reaction of ethanol with water-immiscible organic acids.

Materials

Valeric, hexanoic, octanoic and nonanoic acids, analytical grade were obtained from BDH Chemicals, Ltd. Ethanol, glucose and oleic acid, analytical grade, were form Merck.

k- Carrageenan (Gelcarin CIC) was a gift from FMC, USA. Yeast extract was from Difco. All other reagents used were either Laboratory or analytical grade.

Growth medium and culture conditions

The fermentation and growth medium contained (g/l): glucose, 400; yeast extract, 5; KH_2PO_4, 5; $(NH_4)_2SO_4$, 5; and $MgSO_4.7H_2O$, 1; pH was 4.5.

Extractive and control fermentations of high glucose concentration (400 g/l) were carried out in 100 ml shake flasks, provided with a rubber stopper end a needle, containing 50 ml of both fermentation medium and organic phase, for different periods of time in a rotary shaker (170 rpm) and at 30ºC. Glucose consumption, ethanol production and cell growth were monitored.

Esterification reaction

Equilibrium and kinetics studies of ethanol esterification by a lipase (86 mg protein/l) at pH 4.5 and 30ºC were carried out in 100 ml shake flasks containing 50 ml of both aqueous (fermentation medium) and organic phases, for different reaction conditions and ethanol and organic acids concentrations.

Whole cell immobilization

Saccharomyces bayanus cells were entrapped in k-carrageenan according to the technique of Wada et al. (1979)[9]. The inoculum (0.1 g/l) was mixed with a solution of k-carrageenan(final concentration 4.5%) at 45ºC. The mixture obtained was added dropwise to a 3.5% KCl solution at room temperature. The resulting k-carrageenan gel beads (mean diameter of 4 mm) entrapped the yeast cells and were left to harden for 30 minutes in the saline solution. The entrapped cells were transfered into a shake flask containing the fermentation medium and the extractant, and were incubated as described for the free cell suspensions.

Distribution coefficients for ethanol

The partition of ethanol between the organic solvent and the aqueous phase was carried out in shake flasks in a rotary shaker (170 rpm) at 30ºC and pH 4.5. The aqueous phase was fresh fermentation medium supplemented with variable ethanol concentrations. The residual ethanol in the aqueous phase was determined for the different experimental conditions, time, initial ethanol concentration and solvent/aqueous phase ratio.

Analytical methods

Free cell concentration was estimated from optical density at 640 nm, using appropriate diluitions.

Yeast cell concentration in the immobilized preparation was determined by dissolving the entrapped cell beads of k-carrageenan in a 0.9% NaCl solution at room temperature following by optical density measurements.

Ethanol concentration in the fermentation medium was determined by gas-liquid chromatography, using a HewlettPackard model 5710 A, with flame ionization detector, n-Butanol was employed as the internal standard and samples were injected onto a Porapack Q (5 mm i.d. * 1 m 60-80 mesh) column. The samples were previously prepared, removing the cells by centrifugation, and providing appropriate dilutions , when necessary. The temperatures of injector, detector and oven were kept at 250, 250 and 110°C, respectively. Nitrogen was used as the carrier gas at a flow rate of 45 ml/min.

Glucose was measured by the dinitro salicylic acid method.

RESULTS AND DISCUSSION

Extraction of ethanol by organic acids

Physical liquid-liquid extraction. Water-immiscible organic acids were used as extracting solvents of ethanol from a fermentation medium. This group of compunds provide a higher separation factor, α, than other solvents for the same distribution coefficients[7,8].

TABLE 1

Physical and ethanol extraction properties of organic acids

Organic Acid	Ethanol Distribution Coefficient K_D (30°)	Separation factor α	Solubility Parameter $\delta(cal/cm^3)^{0.5}$
Valeric	1.13	13	9.1
Hexanoic	0.944	15	8.7
Octanoic	0.525	23	8.0
Nonanoic	0.464	25	7.7
Oleic	0.171	52	5.4

Table 1 shows some physical and ethanol extraction properties of four aliphatic carboxylic acids (C_5, C_6, C_8 and C_9) and the unsaturated, oleic acid. The distribution coefficient values shown in Table 1 and Figure 1 were obtained at 30°C using a cell-free fermentation medium at pH 4.5, to which ethanol was added to a final concentration of 100 g/l, the solvent/fermentation medium ratio was equal to 1 and equilibrium time was 24 h. The ethanol distribution coefficients between hexanoic and octanoic acids and fermentation medium agree with those obtained by other authors[8] using water as the aqueous phase, the values being 1.0 and 0.60, respectively.

The equilibrium distribution coefficient for ethanol decreases with the chain length of the organic acid. This may be related with the decrease of

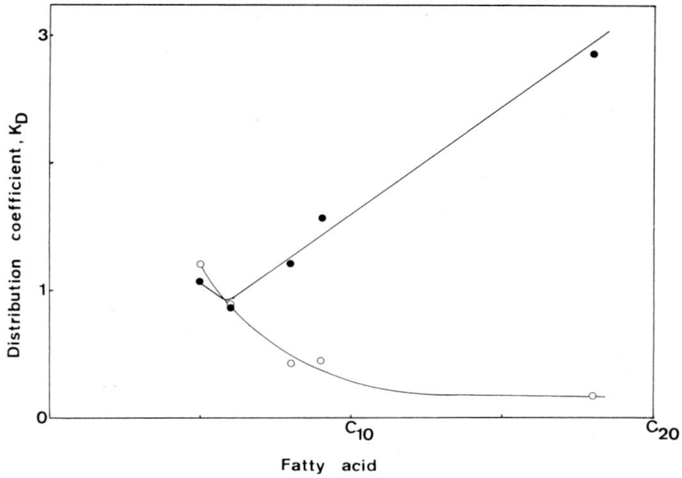

Figure 1. Distribution coefficients for 100 g/l of initial aqueous ethanol between water-immiscible organic acids and aqueous phase (1:1 ratio): (○) physical liquid extraction and (●) liquid extraction with enzyme reaction.

the solvent solubility in water or with the solvent polarity, measured by the Hildebrandt solubility parameter, δ[10]. However, the molar extraction ratio of ethanol with water-immiscible organic acids was found to be 1 mole of ethanol per 8.5 mole of fatty acid. The partition of ethanol between the organic and aqueous phases increases with the enhancement of the ethanol concentration in the aqueous phase[11,12].

Enzymatic Extraction. The extraction of 100 g/l ethanol from the aqueous phase is strongly enhanced at pH 4.5 and 30°C when a lipase (86 mg/l) from the mold *Mucor miehei* is added to an extraction system containing an organic to aqueous phases ratio of 1. (Figure 1). The enzyme exhibits a pronounced substrate specifity which increases with the chain length of the carboxylic acids. Furthermore, valeric and hexanoic acids do not undergo esterification at all.

The *Mucor miehei* enzyme preparation seems to display a relatively high synthetic activity, even in the presence of high water content, when hydrolysis usually prevails[13]. The esterification yields of ethanol, over 24 hours, with excess of octanoic, nonanoic and oleic acids were 35, 43 and 65%, respectively.

Enzymatic synthesis of ethyl oleate

Effect of ethanol and oleic acid. The ethyl oleate synthesis was studied at pH 4.5 and 30°C for 24 h using different ethanol and oleic acid concentrations (Figure 2).

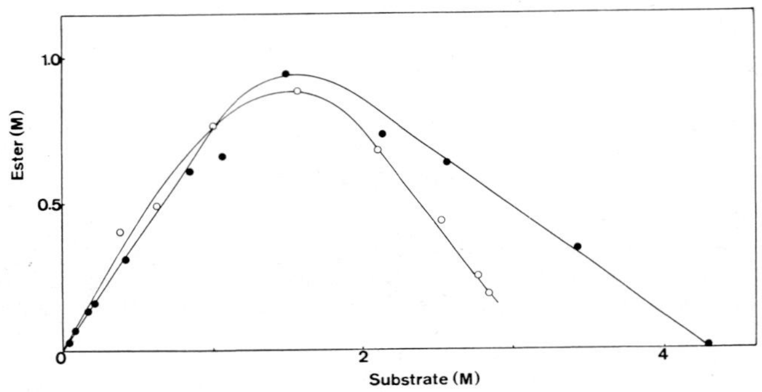

Figure 2. Effect of ethanol (●) and oleic acid (○) on the synthesis of ethyl oleate, catalysed by a *Mucor miehei* lipase.

The influence of ethanol on the rate of the ester formation shows a typical substrate inhibition pattern. The enzymatic synthesis of ethyl oleate is strongly inhibited, at ethanol concentration, in the aqueous phase, higher than 140 g/l, and no activity was detected with 400 g/l ethanol. The inhibition of the enzyme by increasing alkanols concentrations were also reported [13,14], in systems with very low water content. However the ethanol concentration usually found in the fermentation broth is less than 150 g/l, which allows the enzymatic esterification of ethanol without substrate inhibition, by the lipase preparation used in this work. The esterification yields of ethanol with 1.58 M oleic acid decreased from 83% to 63% with the increase of initial ethanol concentrations in the aqueous phase from 4 to 140 g/l, respectively.

The influence of oleic acid on the enzymatic synthesis of ethyl oleate was studied at different solvent/aqueous phase ratios, with 110 g/l ethanol in the aqueous medium. The effect of oleic acid is similar to that of ethanol; the rate of esterification reaction decreased at overall concentration of oleic acid in the reaction mixture higher than 1.58 M, which corresponds to a 1:1 solvent/aqueous phase ratio. The decrease of the ethyl oleate synthesis at high solvent/aqueous phase ratios is even more pronouced than the aqueous ethanol effect (Figure 2). The rate of reaction appears to be dependent on the oleic acid concentration, which tends to either inhibit the catalyst or to slow down the rate of acid diffusion to interface between the two phases, where the reaction seems to occur[14,15]. Oleic acid was completely esterified at low concentration (0.3-0.4 M), the yield of esterification decreasing with the increase of oleic acid phase. At 1:1 oleic acid/aqueous phase ratio the esterification yields of oleic acid (1.58 M) and ethanol (1.22 M) were 56 and 72% respectively.

Effect of reaction time. The time course of the esterification reaction at pH 4.5 and 30°C of ethanol (1.07 M) with oleic acid (1.58 M) in a phase ratio of 1, by the *Mucor miehei* lipase is shown in Figure 3.

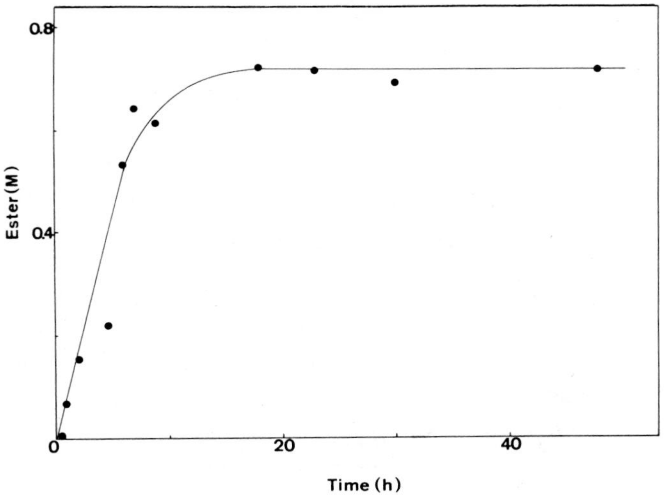

Figure 3. Effect of reaction time on the synthesis of ethyl oleate catalysed by a *Mucor miehei* lipase, from 100 g/l of initial aqueous ethanol in a 1:1 oleic acid/aqueous phase ratio.

As can be seen from this figure an equilibrium concentration of 0.71 M ethyl oleate was reached after 18 hours, leading to a esterification ethanol yield of 66%. Similar conversions were obtained in the synthesis of oleic acid glycerides by *Mucor miehei* and *Chromobaterium viscosum* lipases in the presence of very low (3 - 4%) water content[16], and ethyl oleate by *Mucor miehei* enzyme[13].

The ester concentration could not be increased with new enzyme addition, which confirms the equillibrium concentration obtained. An equilibrium constant of 80 can be calculated for the esterification of ethanol in the biphasic system.

Effect of water activity. The effect of water content on the conversion of 40 g/l ethanol in the aqueous phase, with 1.58 M oleic acid in the biphasic system was examined by keeping the water activity of ethanol-glucose-water solution in the ranges shown in Figure 4. The organic phase/aqueous media ratio was 1:1. Glucose was selected to reduce the water content in the aqueous phase, as it is the substrate in the fermentation studies.

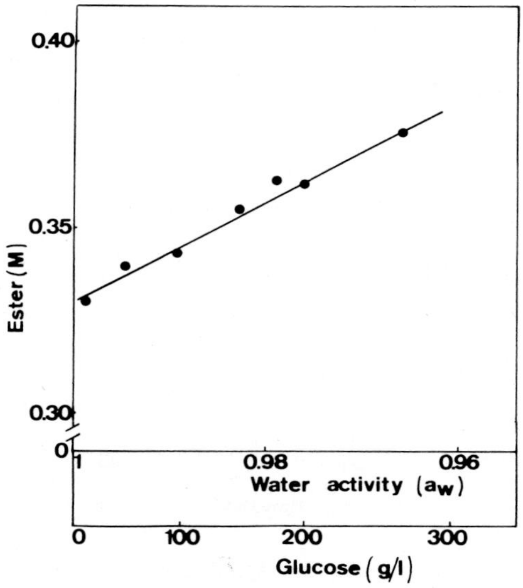

Figure 4. Effect of water activity on the synthesis of ethyl oleate catalysed by a Mucor miehei lipase, from 40 g/l of initial aqueous ethanol in a 1:1 oleic acid/aqueous phase ratio.

The degree of esterification achieved is inversely proportional to the water content of the system, the esterification yields of ethanol being 74% and 87% at water activities of 1 and 0.965, respectively. The higher conversion rate of ethanol, at high glucose concentration, is due to the best performance of ester synthesis by the Mucor miehei enzyme system under low water contents[13,16]. For the experimental conditions used, the following correlation of ethyl oleate concentration, in the reaction mixture, with the water activity, was found: (Ester) (M) = $1.68 - 1.35\ a_W$.

The equilibrium constant increases with the decreasing of water activity and a biphasic constant of 190 was reached at low water activity (0.965).

The distribution coefficient, at high (270 g/l) glucose concentration also increased, by a factor of 2.6 fold relatively to the value in the absence of glucose.

Extractive fermentation of ethanol

The process feasibility for *in situ* physical and enzyme-catalysed extractions of ethanol from a fermentation broth and the protection of cells by immobilization against solvent toxicity was tested in several extractive fermentations of 400 g/l of glucose solutions (Figure 5). Whole cells of

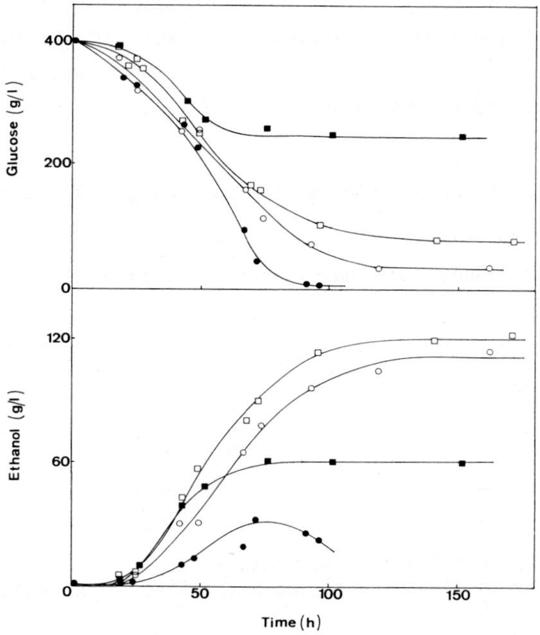

Figure 5. Fermentation of 400 g/l of glucose by *Saccharomyces bayanus*: (■) free cells; (□) k-carrageenan entrapped cells; (○) k-carrageenan entrapped cells in 1:1 oleic acid/fermentation medium ratio; and (●) k-carrageenan entrapped cells in 1:1 oleic acid/fermentation medium ratio in the presence of a *Mucor miehei* lipase.

Saccharomyces bayanus were entrapped in 4.5% w/v k-carrageenan and oleic acid used as the extracting solvent in a 1:1 organic phase/fermentation medium ratio.

Immobilization of whole cells has been used to improve the ethanol fermentation performance under substrate inhibition(12), due to the higher metabolic efficiency than the free cell systems. Figure 5 shows this effect of entrapped cells on the fermentation of 400 g/l of glucose, the substrate residual level being 77 and 245 g/l, for the k-carrageenan entrapped cells and their free counterparts, respectively (Table 2). Ethanol produced by the free cells is lower (62 g/l) than in the entrapped cells (122 g/l), which suggests that the inhibitory effect of substrate is much more pronounced than the end-product inhibition. The very different fermentation behaviour may be due to higher cell viability in the immobilized cell system than in the free cell. At high sugar concentrations, the water activity (osmolality = 3.3 Os/Kg) and oxygen supply are low, which may lead to different metabolic pathways(12,17) with accumulation of by-products other than ethanol. The gel entrapment of yeast cells creates a gradient of glucose, allowing lower micro environmental concentrations of sugar, which would be non-inhibitory to the fermentation by the entrapped cells(12).

TABLE 2

Comparison of ethanol fermentation systems of 400 g/l of glucose

System	Residual Glucose (g/l)	Ethanol in aqueous phase (g/l)	Glucose uptake rate (g/lh)
A	245	61.5	1.7
B	77	122	4.0
C	38	115	4.6
D	7	22.6	5.7

A - Free cells
B - Entrapped cells
C - Entrapped cells in extractive fermentation
D - Entrapped cells in lipase-catalysed extractive fermentation

Physical extractive fermentation occurs by reducing ethanol concentration in the aqueous phase and consequently product inhibition, which leads to the utilization of higher glucose solutions (Figure 5 and Table 2). The residual glucose concentration in the fermentation broth was 38 g/l, when a 1:1 oleic acid/fermentation medium ratio was used. As oleic acid is toxic to the cells at high concentrations (1.58 M)(12), these results suggest that immobilization of cells, by entrapment in k-carrageenan, protects the cell viability probably due to steric hindrance and/or diffusional effects.

The glucose consumption increases, while the level of ethanol in the aqueous phase strongly decreases (26 g/l) when a Mucor miehei lipase (86 mg/l) was added to the liquid-liquid extraction - fermentation system (Table 2).

This fermentation process was carried out with controlled pH at 4.5, in order to maintain the activity of the lipase, which tends to be inactivate at lower pH values, usually obtained in the ethanol fermentation without pH control.

Glucose was almost completely consumed, without increasing the organic extractant/fermentation medium ratio, which is necessary with physical extractant(12).

The total ethyl oleate synthesized during the fermentation was 1.04 M, which corresponds to esterification yields of ethanol and oleic acid, 78.3 and 65.7%, respectively. These yields are higher than those reported previously in aqueous phases without glucose, probably, due to the effect of the low water activities of the fermentation broth, which varies from 0.94 to 1 during the ethanol fermentation.

In this work it was found that glucose can be metabolized by gel entraped yeast cells and the ethanol produced by fermentation can be esterified in situ with the extractant by a Mucor lipase yielding ethyl oleate, and a very low ethanol level being accumulated in the fermentation broth. With this enzyme-catalysed extractive fermentation process it is potentially possible to synthesize esters of water-immiscible fatty acids from sugar feedstoks via ethanol route.

REFERENCES

1 M. Minier and G. Goma, Ethanol production by extractive fermentation, Biotechnol. Bioeng., 24 (1982) 1565-1579.

2 H.Y. Wang, F.M. Robinson and S.S. Lee, Enhanced alcohol production through on-line extraction, Biotechnol. Bioeng. Symp., 11 (1981) 555-565.

3 G.R. Cysewski and C.R. Wilke, Rapid ethanol fermentation using vacuum and cell recycle, Biotechnol. Bioeng., 19 (1977) 1125-1143.

4 T.K. Ghose, P.K. Roychoudhury and P. Ghosh, Simultaneous saccharification and fermentation of lignocellulosics to ethanol under vacuum cycling and step feeding, Biotechnol. Bioeng., 26 (1984) 377-381.

5 B. Mattiasson and M. Larsson, Extractive bioconversions with emphasis on solvent production, Biotechnol. Genetic Reviews, 3 (1985) 137-174.

6 H. Ishikawa, H. Nishida and H. Hikita, Theoretical analysis of a new separations process utilizing extraction and enzyme reaction, J. Chem. Eng. Japan, 19 (1986) 149-153.

7 C.L. Munson and C.J. King, Factors influencing solvent selection for extraction of ethanol from aqueous solutions, Ind. Eng. Chem. Process Des. Dev., 23 (1983) 109-115.

8 J.W. Roddy, Distribution of ethanolwater mixtures in organic liquids, Ind. Eng. Chem. Process. Des. Dev., 20 (1981) 104-108.

9 M. Wada, J. Kato and I. Chibata, A new immobilization of microbial cells, Eur. J. Appl. Microbiol. Biotechnol., 8 (1979) 241-247.

10 L.E.S. Brink and J. Tramper, Optimization of organic solvent in multiphase biocatalysis, Biotechnol. Bioeng., 27 (1985) 1258-1269.

11 M. Matsumura and H. Markl, Application of solvent extraction to ethanol fermentation, Appl. Microbiol. Biotechnol., 20 (1984) 371-377.

12 M.R.A. Barros, J.M.S. Cabral and J.M. Novais, Production of ethanol by immobilized Saccharomyces bayanus in extractive fermentation systems, Biotechnol. Bioeng. (in press)

13 I.L. Gatfield, The enzymatic synthesis of ester in nonaqueous systems, Ann. N.Y. Acad. Sci., 434 (1984) 569-572.

14 T. Knox and K.R. Cliffe, Synthesis of long-chain esters in a loop reactor system using a fungal cell bound enzyme, Process Biochem., October 1984, 188-192.

15 S. Mukataka, T. Kobayashi and J. Takahasi, Kinetics of enzymatic hydrolysis of lipids in biphasic organic-aqueous systems, J. Ferment. Technol., 63 (1986) 461-466.

16 T. Yamane, M.M. Hoq, S. Shimizu, S. Ishida and T. Funada, Continuous synthesis of glycerides by lipase in a microporous membrane bioreactor, Ann. N.Y. Acad. Sci., 434 (1984) 558-568.

17 B. Mattiasson and B. Hahn-Hagerdal, Microenvironment effects on metabolic behaviour of immobilized cells, a hypothesis, Eur. J. Appl. Microbiol. Biotechnol., 16 (1982) 52-55.

C. Laane, J. Tramper and M.D. Lilly (Editors), *Biocatalysis in Organic Media*, Proceedings of an International Symposium held at Wageningen, The Netherlands, 7–10 December 1986.

NICOTINAMIDE COFACTOR-REQUIRING ENZYMATIC SYNTHESIS IN ORGANIC SOLVENT-WATER BIPHASIC SYSTEMS

Chi-Huey Wong, Department of Chemistry, Texas A&M University
College Station, Texas 77843

SUMMARY

This talk illustrates several practical one-pot syntheses of two valuable and separable compounds in a biphasic system using nicotinamide cofactor-requiring enzymes as catalysts. In each system, two synthetic reactions occur in the aqueous phase where the cofactor is recycled, and one product is extracted into the organic phase while another is retained in the aqueous phase. The effective separation of products and elimination of product inhibition during the reactions make the biphasic system practical for large scale operation.

The major factors which have been considered to affect the practicality of nicotinamide cofactor-requiring oxidoreductases are the cost and stability of the cofactors and enzymes, the solubility and stability of substrates and products, and the degree of inhibition of enzyme activities in the presence of substrates or products. The costs of the cofactors range from $700/mol for NAD to $200,000/mol for NADPH. These prices are sufficiently high that they exclude stoichiometric use of cofactors. Instead, it has been necessary to develop schemes for the <u>in situ</u> regeneration of the cofactors. A very large amount of work has been devoted to developing and testing various types of systems for regeneration of nicotinamide cofactors,[1] and several of them have proven very practical for large-scale synthesis.

For a nicotinamide cofactor-dependent reaction to be practical for synthesis of fine chemicals, the cofactors must be recycled for at least 10^3-10^5. To achieve this high turnover of the cofactor, the yield and selectivity in reaction for each cycle (particularly for the stereoselective reduction of NAD(P) to the active NAD(P)H) must be almost 100%, and enzymatic methods seem to be the most selective for the regeneration of NAD(P)H.

The nicotinamide cofactors are intrinsically unstable in solution. The oxidized forms are stable in slightly acidic conditions but unstable in basic conditions. The reduced forms are stable in slightly basic conditions but unstable in acidic conditions. The reduced cofactors are destroyed by several processes, the most important of which is a general-acid catalyzed protonation at the 5 position of the dihydronicotinamide ring, followed by a fast addition of water to the resulting iminium salt and further irreversible conversion to

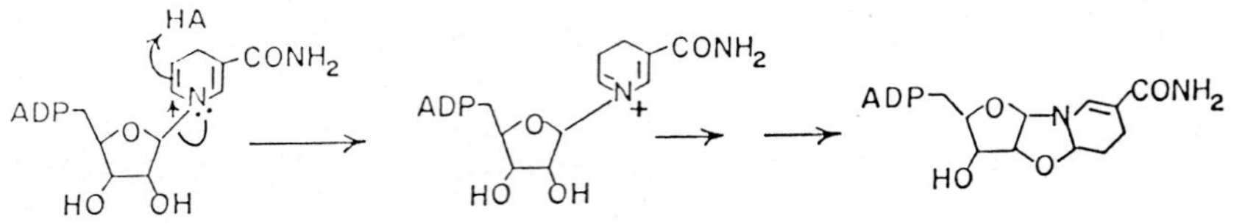

Figure 1. General acid-catalyzed decomposition of NAD(P)H.

$$-d\ln[NADH]/dt = k^{obsd}_{NADH} = (k_H[H^+] + k_{HA}[HA] + k_{H_2O}) \quad (1)$$

$$-d\ln[NADPH]/dt = k^{obsd}_{NADPH} \quad (2)$$

$$k^{obsd}_{NADPH} - k^{obsd}_{NADH} = \frac{k_I[H^+]}{K_a + [H^+]} = \frac{k_I}{1 + 10^{pH-pK_a}} \quad (3)$$

a cyclic ether product (Figure 1).[2,3] The kinetics of the decomposition are shown in eq 1-3. Since NADPH has one extra phosphate group, it is less stable than NADH because the extra phosphate contributes to the intramolecular protonation of the dihydronicotinamide ring. The difference of the stability between NADH and NADPH is shown in eq 3 where k_I is the intramolecular rate constant (0.15 h^{-1}) and $H^+/k_a + H^+$ is the fraction of the monoprotonated 2'-phosphate (pK = 6.2) of NADPH.[2] Figure 2 indicates the Brφnsted plot and the stability of NADH and NADPH in different buffers. These results suggest that under neutral conditions certain buffers, particularly phosphate buffer, catalyze the decomposition of NAD(P)H. The best operation condition is to control the reaction pH by adding NaOH or HCl solution automatically to the reaction mixture containing a small amount of organic buffers (such as Tris, Hepes, or triethanolamine buffer) and an appropirate amount of nonprotic inorganic salt to maintain the appropriate dielectric environment for the enzymes. Although these organic buffers appear to stabilize NAD(P)H to some extent, they cause problems in work-up. It is better to keep them at minimum concentration. Our experience indicates that the compromised pH for NAD and NADH is 7.6 and that for NADP and NADPH is 8.6 (Figure 3).[2]

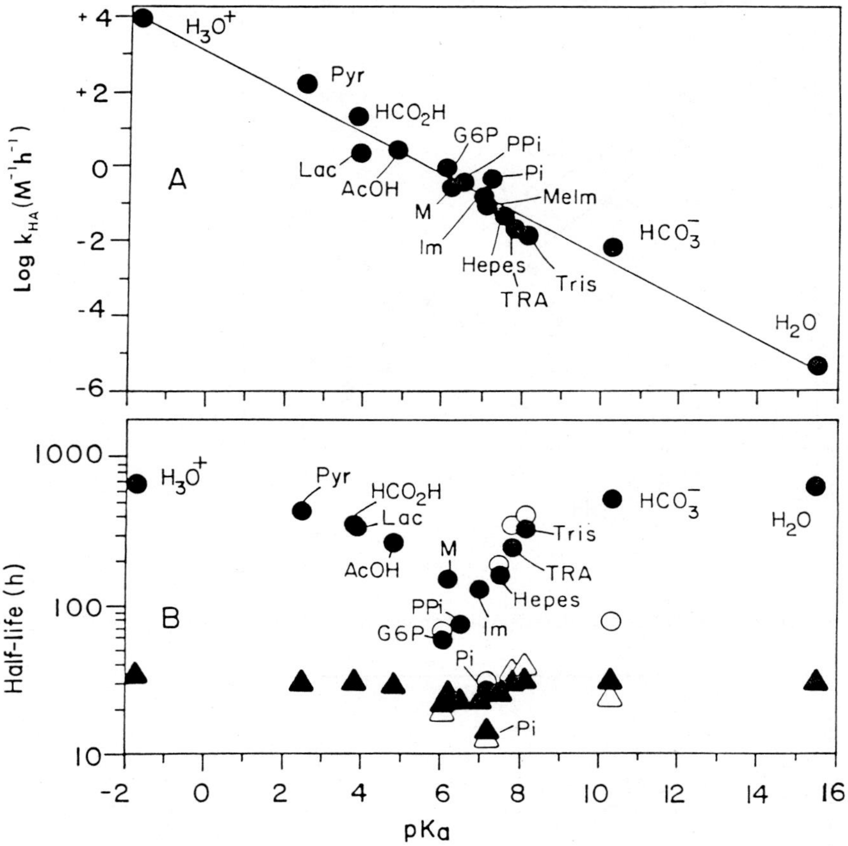

Figure 2. (A) Bronsted plot for acid-catalyzed decomposition of NAD(P)H. (B) Plots of calculated half-life vs pKa for decomposition of NADH (●) and NADPH (▲) in 0.1 M buffer, pH 7.0, 25 °C. The observed values are indicated by (O) for NADH and (Δ) for NADPH.

The problem of product inhibition in the nicotinamide cofactor-requiring enzymatic synthesis is often severe. In particular, the oxidation of alcohols to aldehydes or ketones is not only thermodynamically unfavorable (and the reaction must be driven by a coupled favorable reaction) but also subject to product inhibition (the active site of alcohol dehydrogenases appears to bind the product ketone or aldehyde more strongly than the reactant alcohol). Of different types of product inhibition, the kinetically non-competitive inhibition is the most severe. Figure 4 indicates a theoretical analysis of reactions suffering from non-competitive inhibition. The relative rate

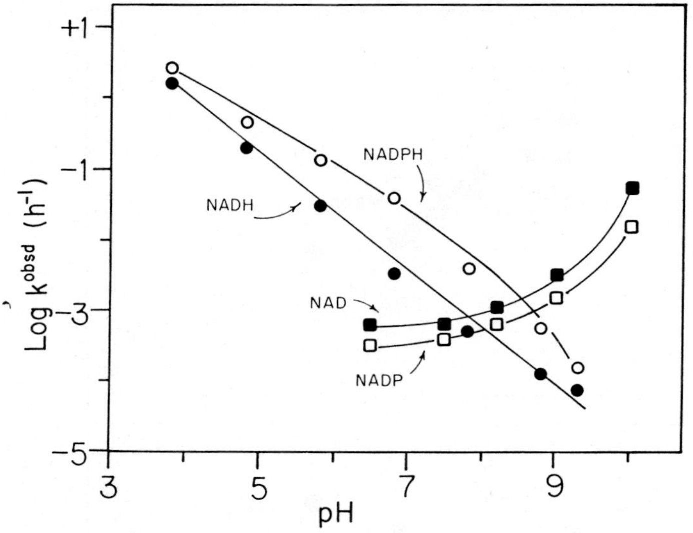

Figure 3. Rate constants for decomposition of NAD(P)(H).

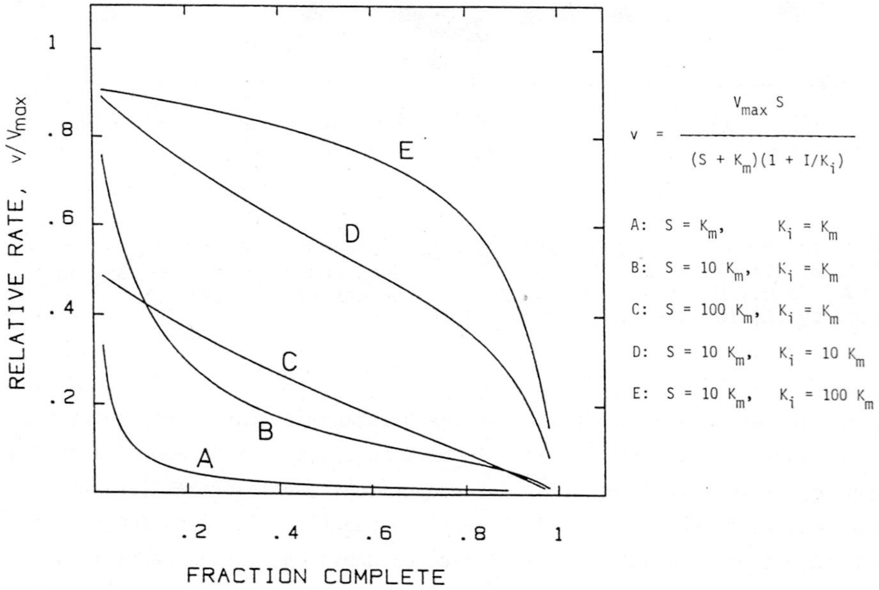

Figure 4. Theoretical predication of relative rate in the presence of a noncompetitive inhibitor during the progress of reaction.

changes during the progress of reaction and is significantly affected by the relative affinity of substrate (K_m) and inhibitor (K_i) for the enzyme. A typical example is shown in curve C where the reaction rate drops to about 20% of the original at 60% completion with an initial substrate concentration of 100 K_m and $K_i = K_m$. It appears to be no general strategy to deal with product inhibition, other than to remove the product as it is formed by extraction or by some other techniques.

Another problem of nicotinamide cofactor-requiring enzymatic reactions is that many interesting substrates and products are insoluble or unstable in aqueous solution. The limitations place constraints on the ultimate concentrations of products attainable and on the rates of reaction. These problems should be solved if NAD(P)-dependent enzymatic synthesis is to be employed in industry for large-scale process, particularly for the synthesis of compounds which are labile or insoluble in aqueous media.

Recent Developments

To deal with the costs of cofactors and enzymes and the problems of solubility and inhibition, we have recently developed improved procedures for enzymatic synthesis requiring nicotinamide cofactors in organic solvent-water biphasic systems.[4] Bioconversion in biphasic or reverse micellar systems has been reported and used for a long time.[5] As shown in Figure 5, an NAD-requiring asymmetric oxidation of a meso-diol catalyzed by horse liver alcohol dehydro-

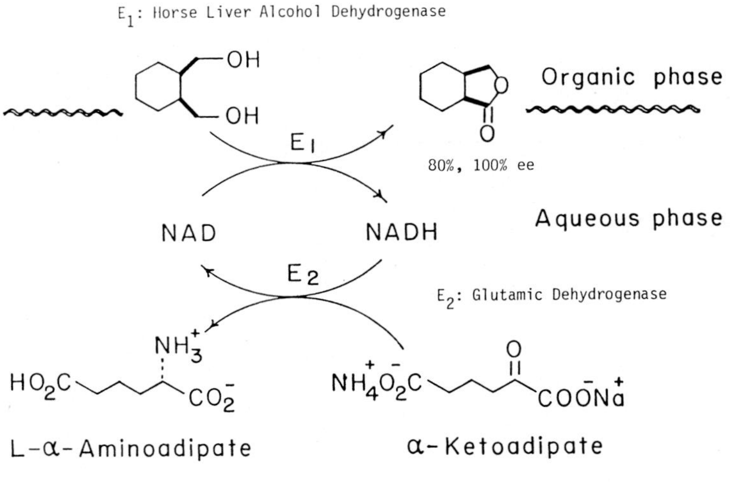

Figure 5. Coupling of two nincotinamide cofactor-requiring enzymatic syntheses in a biphasic system.

genase (HLADP) is coupled with an NADH-requiring asymmetric reductive amination of 2-ketoadipate catalyzed by glutamate dehydrogenase (GluDH). Each of the two enzymatic reactions is synthetically useful and generates the proper form of the cofactor for the other. The reactions are carried out in the cyclohexane-water biphasic system where the chiral lactone produced is extracted from the aqueous phase to separate from the other water-soluble product (L-2-amino-adipate) and to minimize product inhibition. A number of chiral lactones prepared by Jones et. al[6] from meso-diols can be produced on mol scales by using this strategy. The turnover numbers for the enzymes and the cofactors are about 10^{8-10} and 10^{3-4} respectively (these numbers are not optimized). The enzymes can be used as free forms or as water-soluble cross-linked forms prepared by reaction with a hydrophilic polyacrylamide-acryloxysuccinimide polymer as shown in Figure 6. In these enzymatic syntheses, 2-aminoadipate is completely soluble in the aqueous phase and does not inhibit the enzyme activity. Although the lactone is

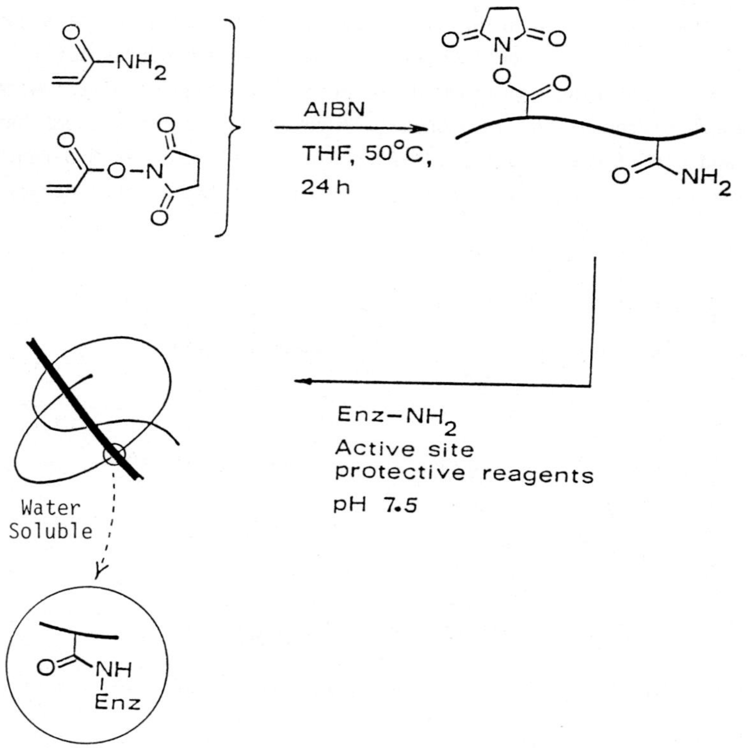

Figure 6. Preparation of cross-linked and water-soluble enzymes.

an inhibitor ($K_i \sim 80$ mM) for HLADH, it is partially removed from the aqueous solution when it is formed (the partition coefficient is about 1); the inhibition from lactone is lessened. Since both cofactors are in a very low concentration (~ 0.1 mM), their inhibition can be neglected. To increase the lifetime of enzymes, the reaction was usually carried out with a slow stirring of the aqueous phase without disturbing the interface.

The direction of the alcohol dehydrogenase-catalyzed reactions can be reversed if the synthetic reaction is coupled with glucose-6-phosphate/glucose-6-phosphate dehydrogenase (G-6-P/G-6-PDH, Figure 7). In this system, the initial product gluconolactone phosphate is hydrolyzed spontaneously to 6-

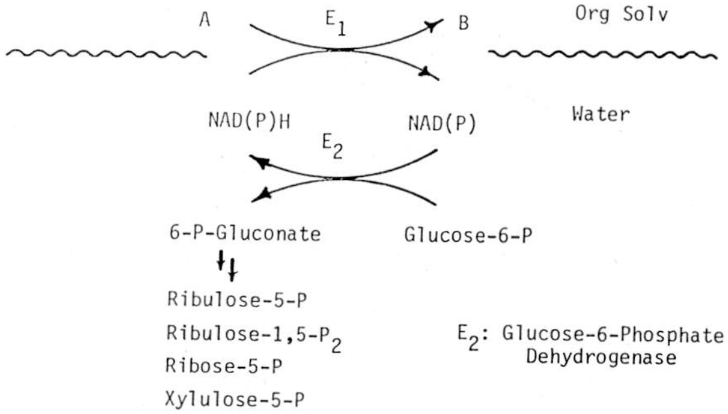

A	Enzyme E_1	B	%ee
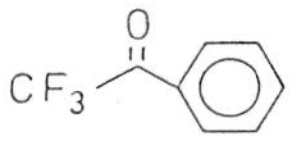	TADH	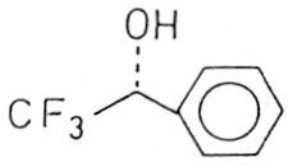	97
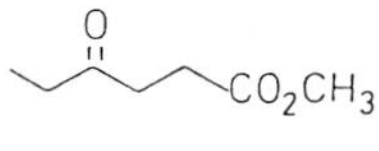	TADH	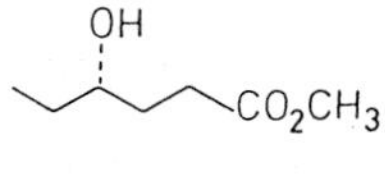	97
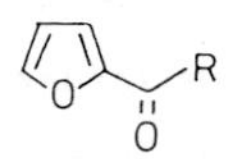	TADH	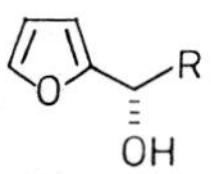	92
Cortisone	Cortisone reductase	3α,20 Hydroxysteroid	97

Figure 7. Coupling of two nicotinamide cofactor-requiring enzymatic synthesis in a biphasic system. TADH = Thermoanaerobium brockii alcohol dehydrogenase

phosphogluconate which is not an inhibitor of G-6-PDH. 6-Phosphogluconate can be easily isolated from the aqueous phase; it is a useful starting material for preparation of a number of valuable sugar phosphates.[7,8] The NAD(P)H generated can be used for enzymatic reduction of different ketones as shown in Figure 7. All the alcohols produced are poorly soluble in the aqueous phase and extracted into the organic phase (butyl acetate) to eliminate the product inhibition and increase the working concentration. Several new compounds produced are synthetically useful. (S)-Methyl 4-hydroxyhexanoate can be easily converted to a lactone pheromone (eq 4) and the alkyl furan carbinol is useful for the synthesis of L-sugars and L-aminosugars (eq 5).[9]

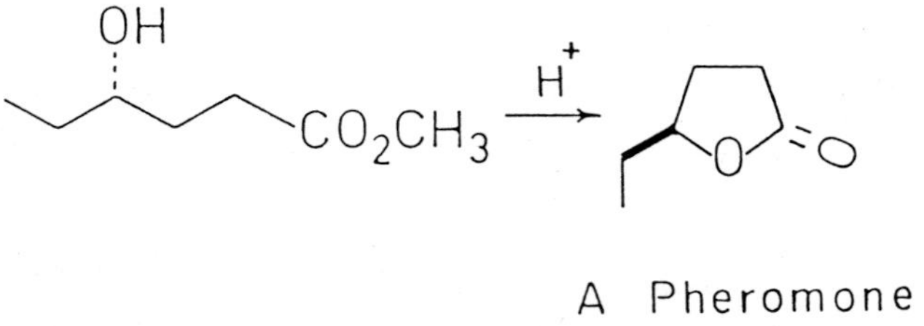

(4)

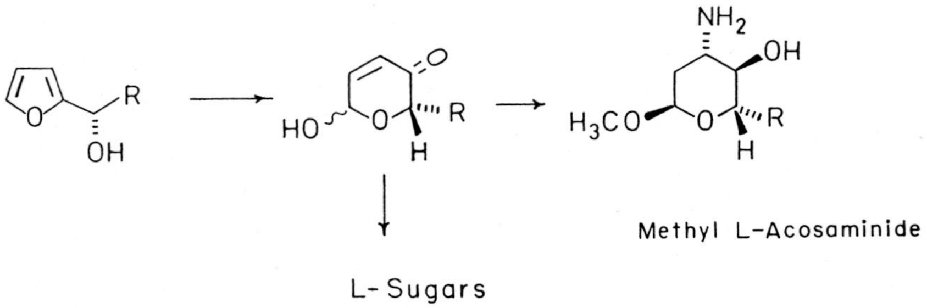

(5)

Although the biphasic systems can easily produce mol quantities of products in laboratory, they are difficult to operate continuously to produce larger quantities of products. In order to make the biphasic reactions possible for operation in a column reactor, we entrap both enzymes and the cofactor in the matrix of a cross-linked neutral polyester, XAD-8 (Figure 8). Entrapment of thermolysin in XAD-7 has been used for peptide synthesis.[10] We simply extend the application to cofactor-requiring enzymes. The preparation was carried out by mixing the polymer with enzymes and cofactors (NAD or NADP) in a buffer solution until equilibrium was reached. The mixture was filtered, washed with buffer, and dried at room temperature overnight until the beads look dry. In

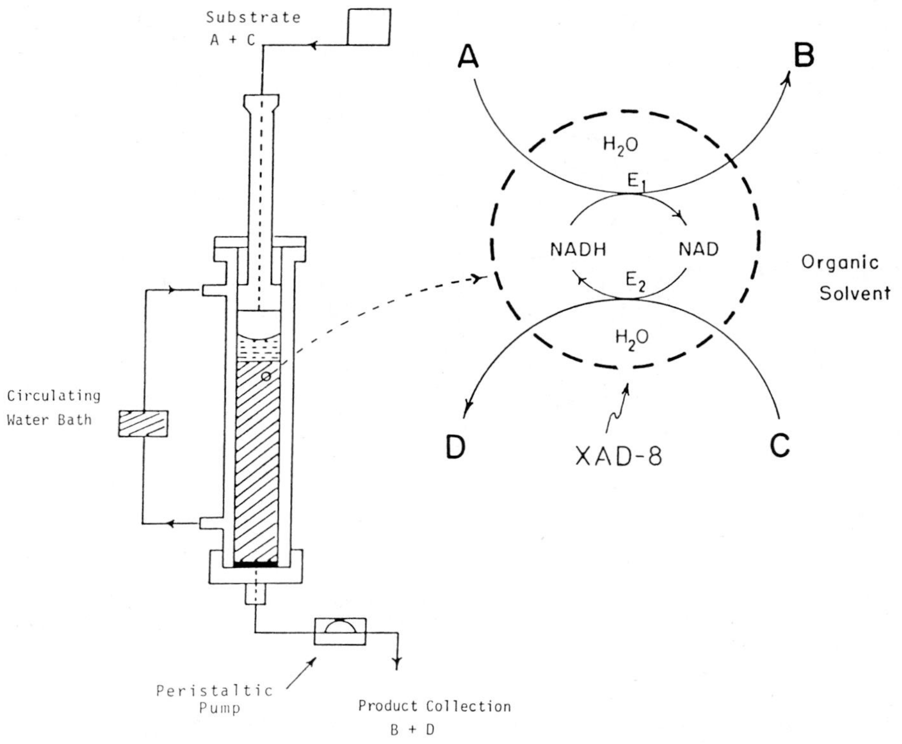

Figure 8. Entrapment of enzymes and NAD in a cross-linked polyester matrix for use in synthesis in organic solvent.

this preparation, both enzyme and cofactor are entrapped inside the polymer matrix containing aqueous solution. The entrapped enzymes and cofactors were then suspended in a water-immicible solvent such as butyl acetate containing substrates (A&C) and shaked at room temperature. The reaction progress was monitored by measuring the concentration of B or C periodically. In a typical reaction system where A = pyruvaldehyde dimethylacetal (50 mM), C = cyclohexanol (0.1 M), and E_1 = E_2 = HLADH (the local concentration of HLADH is 5 mg/mL and that of NAD is 0.5 mM in the aqueous phase inside the matrix), cyclohexanone (D) and L-lactaldehyde dimethyl acetal (B) were produced steadily up to about 30 mM over a period of 2 days and then slowly produced until 40 mM of each product was formed over a total period of 5 days.

A similar system was prepared as shown in Figure 9 where two different enzyme reactions were coupled with each other: the oxidation of diol was catalyzed by

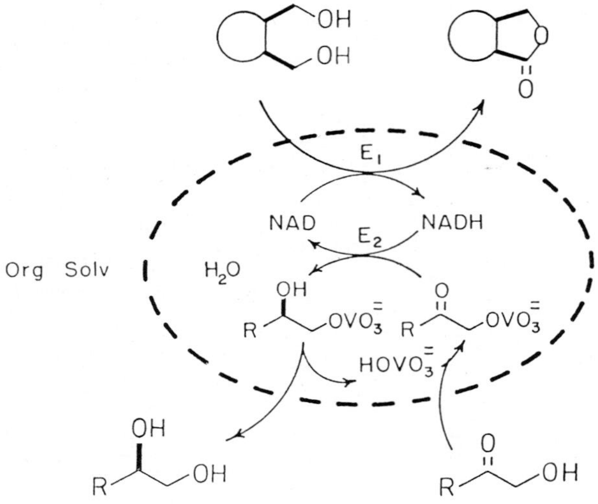

Figure 9. Synthesis in organic solvent containing a suspension of entrapped enzymes and NAD. E_1, horse liver alcohol dehydrogenase; E_2, glycerophosphate dehydrogenase.; R = H.

HLADH (E_1) and the reduction of acetol (R = H) was catalyzed by glycerophosphate dehydrogenase (E_2) in the presence of inorganic vanadate entrapped inside the matrix. When acetol diffuses into the aqueous phase, it reacts spontaneously with vanadate ($k = 1\ M^{-1}s^{-1}$) to acetol vanadate which, an analogue of acetol phosphate, is accepted by glycerophosphate dehydrogenase as a substrate. The product, (R)-1,2-propanediol-1-vanadate, is then hydrolyzed spontaneously to (R)-1,2-propanediol and extracted to the organic phase. This process allows preparation of two compounds simultaneously in organic solvents. Optimization of the reaction conditions is still in progress. Although the last two systems have some advantages from the engineering point of view, they suffer from several drawbacks: first, the reactions are slow due to diffusional problems; second, the enzymes may be subject to interfacial denaturation; third, the reaction pH's are difficult to adjust; finally, the systems appear to generate a mixture of products and unreacted substrates soluble in the organic phase and thus may complicate work-up.

Reactor Configuration

The reactor configuration most convenient for use in laboratory-scale synthesis is the batch reactor. The biphasic systems described in this presentation can be operated in a batch reactor or in a column reactor where both enzymes (free or cross-linked and water-soluble) and cofactors are entrapped in the aqueous phase. Immobilization of enzymes on insoluble support improves their stability and facilitates their recovery from reactor, but may be denatured at the interface due to vigorous stirring. More convenient for large-scale synthesis are column, hollow fiber, and membrane-based reactors. These reactors allow continuous processing of an influent substrate solution while product is removed from the effluent stream. In column or hollow fiber reactors, soluble nicotinamide cofactots may be present in the feed solution. An alternative approach is that shown in Figure 8; both enzymes and cofactors are entrapped inside the polymer matrix with water and suspended in a single-phase, water-immiscible organic solvent.

In summary, nicotinamide cofactor-requiring enzymatic synthesis of either water-soluble or water-insoluble materials can be carried out in a laboratory on mole scales with the cofactor being recycled for 10^3-10^5 times. Besides reducing the operation cost in synthesis, cofactor regeneration can be used to drive the equilibrium toward product formation, to prevent the accumulation of cofactor by-product which may inhibit the reactions, and to simply the reaction wark-up. The most highly developed process up to pilot plant scales is that described by Kula et. al[11] for the contaneous production of amino acids in a membrane reactor containing a homogeneous aqueous solution of enzymes and derivatized NAD. Whether the aforementioned biphasic processes can be scaled up to a full, multi-kilogram process remains to be investigated. One thing for sure is that further improvements of economy and stability regarding the synthetic application of cofactor-requiring enzymes will eventially emerge through screening and through protein engineering and recombinant DNA technology.

REFERENCES

1 G.M. Whitesides, C.-H. Wong, Enzymes as Catalysts in Synthetic Organic Chemistry, Angew Chem. Int. Ed. Engl., 24 (1985) 617-638.
2 C.-H. Wong, G.M. Whitesides, Enzyme-Catalyzed Organic Synthesis: NAD(P)H Cofactor Regeneration by Using Glucose-6-phosphate and the Glucose-6-Phosphate Dehydrogenase from Leuconostoc mesenteroides, J. Am. Chem. Soc., 103 (1981) 4890-9.
3 S.L. Johnson, P.T. Tauzon, Acid-Catalyzed Hydration of Reduced Nicotinamide Adenine Dinucleotide and Its Analogues, Biochemistry, 16 (1977) 1175-83.
4 J.R. Matos, C.-H. Wong, Biphasic One-Pot Synthesis of two Useful and Separable Compounds Using Cofactor-Requiring Enzymatic Reactions: Glutamate Dehydrogenase Catalyzed Synthesis of L-α-Aminoadipate Coupled With Alcohol Dehydrogenase Catalyzed Synthesis of a Chiral Lactone, J. Org. Chem., 51 (1986) 2388-9.

5 P.L. Luisi, C. Laane, Solubilization of Enzymes in Apolar Solvents via Reverse Micelles, Trends in Biotechnology, 4 (1986) 153-60.
6 J.B. Jones, Enzymes in Organic Synthesis, Tetrahedron, 42 (1986) 3351-3403.
7 C.-H. Wong, S.D. McCurry, G.M. Whitesides, Practical Enzymatic Syntheses of Ribulose 1,5-Bisphosphate and Ribose 5-Phosphate, J. Am. Chem. Soc., 102 (1980) 7938-9.
8 O. Ghisalba, H.-P. Schar, G.M. Ramos Tombo, Applications of Microbes and Microbial Enzymes in Environmental Control and Organic Synthesis, in: M.P. Schneider (Ed.), Enzymes as Catalysts in Organic Synthesis, D. Riedel Publishing Co., 1986, pp 233-250.
9 D.G. Drueckhammer, C.-H. Wong, unpublished.
10 K. Oyama, S. Nishimura, Y. Nonada, K. Kihara, T. Hashimoto, Synthesis of an Aspartame Precursor by Immobilized Thermolysin in an Organic Solvent, J. Org. Chem., 46 (1981) 5241-2.
11 R. Wichmann, C. Wandrey, A.F. Buckmann, M.R. Kula, Continuous Enzymatic Transformation in an Enzyme Membrane Reactor With Simultaneous NADH Regeneration, Biotech. Bioeng. 23 (1981) 2789-2802.

C. Laane, J. Tramper and M.D. Lilly (Editors), *Biocatalysis in Organic Media*, Proceedings of an International Symposium held at Wageningen, The Netherlands, 7–10 December 1986.

ENZYMATIC SYNTHESIS OF ASPARTAME IN ORGANIC SOLVENTS

KIYOTAKA OYAMA

Organic Chemistry Laboratories, Chemical Research Center, Toyo Soda Manufacturing Co., Ltd., 4560 Oaza Tonda, Shin-Nanyo, Yamaguchi 746, Japan.

SUMMARY

It was found that N-protected aspartic acid (X-Asp-OH) and phenylalanine methyl ester (PM) react in the presence of a metalloproteinase to give X-APM, a precursor of the synthetic sweetener aspartame. In this reaction, the condensation occurs exclusively at the α-carboxylate of X-Asp-OH, even though the side chain carboxylate is unprotected. Furthermore, only L-isomers react to give the L-L dipeptide when inexpensive racemic substrates are used. Because of the attractive features of the reaction, the basic studies and process development work were carried out with the aim to establish the industrial technology based on this novel enzymatic reaction. The studies were also made on the use of immobilized enzyme in organic solvents and some results were also presented.

INTRODUCTION

It has been known for a long time that peptides can be synthesized by a reverse reaction of enzymatic hydrolysis of peptides, e.g., Mohr and Strohschein (ref. 1) reported in 1909 that the following condensation takes place in the presence of papain (eqn. 1).

$$\text{Bz-Leu-OH} + \text{H-Leu-NH}_2 \longrightarrow \text{Bz-Leu-Leu-NH}_2 \qquad (1)$$

During 1937-1944, Bergamann and his group (ref. 2) carried out an extensive studies on this type of peptide synthesis, chiefly because they wanted to elucidate the mechanism of the biosynthesis of proteins in living systems. However, interest in peptide synthesis by a proteolytic enzyme waned rapidly when the mechanism of in vivo synthesis of proteins was clarified in the 1950s through molecular biology.

This enzymatic reaction of peptide bond formation is also of interest from the viewpoint of a practical synthetic method. However, it failed to develop into a routin synthetic method, as did the widely used chemical methods using acid chloride, mixed anhydride, active ester, etc. (ref. 3). One of the major problems of enzymatic synthesis at that time might have been attributable to protecting groups. Usually a benzoyl substituent was used to protect the amino group, and an amide or an anilide substituent to protect the carboxylic group. Removal of these substituents after the enzymatic reaction requires severe conditions that usually also destroy the peptide linkage. Furthermore, the enzyme's strict substrate specificities limit its application to substrate with specific structures. This might have beenconsidered as a further limitation of the enzymatic method.

Since then, however, these limitations have been gradually lifted by progress both in synthetic organic chemistry of peptides and in enzyme chemistry. New protecting groups capable of being easily introduced and removed have been developed, and a number of proteolytic enzymes having various substrate specificities have been found, especially from microorganisms. About a decade ago, we started the studies to apply this unique method for the synthesis of valuable materials, especially targeted at the synthesis of aspartame, which drew considerable attention at that time as a new sweetener.

ASPARTAME

Aspartame (L-aspartyl-L-phenylalanine methyl ester, APM) is the methyl ester of the C-terminal dipeptide of the digestive hormone gastrin. It is about 200 times sweeter than sucrose and has a pleasant sweetness without a bitter aftertaste. The sweetness was accidentaly discovered by researchers of G. D. Searle & Co. during the synthesis of the hormone (ref. 4). The safety of the sweetener was confirmed by extensive studies conducted for a long period of time using experimental animals, and its use was approved in many countries throughout the world. Because of its intensive sweetness and pleasant taste, the sweetener is well received as a diet sweetener. The sweetener is composed of L-aspartic acid (L-Asp), L-phenylalanine (L-Phe), and methyl alcohol, and its structure is shown below.

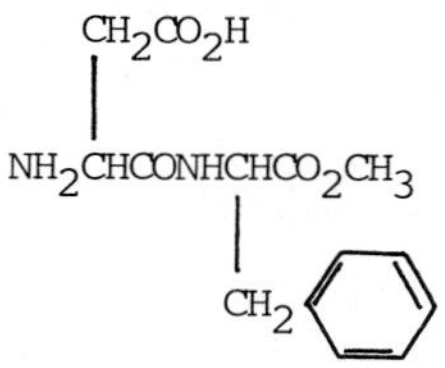

(L-Asp-L-Phe-OMe; APM)

CHEMICAL SYNTHESIS

Many chemical ways have been reported for the synthesis of aspartame (ref. 5). In the classical peptide synthetic method, the side chain carboxylate of Asp is protected in order to avoid the undesired condensation at the side chain. However, in all of the chemical synthetic methods with industrial potential, the side chain carboxylate is not protected and acid anhydrides of Asp are used in order to shorten the process, resulting in the inevitable production of 20-30% of the β-isomer (II) along with the desired α-isomer (I), as shown in eqn. 2 [X=H, benzyloxycarbonyl (Z), and formyl (CHO)].

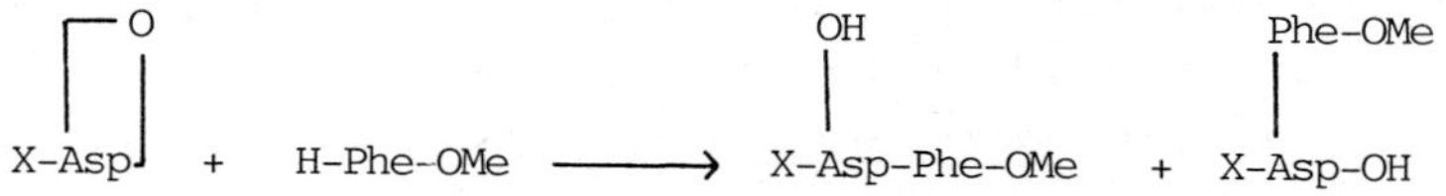

(2)

(I) (II)

Since β-aspartame from (II) has a bittertaste, it must be completely separated from the α-isomer. In the industrial production, recovery of expensive amino acids from (II) is also necessary. In the case of formyl protection, cleavage of the peptide bond also occurs to some extent during the deprotection reaction.

ENZYMATIC METHOD

We started, in 1975, screening of commercially available proteinases, especially focused on the proteinases which have substrate specificities on hydrophobic amino acid residues at the scissible bond of proteins. As a result, it was found that proteinases belonging to a metalloproteinase and a thiol proteinase can catalyze the condensations between N-protected Asp and phenylalanine methyl ester as shown in eqns. 3 and 4 (ref. 6 and 7). However, the latter enzymes were found to also catalyze hydrolysis of the methyl ester

$$\underset{\text{X-Asp-OH}}{\overset{\text{OBl}}{|}} \quad + \quad \text{H-Phe-OMe} \longrightarrow \underset{\text{X-Asp-Phe-OMe}}{\overset{\text{OBzl}}{|}} \qquad (3)$$

$$\underset{\text{X-Asp-OH}}{\overset{\text{OH}}{|}} \quad + \quad \text{H-Phe-OMe} \longrightarrow \underset{\text{X-Asp-Phe-OMe}\cdot\text{H-Phe-OMe}}{\overset{\text{OH}}{|}} \qquad (4)$$

linkages in both the reactants and the products, resulting in lower yields of the desired products. Commercially available crude metalloproteinases often contain serine proteinases, which also hydrolyze the ester linkages, thus usually requiring purification of the enzyme. However, a crude enzyme can be used without extensive purification when applied in the presence of an inhibitor specific to a serine proteinase, e.g., the inhibitor that can be obtained from potatoes. (ref. 8). Among the enzymes studied, thermolysin (E.C.3.4.24), which is obtained from the microorganism found in the Japanese hot spring (Bacillus thermoproteolyticus Rokko), showed a high catalytic activity, marked stability against heat, organic solvents, and extreme pH, etc., and does not show the esterase activity. It was found that thermoase, a much cheaper crude enzyme preparation of thermolysin, could also be used even in the absence of the serine proteinase inhibitor.

In eqns. 3 and 4, the reaction products can be easily led to aspartame by usual catalytic hydrogenation, thus providing unique methods for the synthesis of the sweetener. Especially in eqn. 4, the enzymatic condensation occurs exclusively at the α-carboxylate due to the enzyme's strict regiospecificity, even though the side chain carboxylate is unprotected. In this reaction, the product dipeptide forms an insoluble addition compound with unreacted H-Phe-OMe, thus, shifting the unfavourable equilibrium toward the synthesis side. We also found that when inexpensive racemic substrates are used, only L-isomers react to give the L-L dipeptide, which is then deposited as the addition compound exclusively with D-H-Phe-OMe, thus, one can attain condensation and optical resolution of racemic compounds simultaneously. Because of the outstanding features of the enzymatic method which can not be compared with chemical methods, we thought this might become feasible industrial technology and thus performed basic studies as well as process development.

KINETICS AND REACTION MECHANISM

In order to develop the industrial process, it is important to study the kinetics and to understand the reaction mechanism as much as possible. The kinetic studies were carried out in the following reaction using thermolysin as the catalyst (ref. 9).

$$\underset{\text{(Z-Asp)}}{\overset{\text{OH}}{\text{Z-Asp-OH}}} + \underset{\text{(PM)}}{\text{H-Phe-OMe}} \longrightarrow \underset{\text{(ZAPM·PM)}}{\overset{\text{OH}}{\text{Z-Asp-Phe-OMe·H-Phe-OMe}}} \qquad (5)$$

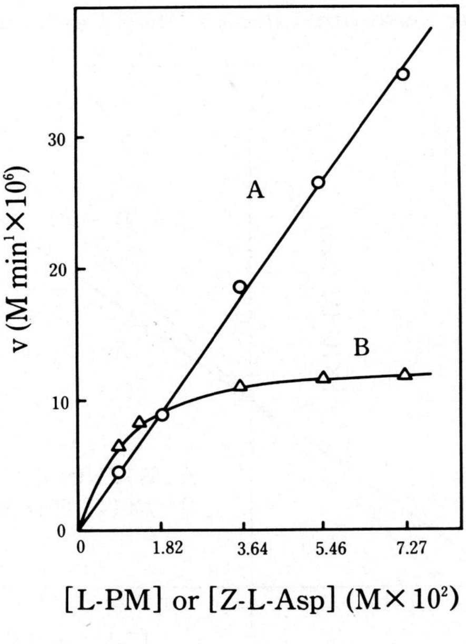

Fig. 1. The effects of the substrate concentrations on the rate of the formation of ZAPM in the thermolysin-catalyzed condensation reaction of Z-L-Asp with L-PM at 40° C ($[E_o] = 4.85 \times 10^{-6}$ M, $\mu = 3.64$). The straight line (A) represents the plot of v versus [L-PM] ([Z-L-Asp] = 1.82×10^{-2} M), and the curve (B) represents the plot of v versus [Z-L-Asp] ([L-PM] = 1.82×10^{-2} M).

It was found that the reaction is first order with respect to the enzyme and L-PM concentrations. On the other hand, with respect to Z-L-Asp, the retardation of the rate, typical Michaelis-Menten behavior, can be seen (Fig. 1). From these facts, the reaction mechanism and the rate law can be expressed as follows.

$$\text{Z-L-Asp} + \text{E} \underset{k_{-1}}{\overset{k_{+1}}{\rightleftharpoons}} \text{Z-L-Asp,E}$$

$$\text{Z-L-Asp,E} + \text{L-PM} \xrightarrow{k_2} \text{ZAPM} + \text{E}$$

$$v = \frac{k_2[E_o][\text{Z-L-Asp}][\text{L-PM}]}{K + [\text{Z-L-Asp}]} \qquad \left(K = \frac{k_{-1}}{k_{+1}} \; ; \; k_{-1} \gg k_2[\text{L-PM}] \right)$$

We also found that presence of D-PM does not affect the reaction rate at all whereas Z-D-Asp retards the reaction. The fact that the two lines of Fig. 2 cross each other on the ordinate indicates that Z-D-Asp acts as a competitive

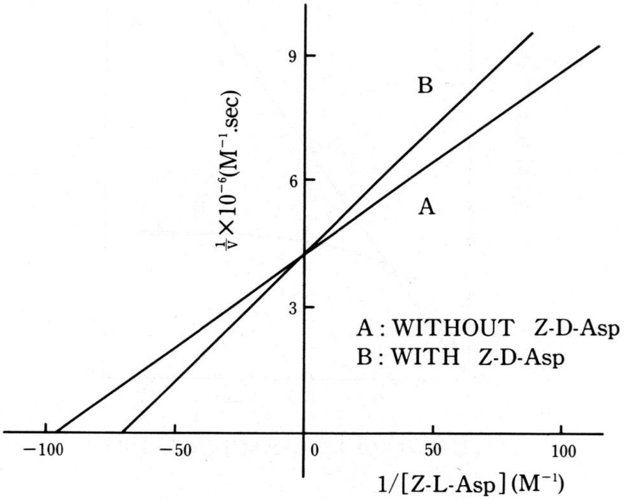

Fig. 2. The effects of Z-L-Asp concentraion on the rate of ZAPM production. Curve A represents the reaction in the absence of Z-D-Asp, and B in the presence of Z-D-Asp ([Z-D-Asp] = 9.09×10^{-3} M, [L-PM] = 1.82×10^{-2} M, $[E_o]$ = 4.85×10^{-6} M)

inhibitor to the reaction with the binding constants of Z-L-Asp (K) and Z-D-Asp (K_i) with the enzyme being 1.03 x 10^{-2} M and 2.81 x 10^{-2} M, respectively. Thermolysin is a widely studied neutral metalloproteinase, but its mode of action is not yet well understood, although many studies have been carried out; especially arguments are being made whether the reaction goes via the amino-enzyme intermediate (ref. 10), the acyl-enzyme intermediate (ref. 11), or attack of a water molecule at a scissible bond (reverse of the general base catalysis route) (ref. 12).

From the above results, it is now evident that the condensation reaction involves initial binding of Z-L-Asp to the enzyme to form the Z-L-Asp-enzyme complex then attack by L-PM on the complex as the rate-determining step to form the condensation product. From the ordered binding of the substrates, the possibility of the amino-enzyme intermediate can be rejected. It is well established that the hydrolysis by α-chymotripsin goes via the acyl-enzyme intermediate, where deacylation rate of a L-form is generally faster than a D-form in the factor of 100-20,000 (ref. 13). On the other hand, binding constants of L- and D-forms to the enzyme do not differ greatly (ref. 14). Although the exact nature of the Z-L-Asp-enzyme complex can not be described yet, the fact that the K values of Z-L-Asp and Z-D-Asp are similar may indicate that the condensation does not involve the covalent bond formation between Z-L-Asp and the enzyme to form the acyl-enzyme intermediate, but takes

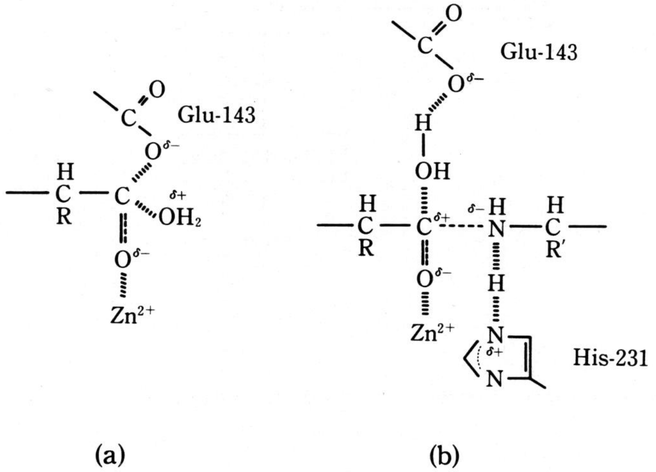

Fig. 3. Possible mechanisms for the synthesis of the peptide: (a) acyl-enzyme route, (b) reverse of general base catalysis route.

place via the reverse of the general base catalyzed hydrolysis where the water molecule is delivered by Glu-143 ((a) and (b) in Fig. 3). This mechanism is also consistent with the result of X-ray crystallographic study of the binding of inhibitors to thermolysin (ref. 15).

From the kinetic results and the facts that L-Asp is available cheaply whereas DL-Phe is much cheaper than L-Phe and racemization of D-Phe is much easier than D-Asp, we chose L-Asp and DL-Phe as the raw materials. The entire process of the enzymatic production of aspartame can be given in Fig. 4.

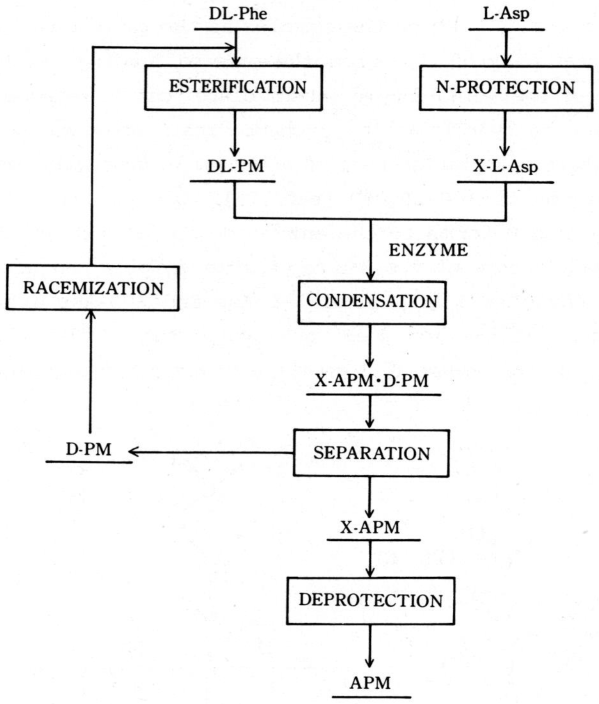

Fig. 4. The enzymatic process for producing aspartame.

EQUILIBRIUM

The peptide synthesis reaction between Z-L-Asp and L-PM to give ZAPM and the hydrolysis reaction of the product to the substrates are in equilibrium as shown below.

Z-L-Asp + L-PM $\rightleftharpoons$ ZAPM

The equilibrium constant (K) in H_2O at pH 7.0 is 1.5 M^{-1} and thus the equilibrium lies well over at the hydrolysis side under normal conditions. Therefore in order to obtain the product in high yield, the equilibrium must be shifted toward synthesis side. The first approach to improve the product yield is to form a product which is sparingly soluble in the reaction medium. In the present reaction, the product ZAPM forms a quite insoluble salt with D-PM in an aqueous solution (solubility of ZAPM·D-PM in H_2O is 5 x 10^{-3} M). In such a case, the follwing equation can be given (eqn. 7), where S is the

$$K = \frac{S}{[A(1-\alpha)-S][B(1-\alpha)-S]} \tag{7}$$

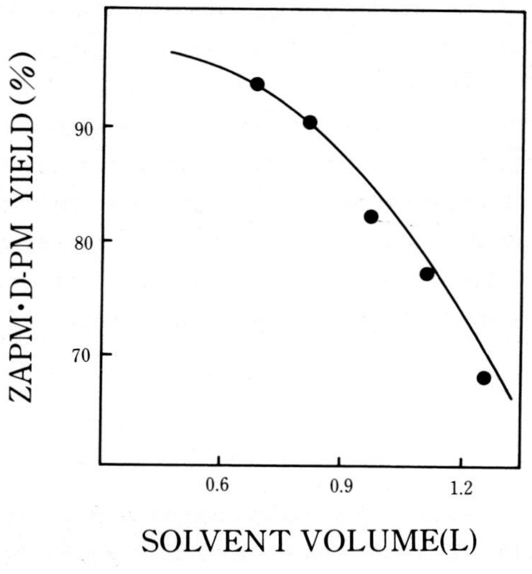

Fig. 5. The dependence of the yield of the equilibrium, Z-L-Asp + DL-PM ZAPM·D-PM, on the solvent volume (K = 1.5 M^{-1}, S = 0.005M, Z-L-Asp = 0.2 mole, DL-PM = 0.5 mole). The line is calculated by eqn. 7, while the dots are obtained experimentally.

solubility of ZAPM·D-PM; A, B are the initial concentrations of Z-L-Asp and L-PM, respectively; and α is the conversion to ZAPM. The equation indicates that the yield is higher with higher initial substrate concentrations and a lower solubility of the product. From eqn. 7 theoretical yields of ZAPM·D-PM with different substrates concentrations are calculable and they were compared with the yields actually obtained (Fig. 5). As can be seen in Fig. 5, these values agree very well (ref. 16).

The second approach is to shift the equilibrium constant toward synthesis side by adding a water-miscible organic cosolvent that can also dissolve the product. For instance, in the following equilibrium catalyzed by α-chymotrypsin, it is known that the equilibrium constant in pure water is increased 84-fold from in 85 % (v/v) aqueous 1,4-butanediol (ref. 17).

$$\text{Z-Trp-OH} \quad + \quad \text{H-Gly-NH}_2 \rightleftharpoons \text{Z-Trp-Gly-NH}_2$$

In the present equilibrium, K is increased 1.5-fold in 50% aqueous methanol solution from in a pure water.

The third approach is the addition of a water-imiscible organic cosolvent, resulting in a two-phase system. If one could find a system in which reactants are soluble in the aqueous layer and the product in the organic layer, transfer of these compounds through the layers may overcome the unfavourable equilibrium. From the practical point of view, use of an immobilized enzyme is of great value. However, it may be rather impractical for enzymatic peptide synthesis in an aqueous solution, since separation of an immobilized enzyme from a deposited product is difficult. Therefore, the immobilized enzyme systems in the second and the third approaches are investigated below.

ENZYME IMMOBILIZATION

When immobilizing the enzyme, following conditions must be considered; (1) high immobilization yield; (2) small enzyme inactivation during immobilization procedures; (3) high catalytic activity per weight of immobilized enzyme; (4) small leakage of the enzyme from supports during usage; (5) cheap and simple immobilization procedure. We investigated various immobilization techniques, including physical adsorption, ionic binding, covalent binding, and holding.

Amberlite XAD is known to adsorb large quantities of proteins by virtue of its hydrophobicity, and often used for recovery of enzymes from a fermentation broth (ref. 18). In fact, it was found that Amberlite XAD-7 and XAD-8 could immobilize a large quantity of the enzyme. Furthermore, the immobilization procedure is so simple that the method is very attractive. However, it was found that , when in the presence of the substrates and/or in organic cosolvents, leakage of the enzyme from the support was too large.

Ionic binding was studied using Amberlite IRC-50 and IRA-94, representative cationic and anionic ion-exchange resins, respectively. However, it was found that amounts of the enzyme bound to the supports was very small. In addition, significant leakage of the enzyme from the supports during the usage was observed. This might be accounted by high ionic strength of the reaction medium due to the ionic nature of the substrates. Therefore, the method was not satisfactory, even though the immobilization procedure is very simple and inexpensive.

Covalent binding was studied using Toyopearl gels as supporting materials. Toyopearl is a hydrophilic polymer gel that was originally developed for use in high performance gel permeation chromatography and now is also used as

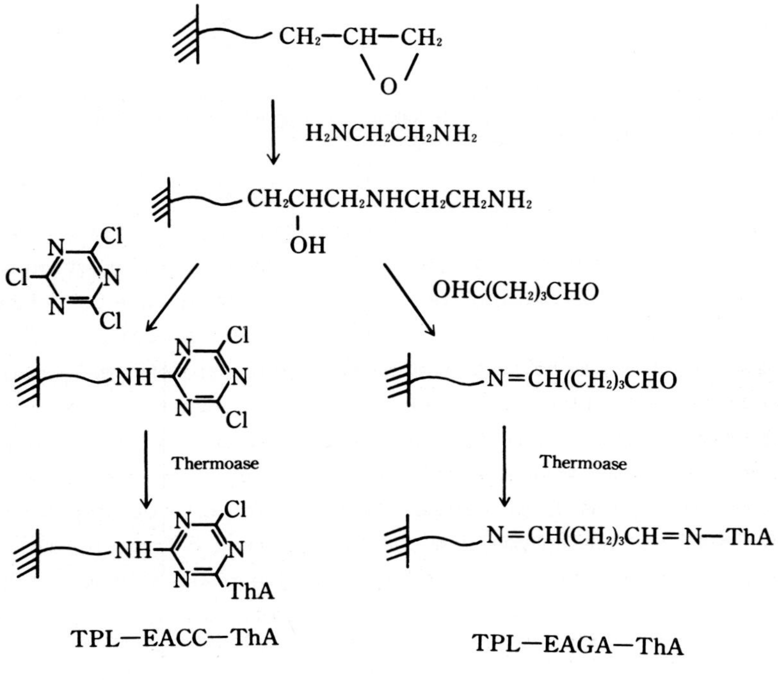

Fig. 6. Immobilization of thermoase on Toyopearl gel by ethylenediamine - glutaldehyde activation method (TPL-EAGA-ThA) and ethylenediamine - cyanuric chloride activation method (TPL-EACC-ThA).

support for affinity chromatography. It has excellent mechanical strength as well as marked resistances to heat, extreme pH, and swelling by organic solvents. Thus, thermoase was immobilized to Toyopearl via several routes. Among them, immobilization with the cyanuric acid activation method (TPL-EACC-ThA) and the glutaraldehyde activation method (TPL-EAGA-ThA) appear to be advantageous from the view points of high immobilization yield, easy immobilization procedures, and high catalytic activity (Fig. 6).

In addition to the conventional immobilization methods, the enzyme was immobilized by holding inside the aqueous pore of supporting material using an apparent single phase of a water-immiscible organic solvent (Fig. 7). In the

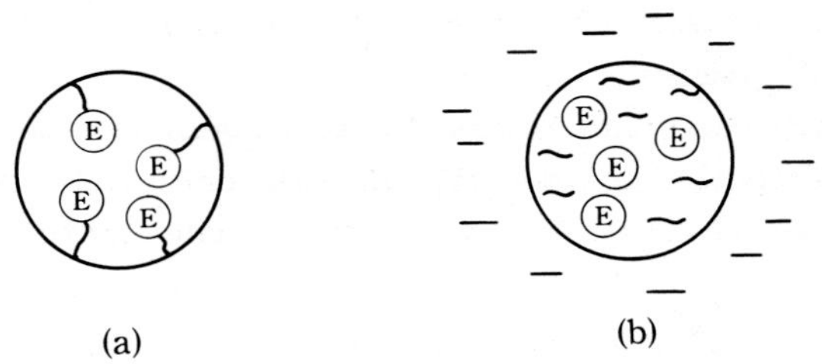

Fig. 7. Conventional immobilized enzyme (a) and "immobilization" by holding aqueous enzyme solution inside the pore of supporting material in an apparent single phase of water-immiscible organic solvent.

apparent single phase system, the enzyme can be "immobilized" in such a support as glass beads even though it has no special interaction with the enzyme. Such "immobilization" is impossible in an aqueous solution or in a biphasic system. The advantages of this type of "immobilization" are that the procedures for immobilization and regeneration are extremely simple; the "immobilization" can be done by mere soaking of supporting material in an aqueous enzyme solution, and the regeneration can be done by washing out any used inactive enzyme from supporting material, which is then soaked again in an aqueous solution of fresh enzyme. However, in spite of simplicity of the method, leakage of the enzyme from the support was too large for the reaction between Z-L-Asp and PM in the apparent single phase system, although in principle the idea was proven to be workable (ref. 19).

EFFECTS OF ORGANIC SOLVENTS

When choosing organic solvents, one must consider the following aspects; (1) effects on the activity and stability of the enzyme; (2) effect on the equilibrium; (3) solubility of the product; (4) partition of the reactants between the aqueous and the organic layers when a water-imiscible organic solvent is used. We investigated a number of organic solvents and found that, in the presence of a water-miscible solvent, the activity as well as stability of the enzyme were markedly impaired, especially such solvents as DMSO, methanol, THF, etc. (Fig. 8), although some organic solvents, e.g., methanol could shift the equilibrium toward the synthesis side as stated before. On the other hand, with an water-imiscible solvent, the impairing effects are generally smaller than with a water-miscible solvent, since in the former

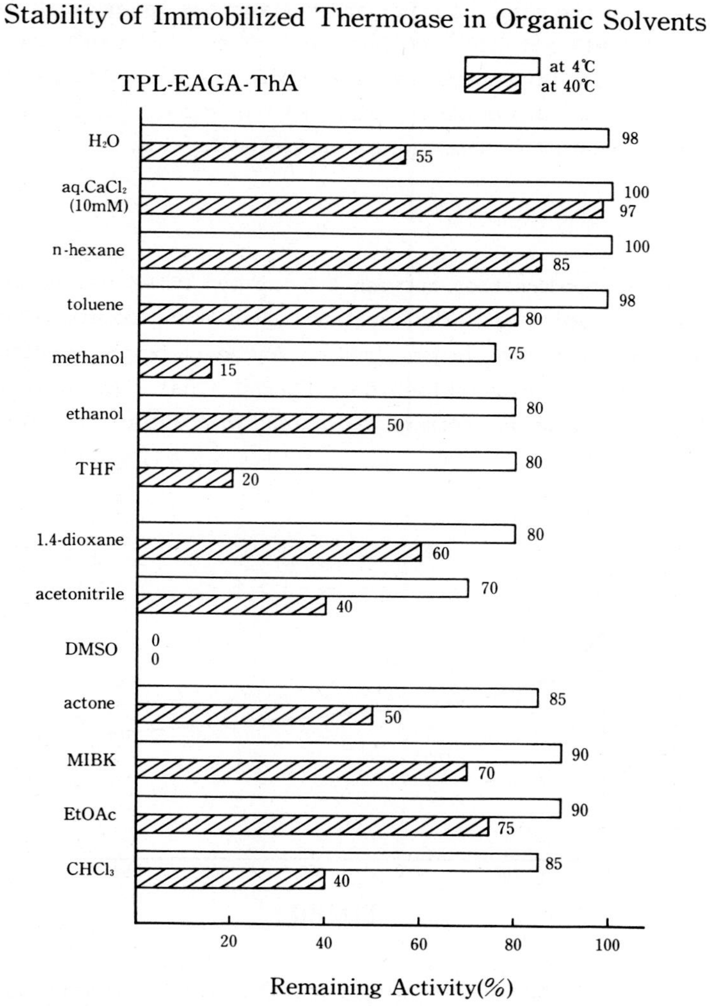

Fig. 8. Stability of immobilized thermoase (TPL-EAGA-Tha) in organic solvents.

case the enzyme can stay in the aqueous phase. The water-wet immobilized enzyme is stable in such solvents as aliphatic and aromatic hydrocarbons, but solubilities of the reactants and the product are very small, thus they are unsatisfactory. Chlorinated hydrocarbons like chloroform can dissolve the product, but the enzyme is not so stable when contacted with these solvents. We finally selected ethyl acetate for the following reasons; (1) the impairing effects on the enzyme are smaller than with other water-imiscible solvents; (2) the solubility of the product in ethyl acetate is high, and the partitions of the reactants into the organic layer are rather low (the partition constants between the organic layer and the aqueous layer for Z-Asp and PM were found to be 0.13 and 0.30, respectively). These facts indicate that most of the reactants stay in the aqueous layer where the enzymatic reaction takes place, and the product dipeptide transfers to the organic layer, thus shifting the equilibrium toward the less favourable synthesis side.

OPERATION

The enzymatic condensation between Z-L-Asp and DL-PM was studied in the biphasic system of water and ethyl acetate. Continuous columm operation was carried out using thermoase immobilized on Toyopearl by the covalent binding through ethylenediamine-glutaraldehyde (TPL-EAGA-Tha). The column reactor packed with the catalyst is thermostated at 40° by a water-circulating jacket,

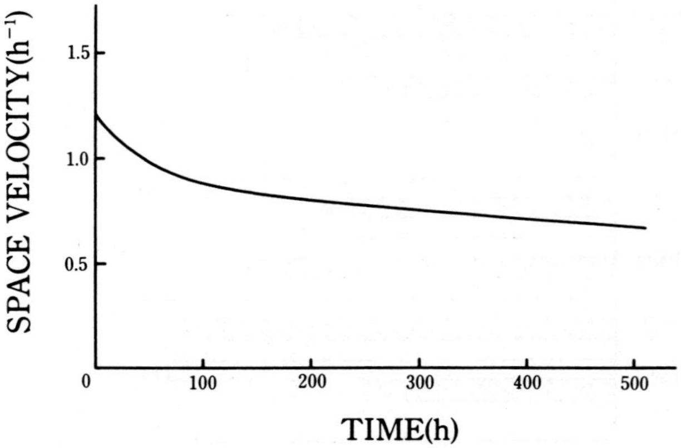

Fig. 9. Results of continuous column operation of the reaction of Z-L-Asp and DL-PM in the ethyl acetate - water system using immobilized thermoase on Toyopearl by the ethylenediamine - glutaldehyde activation method (TPL-EAGA-ThA).

and the reactants mixture solution, consisting of 0.12 M Z-L-Asp, 0.31 M DL-PM, and 0.012 M $Ca(OAc)_2$ in ethyl acetate - water (7:3 in weight ratio), was continuously fed into the reactor from the top, and the product solution mixture exited via the bottom of the reactor by overflow. The result of the operation is illustrated in Fig. 9. At the beginning of the operation, rapid decrease of the activity was observed, which gradually settled down. After 20 days of the operation, the operation was stopped and the packed immobilized enzyme was recovered from the column and analyzed. It was found that along the wall of the column became hard and plugged by deposition of the product and the reaction mixture could run through only the column center. The analysis showed that some activity was still remained in the plugged parts but most of the activity was lost in other parts where the mixture could flow, as shown in Fig. 10.

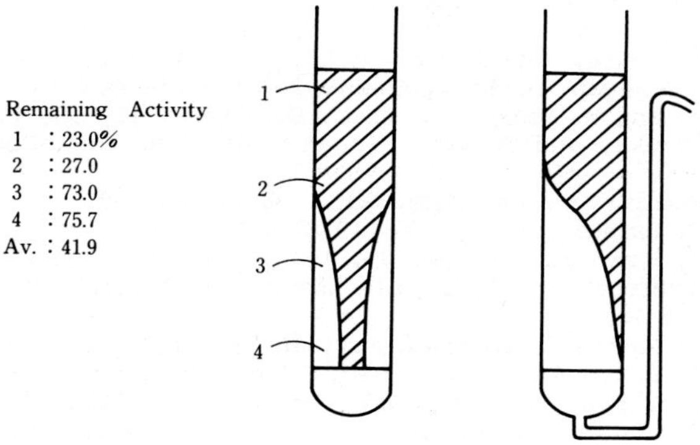

Fig. 10. Remaining activity of TPL-EAGA-ThA in the column reactor after 20 days of operation.

CONCLUSION

The enzymatic method for producing aspartame is an outstanding example of enzyme technology even in general, since it contains many attractive features, especially concerning such points as simultaneous peptide synthesis and optical resolution, and specific peptide-bond formation at the α -carboxylate. On the other hand, the disadvantage of the method as compared with the chemical ones may be that the enzyme is expensive. Therefore the use of immobilized enzyme may be of some value. When using the immobilized enzyme in the synthesis of aspartame precursor, the use of an organic solvent is required. However, the enzyme is not so stable in most of organic solvents.

Therefore it is very important to clarify the mechanism of the enzyme deactivation by an organic solvent, and to establish the way to overcome it, perhaps by development of more sophisticated immobilization methods. Furthermore, development of a good reactor system, that can be suited for such peptide synthesis as the one described here, may be necessary.

REFERENCES

1 E. Mohr and F. Strohschein, Ber., 42 (1909) 2521.
2 M. Bergmann and Frankel-Conrat, J. Biol. Chem., 124 (1937) 321; M. Bergmann and J.S. Fruton, Adv. Enzymol., 1 (1937) 63.
3 J.S. Fruton, Adv. Protein Chem., 5 (1949) 1.
4 R.H. Mazur, J.M. Schlatter, and A.H. Goldkamp, J. Am. Chem. Soc., 91 (1969) 2684.
5 K. Oyama, Bio Industry, 2 (1985) 693.
6 Y. Isowa, K. Oyama, and T. Ichikawa, 35th Annual Meeting of the Chemical Society of Japan, 4B23, Sapporo, Japan, 1976.
7 Y. Isowa, M. Ohmori, T. Ichikawa, K. Mori, Y. Nonaka, K. Kihara, K. Oyama, H. Satoh, and S. Nishimura, Tetrahedron Lett., 2611 (1979).
8 J.C. Melville and C.A. Ryan, J. Biol. Chem., 247 (1972) 3445.
9 K. Oyama, K. Kihara, and Y. Nonaka, J. Chem. Soc. Perk. II, 356 (1981).
10 K. Morihara, H. Tsuzuki, and T. Oka, Biochem. Biophys. Res. Commun. 84 (1978) 95.
11 R. Breslow and D. Wernick, J. Am. Chem. Soc., 98 (1976) 259.
12 T. Sugimoto and E.T. Kaiser, ibid., 100 (1978) 7750.
13 D.W. Ingles and L.J. Knowles, Biochem. J., 104 (1967) 396; 108 (1968) 561.
14 M.L. Bender and L.J. Brubacher, "Catalysis and Enzyme Action", McGraw-Hill, New York, 1973, ch. 6-1.
15 W.R. Kester and B.W. Matthews, Biochemistry, 16 (1977) 2506.
16 K. Oyama, S. Irino, T. Harada, and N. Hagi, "Enzyme Engineering 7", A.I. Laskin, et al. ed., p. 95, New York Academy of Sciences, 1984, New York.
17 G.A. Homandberg, J.A. Matis, and M. Laskowski, Jr., Biochemistry, 17 (1978) 5220.
18 H.Y. Ton, R.D. Hughes, D.B.A. Silk, and R. Williams, J. Biomed. Mater. Res., 13 (1979) 407.
19 K. Oyama, S.Nishimura, Y. Nonaka, K. Kihara, and T. Hashimoto, J. Org. Chem., 46 (1981) 5241.

POSTER PAPERS

C. Laane, J. Tramper and M.D. Lilly (Editors), *Biocatalysis in Organic Media*, Proceedings of an International Symposium held at Wageningen, The Netherlands, 7-10 December 1986.

NATURAL FLAVOUR ESTERS: PRODUCTION BY CANDIDA CYLINDRACAE LIPASE ADSORBED TO SILICA GEL

B. GILLIES[1], H. YAMAZAKI[1] and D.W. ARMSTRONG[2]

[1]Department of Biology, Carleton University, Ottawa, Canada, K1S 5B6

[2]Division of Biological Sciences, National Research Council of Canada, Ottawa, Canada, K1A 0R6

SUMMARY

Candida cylindracae lipase was adsorbed to silica gel. Performance of this immobilized enzyme in n-heptane was examined for ethyl butyrate production. It showed stable activity in repeated use provided that the enzyme was hydrated by passage of water at the conclusion of each batch cycle. This system also catalysed esterification of ethanol with higher fatty acids and esterification of isoamyl and isobutyl alcohols with acetic and butyric acids.

INTRODUCTION

"Natural" esters are in high demand for flavouring in the food, dairy and beverage industries. Esters produced from natural substrates by cells or enzymes are considered to be natural (ref. 1).

Non-polar solvents mixed with lipases (free or immobilized) in an aqueous phase have been used to catalyse interesterification of triglycerides (ref. 2) and esterification of alcohols (ref. 3,4). Some lipases have been immobilized onto a variety of supports and used to catalyse esterification in non-polar solvents (ref. 5). However, long term performance of these systems is not fully documented.

Candida cylindracae lipase has a wide specificity (ref. 6) and is stable in acidic pH (ref. 7). This paper describes the immobilization of this lipase and the requirement of hydration for its long term use.

METHODS

Enzyme immobilization

Candida cylindracae triacylglycerol lipase (Sigma No. L-1754; 700 units/mg solid) was suspended in water to a final concentration of 5% (w/v). Three mL of the supernatant (recovered by decantation) was mixed with 1 g of silica gel (Davisil 646 from Grace Industrial Chemicals; 60 mesh; pore diameter of 150 Å) and left at room temperature (25°C) for 3 h. The gel was then washed three times with water on a glass-sintered funnel and excess liquid was removed by suction with a water aspirator for a few minutes.

Esterification

An alcohol and an organic acid were mixed with n-heptane to a final concentration of 0.4 M and 0.25 M respectively. Ten mL of this substrate solution was mixed with wet immobilized lipase preparation which was derived from 1 g of silica gel. The mixture was shaken at 200 rpm and 30°C. Samples were periodically withdrawn and mixed with 2 volumes of n-heptane containing 0.05% hexanol (internal standard) and analysed by gas chromatography.

RESULTS AND DISCUSSION

Natural ethyl butyrate is in high demand for its pineapple-banana flavour in food industries. Therefore the production of this ester was studied as a model system to determine optimal reaction conditions.

Candida cylindracae lipase was immobilized by adsorption to large-pore silica gel. The air-dried immobilized lipase was reacted with a mixture of 0.25 M ethanol and 0.25 M butyric acid dissolved in n-heptane. After 24 h, 0.15 M ethyl butyrate was produced. When this batch production was repeated with the same immobilized lipase, the activity rapidly decreased; 40% and 90% decreases after 1st and 2nd repetitions. Water is required by all the forces that maintain the catalytically active structure of an enzyme, so it is possible that dehydration of the enzyme was partly responsible for the decrease in activity. To test this possibility, the effect of enzyme hydration was studied.

Effect of enzyme hydration

Different amounts of water were added to air-dried lipase. Figure 1 shows that wet gel containing 10% water gave the highest rate of ester production. Ester production decreased as water concentration increased greater than 20%. The activity of immobilized lipase with 10-20% water content was obtained when the gel was washed with water and excess water was removed by suction on a sintered funnnel with a water aspirator. When the gel was washed in this manner between the repetitions, there was no significant decrease in ester production after 3 repetitions of 24 h batch production.

Effect of substrate concentration

When the effect of substrate concentration was studied, it was found that organic acid concentrations above 0.25 M were inhibitory and increasing ethanol concentration was stimulatory up to 0.4 M beyond which it had inhibitory effects. Ethanol concentration less than 0.4 M resulted in ethanol depletion from the n-heptane phase before 24 h, limiting the extent of ester production. Adsorption of ethanol to the silica gel may account for some of this depletion. Figure 2 shows the kinetics of ethyl butyrate production in 0.25 M butyric acid

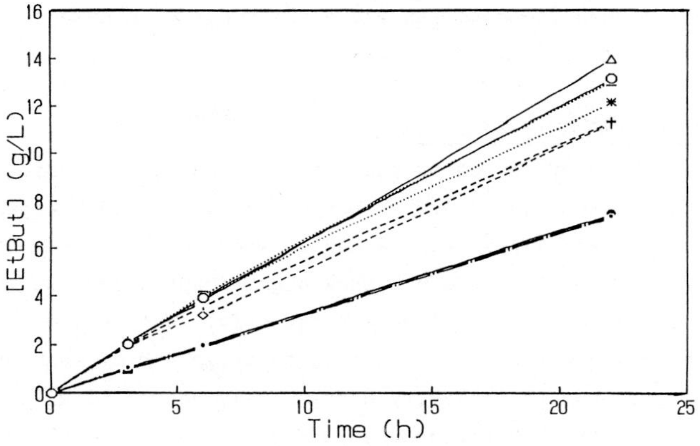

Fig. 1. Effect of hydration of the immobilized lipase on production of ethyl butyrate. Water was added to dry immobilized enzyme and then the substrate of 0.25 M ethanol and 0.25 M butyric acid in n-heptane was reacted. o:dry; Δ:10%; □:20%; ✱:30%; +:40%; ◇:50%: •:60%; ■:70%; ●:80% water added.

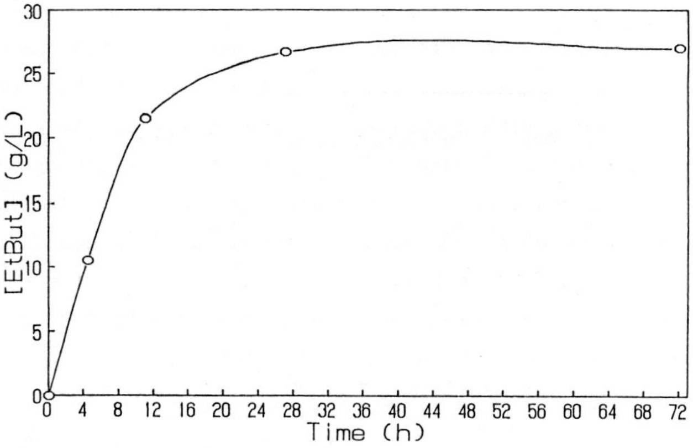

Fig. 2. Kinetics of ethyl butyrate production using immobilized C. cylindracae lipase in 0.25 M butyric acid and 0.4 M ethanol in n-heptane with shaking at 200 rpm, 30°C.

and 0.4 M ethanol. When the reaction was carried out in n-hexane (in place of n-heptane), similar kinetics were obtained.

Repeated use of the immobilized lipase

Figure 3 shows the performance of the present immobilized system in repeated use. Substrate concentration of 0.25 M butyric acid and 0.4 M ethanol were used, and the immobilized enzyme was washed between repetitions. Ethyl butyrate production after 24 h did not show significant change during the first 12 repetitions, but gradually declined subsequently. When the enzyme activity was determined by measuring ester concentration after 4 h of each of the 24 h cycles, it was found that the activity continuously decreased with number of repetitions. Between the repetitions, a portion of the gel was lost (37% after 23 repetitions). However the loss of the gel cannot totally account for the loss of enzyme activity, so enzyme inactivation must have taken place.

Since crosslinking of adsorbed enzyme with glutaraldehyde has been shown to stabilize enzymes, the effect of glutaraldehyde treatment was studied. Glutaraldehyde treatment was not beneficial because it caused significant reduction of the enzyme activity, though it slowed the rate of enzyme decay (Fig. 3).

Enzyme specificity

The substrate specificity of this system will determine its commercial potential for producing a variety of flavour esters. Figure 4 shows esterification of alcohols with a variety of organic acids. After 24 h, the levels of ethyl esters produced ranged from 13 to 45 g/L, although ethyl isovalerate and ethyl lactate were not produced appreciably. Isoamyl and isobutyl alcohol acetates were produced. A large peak corresponding to isoamyl butyrate was detected by GC but not quantitated due to lack of pure standard. The differences in ester production can be ascribed to differences in substrate solubility, specificity of enzyme and rate of reaction.

CONCLUSIONS

Simple immobilization of *C. cylindracae* lipase has been described. The immobilized lipase produced a variety of esters in non-polar solvent. The system can be used repetitively and shows good stability provided the enzyme is hydrated between each repetition. Although shaking was used in the present work, the immobilized lipase can be packed into a column and esters can be produced by circulating substrate solutions in non-polar solvents.

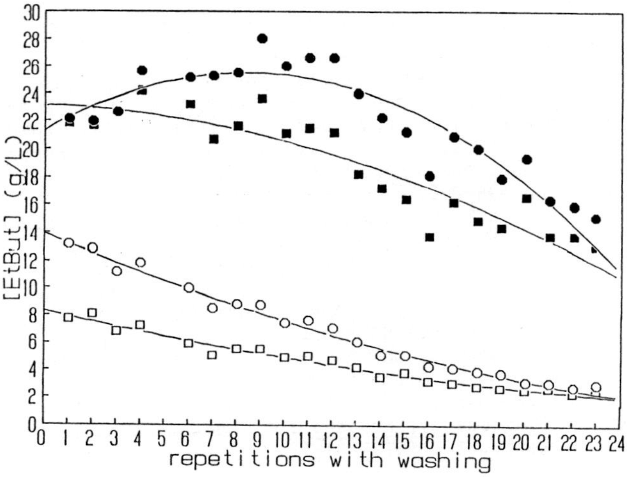

Fig. 3. Repeated use of immobilized lipase. Between each repetition the gel was washed with water and fresh substrate was added. Glutaraldehyde treatment: during the immobilization procedure glutaraldehyde was added to the slurry of silica gel in lipase solution to a final concentration of 1% (w/v) and incubated with occasional stirring. ●:adsorbed lipase (24 h sample); ■:glutaraldehyde treatment (24 h sample); o:adsorbed lipase (4 h sample); □:glutaraldehyde treatment (4 h sample).

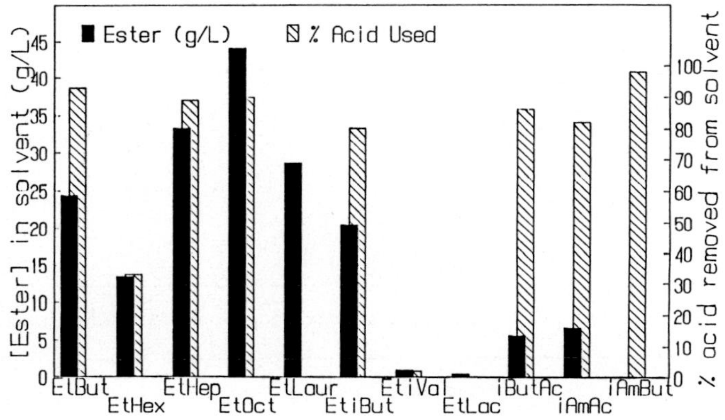

Fig. 4. Ester production by immobilized C. cylindracae lipase using a range of substrates (0.25 M acid, 0.4 M alcohol). The concentration is that of ester in the solvent after 24 h.

REFERENCES

1 D.W. Armstrong and H. Yamazaki, Natural flavours production: A biotechnological approach, Trends in Biotechnol., 4 (1986) 254-265.

2 A.R. Macrae, Interesterification of fats and oils, in: J. Tramper, H.C. van der Plas and P. Linko (Eds.), Biocatalysis in Organic Syntheses, Proc. Int. Symp., Noordwijkerhout, The Netherlands, April 14-17, 1985, Elsevier, Amsterdam, 1985, pp. 195-208.

3 M. Iwai, S. Okumura and Y. Tsujisaka, Synthesis of terpene alcohol esters by lipase, Agric. Biol. Chem., 44 (1980) 2731-2732.

4 I.L. Gatfield and T. Sand, U.S. patent 4,451,565 (1984).

5 C. Marlot, G. Langrand, C. Triantaphylides and J. Baratti, Ester synthesis in organic solvent catalysed by lipases immobilized on hydrophilic supports, Biotechnol. Lett., 7(9) (1985) 647-650.

6 G. Benzonana and S. Esposito, On the positional and chain specificities of _Candida cylindracae_ lipase, Biochim. Biophys. Acta, 231 (1971) 15-22.

7 R.S. Shifreen and P.W. Carr, An investigation of the kinetic characteristics of the lipase from _Candida cylindracae_ for its potential in triglyceride analysis, Analyt. Lett., 12 (1979) 47-69.

C. Laane, J. Tramper and M.D. Lilly (Editors), *Biocatalysis in Organic Media*, Proceedings of an International Symposium held at Wageningen, The Netherlands, 7–10 December 1986.

PREPARATION OF ACYL DERIVATIVES OF 1-HYDROXY ALDOSE BY LIPASE-CATALYZED HYDROLYSIS OR ALCOHOLYSIS OF FULLY ACYLATED ALDOSE IN ORGANIC SOLVENT

Jei-Fu Shaw and Ean-Tun Liaw

Institute of Botany, Academia Sinica, Taipei, Taiwan, Republic of China

ABSTRACT

The acyl derivatives of 1-hydroxy aldose can be prepared with high yield by lipase catalyzed hydrolysis in water-organic solvent two-phase system or alcoholysis in dry organic solvent. The yield of 2,3,4,6-glucose tetraacetate was 93% when the β-D-glucose pentaacetate was hydrolyzed in ethylacetate/0.1 M KCl (pH 7.9) system with A. niger lipase as catalyst. The yield of 2,3,4,6-glucose tetraacetate was 100% when it was prepared by the lipase catalyzed alcoholysis of β-D-glucose pentaacetate in either n-butanol, isopropanol or t-butanol. The order of the reaction rate was as follows: isopropanol > n-butanol > t-butanol. These processes offered many advantages over the chemical process disclosed by Japanese Patent (2).

INTRODUCTION

Esters of mono- and di-saccharides have a number of interesting and potentially useful properties including surface activity, antitumor activity, and plant growth inhibiting activity (3-6). The acyl derivatives of 1-hydroxy aldose are useful intermediates for medicines and pesticides (2). The 2,3,4,6-glucose tetraacetate can be prepared with 75% yield by a rather tedious and energy consumming chemical method (2). In the previous report, we have discovered that this compound can be efficiently prepared with 74% yield by lipase catalyzed hydrolysis of β-D-glucose pentaacetate at room temperature (1). In the present study, we found that the yield can be greatly increased by controlling the lipase-catalyzed reaction in organic solvent.

MATERIALS AND METHODS

Lipase from Aspergillus niger (AP 6), which had a specific activity of 63.7 Amano units per mg solid, was purchased from Amano Pharm. Co. (Nagoya, Japan). β-D(+)-Glucose pentaacetate was obtained from Sigma Chemical Co. (St. Louis, MO). All other biochemicals and chemicals were of reagent grade.

Time courses of lipase-catalyzed hydrolysis of glucose pentaacetate in a water-organic solvent two-phase system were followed potentiometrically using a Radiometer pH-stat. In a typical experiment, 10 ml of 0.1 M glucose pentaacetate in water immiscible organic solvent/0.1 M KCl (1:1 ratio) at 30 C were hydrolyzed at pH 7.9 and 1 M KOH was used for titration. Periodically, the reaction mixture was extracted with ethylacetate (a 10-fold excess by volumes) and assayed by gas chromatography.

Lipase-catalyzed alcoholyses of glucose pentaacetate were initiated by addition of an Celite immobilized lipase to a 0.1 M glucose pentaacetate in 2 ml dry alcohols (by shaking 3A molecular sieves, E. Merck Co., West Germany). The mixture was placed in a stoppered glass vials and shaken on an orbital shaker at 250 rpm and 28 C. Since the glucose pentaacetate was only partially soluble in the alcohols, 1 ml of ethylacetate was added to completely dissolve the substrate and then 1 ml of mixture was dried by rotary evaporator and assayed by gas chromatography. The lipase was immobilized on Celite as follows: 40 g of Celite 545 (Hyashi Pure Chemical Industries. Japan) washed with 10 mM phosphate buffer (pH 7) was mixed with A. niger lipase (20 g dissolved in 120 ml of 10 mM phosphate buffer, pH 7) with stirring at 4 C. Five fold excess of cold (-20 C) acetone was added and was stirred at 4 C for 30 min . The immobilized enzyme was filtered and dried first under air and then under vaccum at room temperature.

To identify the reaction products, the ethylacetate extract was dried by rotary evaporator and derivatized with 1,1,1,3,3,3 hexamethyldisilazane according to the procedure of Brobst and Lott (7) and assayed by gas chromatography (a Hitachi 263-30 gas chromatography equipped with a 2 m glass column packed with Supelco's 3% OV-17 on 80/100 Chromosorb W HP). Then 5 μl of that derivatized solution was injected into the gas chromatograph (N_2 as a carrier gas, 30 ml/min; detector and injector port temperature were 270 C), and the temperature of the column was increased from 180 to 220 C at 4 C/min. The retention times for glucose pentaacetate, glucose tetraacetate and glucose triacetate after derivatization were 9.0, 7.0 and 6.5 min respectively. The derivatized products must be analyzed quickly since glucose pentaacetate gradually transformed into derivatized glucose tetraacetate under the derivatizing condition.

The position of acylation in glucose tetraacetate was analyzed using ^{13}C nuclear magnetic resonance spectroscopy as described in ref. 8, except that deuterated dimethylformamide was used as a solvent.

RESULTS AND DISCUSSION

In the previous work, we established that a suspension of β-D-glucose pentaacetate (0.1 M) in aqueous solution of 0.1 M KCl can be hydrolyzed in a stepwise mode by several lipase catalyzed reaction at pH 7.9 (1). The enzymatic deacylation of glucose pentaacetate affords significant accumulations of glucose tetra-, tri-, and di-acetates depending on the degree of conversion. In the the case of <u>Aspergillus niger</u> lipase, at 20% conversion glucose tetraacetate constitutes 75% of total material, at 40% conversion glucose triacetate constitutes 66%, and at 60% conversion glucose diacetate constitutes 30% of total material. Since the yield is not satisfactory, we then endeavored to find the condition which would improve the yield. Table 1 shows the product composition of <u>A. niger</u> lipase-catalyzed hydrolysis in various organic-aqueous solution two-phase system. The best system for the glucose tetraacetate production was the ethylacetate/0.1 M KCl (7:3) system when the catalysis was carried out at pH 7.9 and 30 C. Fig 1. depicts the concentration <u>vs</u> time profiles for various glucose esters in the course of <u>Aspergillus niger</u> lipase-catalyzed hydrolysis of glucose pentaacetate in the ethylacetate/0.1 M KCl (7:3) system. The data indicate that this reaction system apparently favored the accumulation of glucose tetraacetate. The relative rate for glucose triacetate production was lower in the two-

TABLE 1

Product composition of β-D(+)-glucose pentaacetate hydrolysate of lipase-catalyzed reaction in organic solvent-aqueous solution two-phase system.

Hydrolysis system [(a)]	Reaction time (min) [(b)]	Product composition [(c)], mM GPA	GTEA	GTRA	GDA
0.1 M KCl	15	16	75	9	0
Toluene/0.1 M KCl (1:1)	47	13	35	48	4
Chloroform/0.1 M KCl (1:1)	80	10	46	25	9
Ethylacetate/0.1 M KCl (3:7)	23	10	88	2	0
Ethylacetate/0.1 M KCl (1:1)	30	7	91	2	0
Ethylacetate/0.1 M KCl (7:3)	33	2	93	5	0

(a) 0.5 g of soluble <u>Aspergillus niger</u> lipase was used. 0.1 M of glucose pentaacetate in different system was hydrolyzed at 30 C and pH 7.9.

(b) Reaction time for 20% conversion of glucose pentaacetate.

(c) GPA: glucose pentaacetate; GTEA: glucose tetraacetate; GTRA: glucose triacetate; GDA: glucose diacetate.

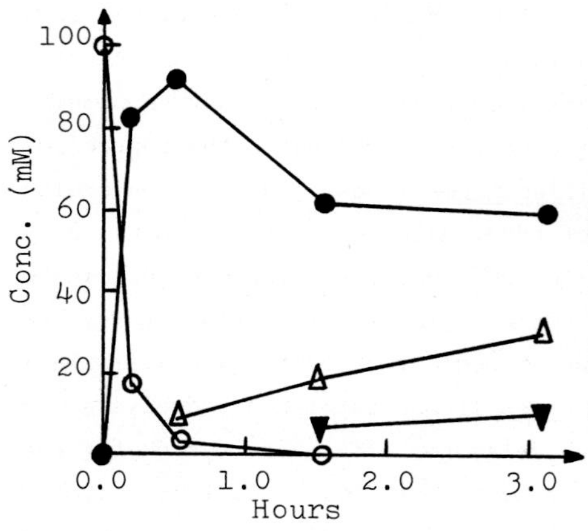

Fig. 1. The concentration vs. reaction time profiles for various glucose esters in the course of Aspergillus niger lipase-catalyzed hydrolysis of β-D(+)-glucose pentaacetate in ethylacetate/0.1 M KCl (7:3) system. The reaction conditions are as described in Methods and Table 1. (○)-glucose pentaacetate; (●)-glucose tetraacetate; (Δ)-glucose triacetate; (▼)-glucose diacetate.

phase system than that in aqueous solution (1). When the reaction was stopped after 29 min by extraction with ethylacetate (10 fold excess by volume) and followed by evaporation of the solvent, the isolated product gave one major spot in TLC [acetone/chloroform (1:1) mixture as a solvent].

Later, it occurred to us that lipase might be able to catalyze alcoholysis of glucose pentaacetate and might exhibit selectivity in different alcohols. The alcoholyses also eliminate the need for controlling the pH during the reaction and extraction of products from water after the reaction. As shown in Table 2, all the tested alcohols reacted in lipase-catalyzed alcoholysis of glucose pentaacetate. It is clear that

TABLE 2

Product composition of lipase catalyzed alcoholysis in various alcohols[(a)].

System	Reaction time (h)[(b)]	Product composition, mM		
		Glucose pentaacetate	Glucose tetraacetate	Glucose triacetate
Methanol	30	48	38	14
Ethanol	23	2	94	4
n-Propanol	28	0	91	9
Isopropanol	20	0	100	0
n-Butanol	21.5	0	100	0
t-Butanol	70	0	100	0
Amyl alcohol	21.5	0	95	5

(a) 0.2 g of celite immobilized Aspergillus niger lipase was incubated with 0.1 M glucose pentaacetate and 2 ml of various alcohols at 28 C.
(b) Reaction time for maximal production of glucose tetraacetate.

lipase indeed exhibited selectivity in different alcohols. In the case of isopropanol and n-butanol, the yield of glucose tetraacetate was 100%. t-Butanol also produced 100% glucose tetraacetate but at a much slower rate. Fig 2 depicts the concentration vs time profiles for various glucose ester in the course of Aspergillus niger lipased-catalyzed alcoholysis of glucose pentaacetate. This apparently indicated that methanol, being the smallest molecule among alcohols, had less selectivity toward glucose ester substrate.

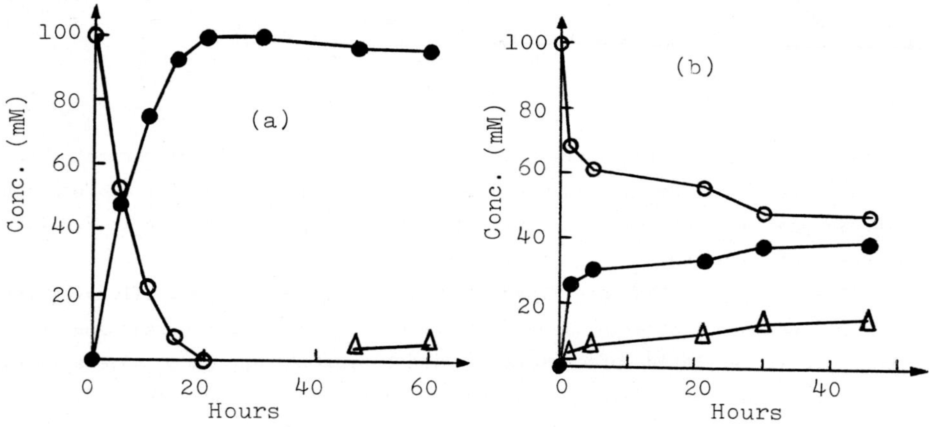

Fig. 2. The concentration vs. reaction time profiles for various glucose esters in the course of A. niger lipase-catalyzed alcoholysis of β-D(+)-glucose pentaacetate in isopropanol (a) and methanol (b). The reaction conditions are as described in Methods and Table 2. (○)-glucose pentaacetate; (●)-glucose tetraacetate; (Δ)-glucose triacetate.

In the case of isopropanol, no significant amount of glucose triacetate appeared 10 h after the glucose tetraacetate reached 100%. As shown in Fig 3, the production of glucose tetraacetate followed Michaelis-Menten kinetics. The Km values of glucose pentaacetate in isopropanol and butanol was 0.047 and 0.055 M respectively. This indicated that lipase had higher affinity for glucose pentaacetate in isopropanol but the Vm was about the same.

The lipase catalyzed hydrolysis or alcoholysis was possibly a kinetic process as follows: glucose pentaacetate $\xrightarrow{k_1}$ glucose tetraacetate $\xrightarrow{k_2}$ glucose triacetate $\xrightarrow{k_3}$ glucose diacetate $\xrightarrow{k_4}$ glucose monoacetate $\xrightarrow{k_5}$ glucose. By choosing conditions which have a high k_1/k_2 ratio, we can selectively prepare glucose tetraacetate in nearly 100% yield.

In the ^{13}C NMR spectrum of the enzymatically prepared glucose tetraacetate, the only change compared to that of the initial β-D(+)-glucose penta-

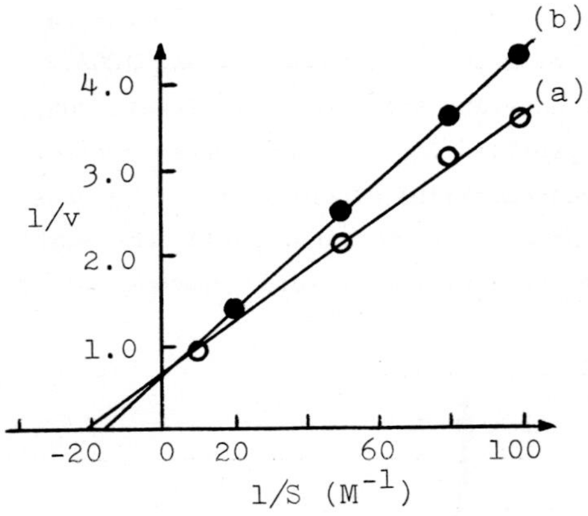

Fig. 3. The double reciprocal plot of lipase-catalyzed alcoholysis of glucose pentaacetate in isopropanol (a) and butanol (b). Initial velocities of A. niger lipase catalyzed reactions were measured at 28 C. 0.1 g Celite immobilized lipase was used and the velocity was expressed as μ mole/min.

acetate was in the peak corresponding to C-1 (Fig 4). The anomerization of glucose tetraacetate provides the additional evidence that C-1 carbon in glucose pentaacetate was deacetylated. Therefore, the product is 2,3,4,6-glucose tetraacetate.

It appears that the same lipase catalyzed process may be applicable for the production of other acyl derivatives of 1-hydroxy aldose starting with their fully acylated precursors . The enzymatic approach offers many advantages over the chemical process disclosed by Japanese patent (2). The

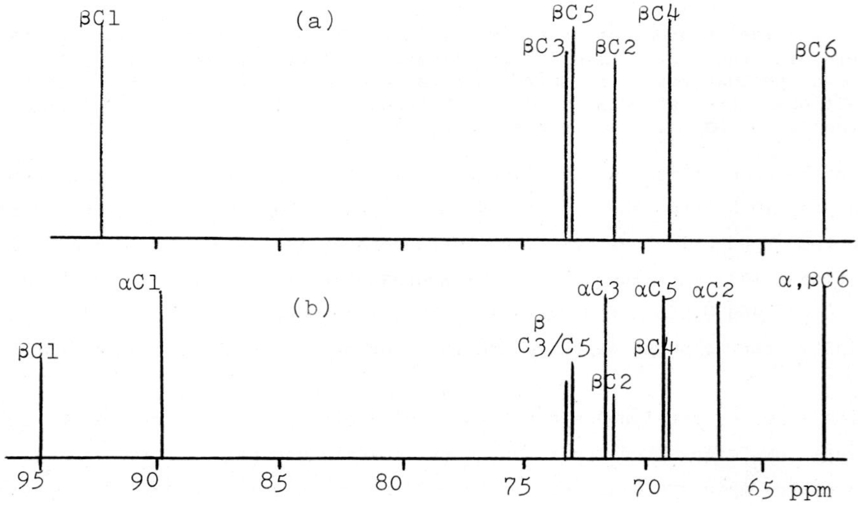

Fig. 4. ^{13}C-NMR spectrum of glucose esters. (a)-β-D(+)-glucose pentaacetate; (b)- glucose tetraacetate.

Japanese patent disclosed that refluxing glucose pentaacetate in CF_3CO_2H and $(CF_3CO)_2O$ for 16.5 h followed by methanolysis gave 75% 2,3,4,6-glucose tetraacetate. In the enzymatic process described in this paper, we can obtain 100% yield of 2,3,4,6-glucose tetraacetate in room temperature by simple lipase-catalyzed alcoholysis. This process has four advantages: 1. high yield and no need for product purification, 2. save energy, 3. simple, and 4. save additional reagents such as CF_3CO_2H and $(CF_3CO)_2O$.

ACKNOWLEDGEMENT. This work was supported by National Science Council, R.O.C. and Academia Sinica, R.O.C.

REFERENCES

1. J. F. Shaw and A.M. Klibanov, Preparation of various glucose esters via lipase-catalyzed hydrolysis of glucose pentaacetate, Biotechnol. Bioeng., in press.
2. Noguchi Research Institute, Acyl derivatives of 1-hydroxy aldoses, JPn. Kokai Tokkyo koho. JP 60 69,093. [8569,093], 19 Apr 1985.
3. Y. Nishikawa, M. Okabe, K. Yoshimoto, G. Kurono, and F. Fukuoka, Chemical and biochemical studies on carbohydrate esters. 3 . Antitumor activity of unsaturated fatty acids and their ester derivatives, Chem. Pharm. Bull. 24(4) (1976) 756-762.
4. Y. Nishikawa, K. Yoshimoto, M. Okada, T. Ikekawa, N. Abiko, and F. Fukuoka, Chemical and biochemical studies on carbohydrate esters.5. Anti Ehrlich Ascites tumor effect and chromatographic behaviors of fatty acyl monoesters of sucrose and trehalose, Chem. Pharm. Bull. 25(7) (1977) 1717-1724.
5. Y. Nishikawa, K. Yoshimoto, and M. Ohkawa, Chemical and biochemical studies on carbohydrate esters.7. Plant growth inhibition by an anomeric mixture of synthetic 1-o-lauroyl-D-glucose, Chem. Pharm. Bull. 27(9) (1979) 2011-2015.
6. H. Seino, T. Uchibori, T. Nishitani, and S. Inamasu, Enzymatic synthesis of carbohydrate esters of fatty acid (1) Esterfication of sucrose, glucose, fructose and sorbitol, J. Am. Oil. Chem. Soc, 61 (1984) 1761-1765.
7. K. M. Brobst and C. E. Lott, Determination of some carbohydrates in corn syrups by gas liquid chromatography of the trimethylsilyl derivatives, Cereal Chem, 43 (1966) 35-42.
8. T.E. Walker, R.E. London, T.W. Whaley, R. Barker, and N.A. Matwiyoff, Carbon-13 nuclear magnetic resonance spectroscopy of [1-^{13}C] enriched monosaccharides. Signal assignments and orientational dependence of germinal and vicinal carbon-carbon and carbon-hydrogen spin-spin coupling constants, J. Am. Chem. Soc., 98 (1976) 5807-5813.

C. Laane, J. Tramper and M.D. Lilly (Editors), *Biocatalysis in Organic Media*, Proceedings of an International Symposium held at Wageningen, The Netherlands, 7–10 December 1986.

THE ANAEROBIC TRANSFORMATION OF LINOLEIC ACID BY ACETOBACTERIUM WOODII

H. Giesel-Bühler*, O. Bartsch, H. Kneifel, H. Sahm und R. Schmid*

Institut für Biotechnologie, Kernforschungsanlage Jülich GmbH, 5170 Jülich 1, FRG

* HA-Biotechnologie, Henkel KGaA, 4000 Düsseldorf, FRG

Summary

Resting cells of Acetobacterium woodii transformed linoleic acid under anaerobic conditions into a more polar product. GC-MS data of the isolated product revealed that linoleic acid was transformed into 10-hydroxy-octadecenoic acid, an isomere of ricinoleic acid. A similar product was obtained with linolenic acid as substrate. The organism can be used for the preparation of 10-hydroxy long chain fatty acids.

Introduction

Acetobacterium woodii is an anaerobic homoacetogenic bacterium with interesting metabolic features: it was for example the first anaerobic bacterium reported to cleave phenylmethylether bonds and to ferment methoxyl groups or methyl residues to acetate in the same way as methanol (1).
We used this versatile organism, among several others, in our studies on the anaerobic biotransformation of unsaturated long chain fatty acids. Our main interest was focused on the transformation products of linoleic acid.

Materials and Methods

Acetobacterium woodii ATCC 29683 was grown on a medium containing glucose and harvested in the early stationary growth phase. Transformations were performed in Warburg respirometers under an atmosphere of argon or hydrogen with 3 ml cell suspensions (0,3 g wet packed cells and 2,7 ml 0,1 M phosphate buffer pH 7,0 containing tetracycline).

The water immiscible substrates (10 mM) were added in emulsified form (BSA). The incubations were carried out at 30°C for several hours. For thin layer chromatography the supernatants of the transformation mixtures were acidified and extracted with chloroform or ethylacetate and analyzed on rp18-HPTLC-plates in ethanol/H_2O or methanol/H_2O 9/1 (v/v). The HPTLC-plates were either sprayed with an anisaldehyd/ H_2SO_4-solution and dryed by 110°C for 10 min. (2) or treated with iodine. Methyl esters were prepared according to (6). For HPLC and GC/MS conditions see Fig. 4 and 5.

Results and Discussion

The growth of A. woodii on glucose was affected by the addition of unsaturated long chain fatty acids; in the presence of Tween 80 (3 %), linoleic acid (1 mM) inhibited growth completely (results not shown). Tween 80 itself had no effect. Resting cells in contrast were able to transform linoleic- and linolenic acid into a more polar product (Fig. 1). The analogous unsaturated long chain fatty alcohols, however, were not transformed in this way.

The transformation product of linoleic acid was obtained under an atmosphere of hydrogen or argon with whole cells and with the supernantant of cell extracts. Heat treated cells showed no transformation activity (Fig. 2). The chromatographic behavior of the transformation product was very similar to ricinoleic acid (Fig. 3).

In order to characterize it further, the methyl ester of the transformation product was processed by preparative HPLC (Fig. 4). The substance eluted after 9.13 min. was found to be responsible for the HPTLC-spot resulting from the conversion of linoleic acid. It was further investigated by UV-spectroscopy (not shown) and by mass spectrometry (Fig. 5). The interpretation of the mass spectrum led to the conclusion that the unknown transformation product was 10-hydroxy-12-octadecenoic acid, an isomere of ricinoleic acid. Mass spectrum data from literature for this compound were in accordance with our findings (3).

Our results show that A. woodii is able to transform linoleic- and probably other Δ9-unsaturated long chain fatty acids by adding water across the double bond. A similar system acting under aerobic and anaerobic conditions has been characterized from Pseudomonas spec. (4). Our results also indicate that the ability to hydrate Δ9-unsaturated long chain fatty acids seems not to be restricted to anaerobic enteric bacteria, which are so far the only anaerobes reported to transform oleic acid into 10-hydroxy-stearic acid (5).

A. woodii thus provides a good instrument for the preparation of 10-hydroxy long chain fatty acids, especially from multiply unsaturated fatty acids.

Fig. 1: Biotransformations with resting A. woodii cells: HPTLC-analysis of extracted products (details in Materials and Methods).

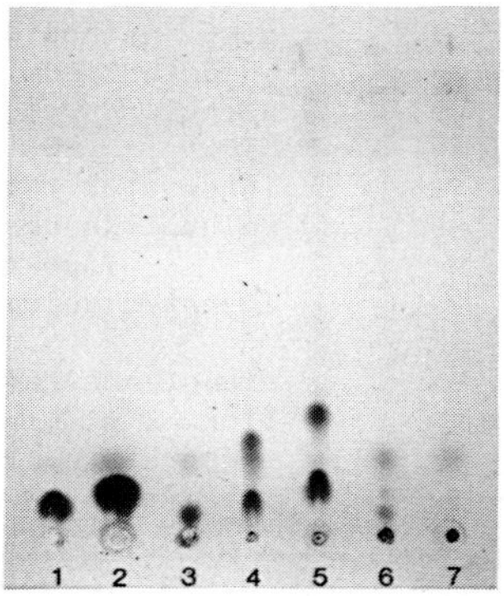

Standards: 1) linoleic acid (R_f 0,063); 2) linolenic acid (R_f 0,089); 3) linoleoyl alcohol (R_f 0,038); Biotransformation with: 4) linoleic acid, R_f-product 0,166; 5) linolenic acid, R_f-product 0,243; 6) linoleoyl alcohol; 7) no substrate

Fig. 2: Biotransformation of linoleic acid with resting cells of A. woodii: HPTLC-analysis

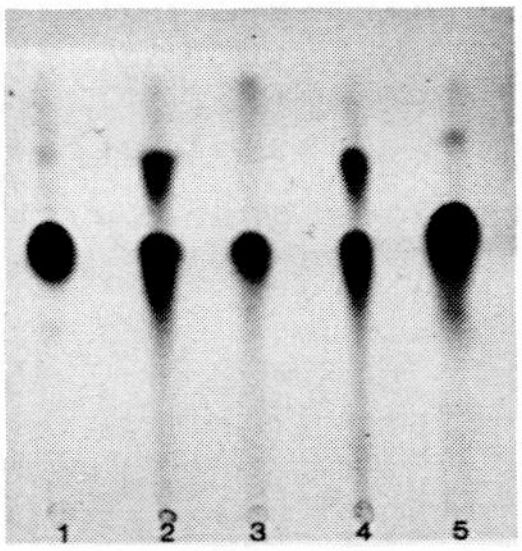

Standard: 1) linoleic acid (R_f 0,518) Biotransformations: 2) under H_2 (R_f-product 0,68) 3) with heat treated cells 4) under Ar (R_f-product 0,675) 5) without cells

Fig. 3: Comparison of the chromatographic behavior of the transformation product and ricinoleic acid by HPTLC

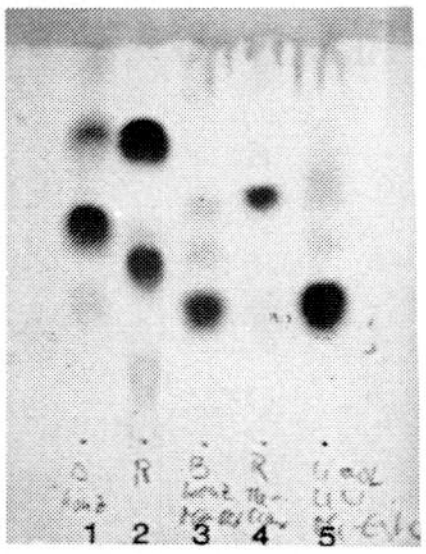

1) Transformation product R_f 0,65, substrat: linoleic acid R_f 0,46
2) Ricinoleic acid standard (impure) R_f 0,65
3) Transformation product methyl ester R_f 0,52, substrate: linoleic acid methyl ester R_f 0,28
4) Ricinoleic acid standard methyl ester R_f 0,53
5) Linoleic acid methylester UV-treated R_f 0,29 UV-product R_f 0,55

Fig. 4: HPLC separation of the extract after microbial conversion of linoleic acid.

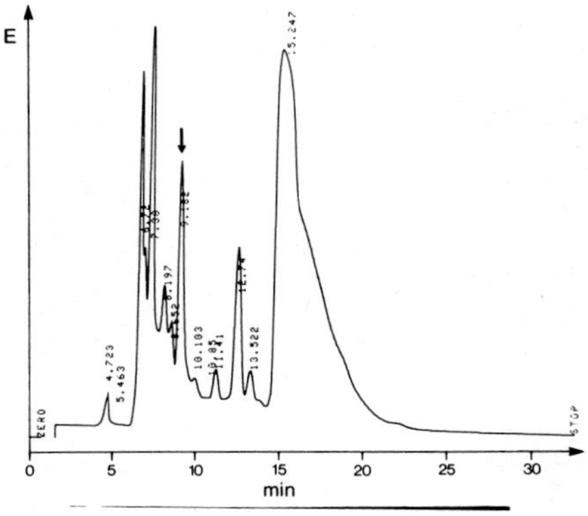

HPLC conditions:
pump: Waters 6000A
injector: U6K
detector: Uvicord S II, LKB set at 206 nm
integrator: Shimadzu C-R 3A
column: Knauer RP 18, 250 x 16 mm, 7 µm particles
injected volume: 200 µl, concentration unknown
eluant: methanol
flow: 6 ml/min
temperature: 20°C
pressure: 4 MPa

9.132 min: transformation product
15.247 min: linoleic acid

Fig. 5: Mass spectrum of the methyl ester of the conversion product (GC/MS, not corrected)

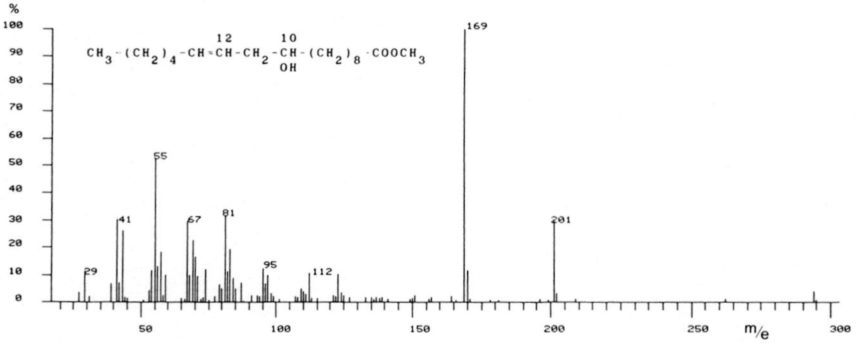

Masses:

312:M^+ (not found)
294:M - H_2O
201:$\bullet$CH-$(CH_2)_8$-CO-OCH_3 (with OH on the CH)
170:201 - OCH_3
169:201 - CH_3OH
112:CH_3-$(CH_2)_4$-CH=CH-$CH_2^{\bullet}$+H^+

GC/MS conditions:

Spectra were obtained using a KRATOS MS 25 mass spectrometer connected to a CARLO ERBA 4200 gas chromatograph
GC: GC column: CP Sil 5 (Chrompak), l=25 m,i.d.: 0,32 mm; temperature: 40°C - 240°C at 10°C/min; injection: 1 µl, split ratio: 33:1; carrier gas: helium
MS: trap current:100 µA; filament: 4.4 A; emission: 1 mA; ion source: 200°C;ionization energy: 70 eV

Acknowledgement

We thank Frau Ing. C. Jebsen for performing the GC/MS experiments.

Literature

1 R. Bache, N. Pfennig: Selective isolation of Acetobacterium woodii on methoxylated aromatic acids and determination of growth yields. Arch. Microbiol. 130 (1986) 255-261
2 J. C. Kohli, A. Arora: A general spray reagent for naturally occuring compounds. Ann. Chim. 2 (1977), 21-23
3 T. Takatori, K. Terazawa, K. Nakano, H. Matsumiya: Identification of 10-hydroxy-12-octadecenoic acid in adipocere. Forensic Sci. Internat. 23 (1983) 117-122
4 W. G. Niehaus, A. Kisic, A. Torkelson, D. J. Bednarczyk, G. J. Schroepfer jr.: Stereospecific hydration of the Δ9 double bond of oleic acid. JBC 245 (1970) 3790-3797.
5 J. P. Thomas: Identification of some enteric bacteria which convert oleic acid to hydroxystearic acid. Gastroenterology 62 (1972) 430-435
6 C. J. W. Brooks, E. C. Horning: Gaschromatographic studies of catecholamines, tryptamines and other biological amines. Part 1. Analyt. Chem. 36 (1964) 1540-1545

C. Laane, J. Tramper and M.D. Lilly (Editors), *Biocatalysis in Organic Media*,
Proceedings of an International Symposium held at Wageningen,
The Netherlands, 7–10 December 1986.

STABILITY OF ENZYMES IN LOW WATER ACTIVITY MEDIA

V. LARRETA - GARDE , Z. F. XU , J. BITON and D. THOMAS
Laboratoire de Technologie Enzymatique , Université de Technologie , B.P. 233,
60206 Compiègne (France)

SUMMARY

The stability of invertase, lysozyme and alcohol dehydrogenase stored in the presence of several additives modifying the water activity of the storage medium has been measured. For invertase, inactivation is delayed when A_w is decreased either by addition of polyols or by immobilization within starch beads. Lysozyme exhibits different behaviours depending on the nature of the additives ; there is no direct relation between A_w and $t_{1/2}$ whereas initial activity is an exponential function of A_w. Sorbitol, at high concentrations, efficiently preserves ADH activity for both yeast and thermostable enzyme, this protective effect is additive to the observed stabilization by immobilization.

INTRODUCTION

The further development of enzyme technology needs, at first, to make use of stable enzymes. In industrial processes, biocatalysts would often act in hot or non- conventional media.To be used, enzyme activities must be stable over long period under operating conditions and this appears not to be the case for most presently available enzymes. It is thus of interest to determine what is limiting enzymatic activity in order to overcome this limitation.

During the ten past years, several authors have proposed explanations to the enzyme stability problem. The mechanism of ribonuclease thermoinactivation has recently been elucidated (1). Immobilization appears to be a good tool for maintaining enzyme activity by avoiding protein aggregation and unfolding. Another solution consists in modifying the molecule microenvironment in aqueous solutions, mainly by addition of polyols or sugars which have been shown to preserve enzyme activity. With this method, enzymes are used in conditions similar to those of industrial processes (heterogeneous, with low water content, non-newtonian media). Enzyme stability may often be correlated to water activity (2).

We have focussed on the effect of additives such as polyols, alcohols and sugars on the activity and stability of different enzymes : invertase, egg-white lysozyme and yeast alcohol dehydrogenase .

METHODS

Invertase, lysozyme and alcohol dehydrogenase activities were measured spectrophotometrically. Thermodynamical water activity (A_w) was obtained with a Novasina unit equipped with a EnBS-4 sensor. Stability was expressed through the half-life time ($t_{1/2}$), where the enzyme exhibits 50% of its initial activity.

RESULTS AND DISCUSSION

Assays on invertase :

Invertase rapidly inactivated when stored in aqueous solution. We investigated the influence of three polyols (sorbitol, mannitol and glycerol) on storage stability of invertase at different temperatures. The enzyme was incubated for 20 min at 55, 60, 63, and 71°C, then the residual activity was measured. The nature of the polyol appeared to be very important for enzyme stability. The decrease of Aw in the storage medium lead to a thermostabilizing effect, but the different polyols showed different effects. There was no residual activity, whatever the temperature when the water activity was lower than 0.8. Only sorbitol showed a good protective effect . As shown in figure , in the presence of sorbitol, inactivation was delayed.

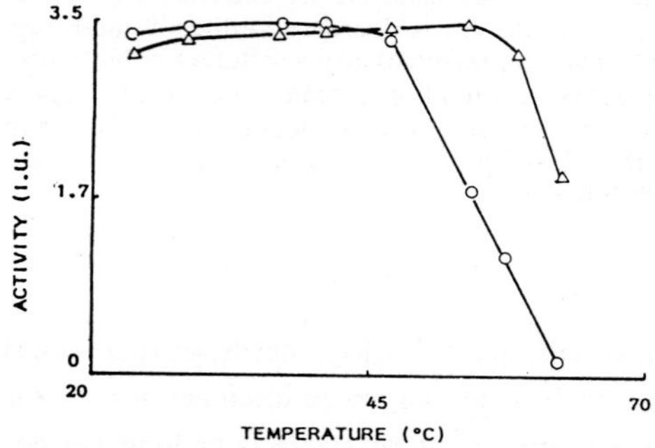

Figure 1 : Residual activity of invertase after 20 minutes incubation in absence (o) or presence of 55 % sorbitol (Δ) as a function of incubation temperature.

Invertase was immobilized within starch beads and the activity measured as a function of time, Aw and temperature (figure 2).

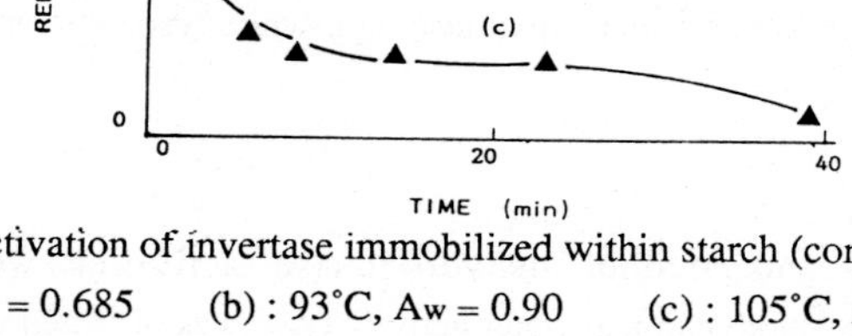

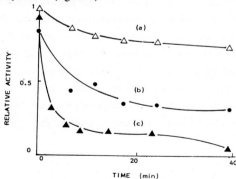

Figure 2 : Inactivation of invertase immobilized within starch (corn / wheat, 1/1, w / w) beads.
(a) : 93°C, Aw = 0.685 (b) : 93°C, Aw = 0.90 (c) : 105°C, Aw = 0.685

As a reference, native invertase completely inactivated at 71°C within 18 minutes. From the results obtained with immobilized invertase, we may conclude that decreasing Aw increases the enzyme stability.

Assays on lysozyme :

We chose to use lysozyme as a model as it is a monomeric and well studied enzyme. With a transition temperature of 77°C at neutral pH, lysozyme is a moderately thermostable enzyme ; this capacity allowed us to realize activity assays at temperatures as high as 100°C with good precision.

Several authors have previously discussed the effect of polyol on enzyme stability. Gekko (3) has shown that the stability factor due to polyol addition increases with the polyol concentration. This result has been obtained for egg-withe lysosyme (3,4), invertase (5), and collagen (6). The stabilizing effect as a function of temperature had been hardly studied untill now.

We thus measured the effect of polyols and alcohol on enzyme stability as a function of storage medium temperature and as a function of additive concentration. The results are presented in Table 1.

Temperature (°C)		25	60	80	100
buffer		960h	265h	38min	7min
	20%	950h			
ethanol	50%	560h			
	80%	150h			
	20%	>1120h	235h	7min	
polyethylenglycol	50%	860h	130h	8min	
	80%	800h	20h	6min	
	20%	720h	>300h	60min	
glycerol	50%	1120h	230h	145min	
	80%	880h	55h	60min	
	20%	1000h	130h	100min	14min
sucrose	50%	880h	225h	132min	15min
	80%	>1200h	280h	>132min	15min
	20%	1200h	>280h	57min	
sorbitol	50%	>1200h	>300h	>>4h	11min
	80%	760h	>>300h	>>4h	12min

Table 1 : $t_{1/2}$ as a function of polyol concentration and storage temperature .

The highest stabilization effect was obtained with sorbitol and sucrose, mainly for high additive concentrations and high temperatures. Glycerol also stabilized lysozyme activity except when concentrated (80 %). The other tested additives did not show any positive influence on lysozyme thermostability.

Lysozyme inactivation with additives present in the storage medium, for example with sucrose at 80°C (fig 3), did not follow first-order kinetics (7). Considering the complexity of an enzyme molecule, this appears quite normal (8).

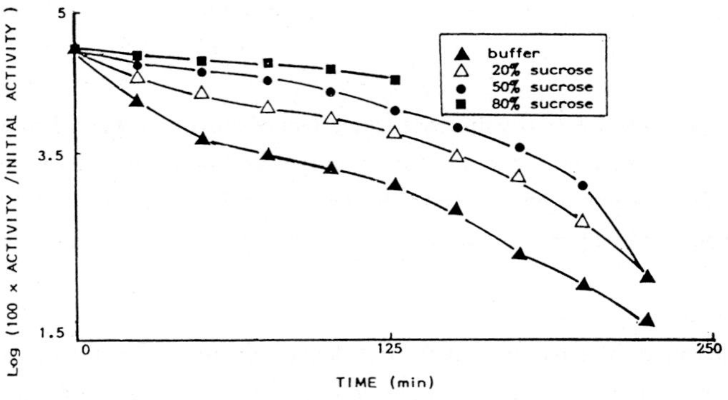

Figure 3 : Inactivation of lysozyme as a function of time of storage at 80°C in the presence of various sucrose concentrations .

The relationship between the enzyme behaviour and water activity has recently been reviewed (2). Enzymes are generally stabilized when their microenvironnemental water activity is decreased. Polyols and sugars are often used to decrease Aw.

Water activity of lysozyme stored in different media and both activity and stability of the enzyme as a function of Aw were measured. Evidently, there was no direct relation between t 1/2 and Aw, when initial activity was an exponential function of Aw (except for glycerol).

All the results obtained with lysozyme showed that enzyme thermostability cannot be directly related to water activity. Enzymatic stability also depends on the nature of Aw depressors. We thus tried to ascertain the importance of the chemical nature of the additive on the activity and stability of another enzyme : yeast alcohol dehydrogenase.

Assays on alcohol dehydrogenase :

We chose to explore further the behaviour of additives using yeast alcohol dehydrogenase (YADH) as it is an oxydoreductase (water does not participate to the enzymatic reaction). The enzyme presents a high specificity : ethanol being the only substrate, polyols may be added to the medium without interferring in the measurements. The enzyme is known to be unstable at high temperatures and over long periods in aqueous solutions.

The enzyme was stored in the presence of sorbitol at 25 and 40°C and its residual activity measured as a function of storage time (Table 2).

t1/2 (min) \ temperature		25°C	40°C
buffer		84	8
sorbitol	20%	131	38
	50%	160	79
	80%	242	150

Table 2 : Effects of various sorbitol concentrations on YADH stability at 25 and 40°C .

The stabilizing effect of sorbitol increased with the polyol concentration in the storage medium. We measured the effect of sorbitol on YADH immobilized within porous albumin foam (figure 4).

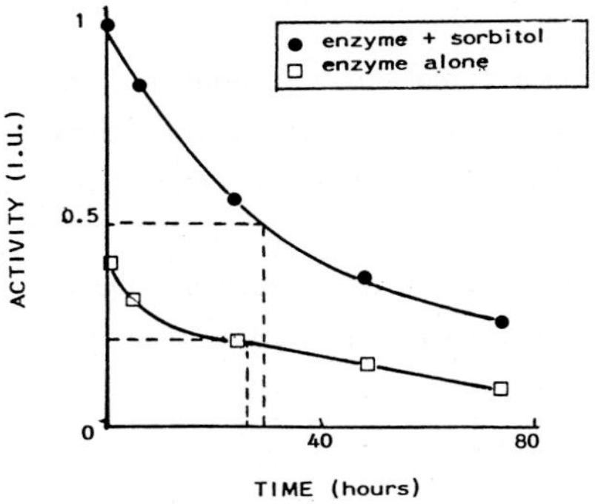

Figure 4 : Effect of sorbitol on immobilized YADH stability.

Sorbitol (0. 1 M) was added to the immobilization mixture before insolubilization. There was a sensitive effect of polyol mainly on initial activity, but also on stability. Immobilization stabilized the enzyme activity. (t 1/2 = 84 min for the native enzyme and 27h for the immobilized one), its effect was additive to that of the polyol .

We tested the effect of sorbitol on a thermostable alcohol dehydrogenase from *Sulfolobus solfataricus* (a kind gift from Pr. Rossi , Naples). At 25°C , the enzyme was little active,but was also well protected against time denaturation by the presence of sorbitol (after a 6 days storage, 25% of the initial activity retained in presence of the polyol while the native enzyme was completely denaturated).

CONCLUSION

We studied the effect of polyols, sugars and alcohols on enzymatic stability as a function of storage medium temperature , polyol concentration and storage medium water activity .

For invertase, thermoinactivation kinetics were modified when the medium water activity was decreased either by addition of polyols or by using non-newtonian media (starch powder). The protective effect of decreasing A_w was shown. The chemical microenvironment of the enzyme was also shown to control the protein thermoresistance.

For lysozyme, sorbitol and sucrose exhibited the best protective effect against thermoinactivation ; thermoresistance increased with the additive concentration . There was no direct relation between $t_{1/2}$ and A_w while initial activity is an exponential function of water activity .

For alcohol dehydrogenase, addition of sorbitol in the storage medium increased stability of the enzyme activity at different temperatures. This result was obtained with the enzyme in its native form, immobilized within porous albumin foam and for a thermostable enzyme.

From all these results, it appears that enzyme stability is directly controlled by the microenvironment of the protein. Water activity is one of the main parameters for enzyme thermoresistance . The chemical nature of the additive is shown to have a great influence on the tested enzyme stability .

In the next future, we shall explore further the influence of the additive chemical structure on enzyme stability. We shall then focus on stability of enzymes when functioning in low-water activity media .

REFERENCES

1 S. Zale and A. Klibanov , Why does ribonuclease irreversibly inactivate at high temperatures ?,Biochemistry, 25 , (1986), 5432-5444.
2 B. Hahn-Hagerdal , Water activity : a possible external regulator in biotechnological processes , Enzyme Microbiol. Technol. , 8-6, (1986), 322-327.
3 K. Gekko, Calorimetric study on thermal denaturation of lysozyme in polyol-water mixtures, J. Biochem.,91, (1982), 1197-1224 .
4 Y. Fujita , Y.Iwasa and Y. Noda , The effect of polyhydric alcohols on the thermal denaturation of lysozyme as measured by differential scanning calorimetry , Bull. Chem. Soc. Japan , 55 , (1982), 1896-1900 .
5 D. Combes and P. Monsan , Effect of water activity on enzyme action and stability , Enzyme Engineering 7 ,(1984), Ann. Acad. Sci. N.Y., 434, 61-63 .
6 K. Gekko and S. Koga , Increased thermostability of collagen in the presence of sugars and polyols , J. Biochem.,94 , (1983) ,199-2057
7 I. Segel , Enzyme Kinetics , (1975) , J. Wiley and Sons, Ed. ,New-York.
8 A. Sadana and J. Henley , Mechanistics analysis of complex enzyme deactivation , Biotechnol. Bioeng.,28, (1986) , 977-987.

C. Laane, J. Tramper and M.D. Lilly (Editors), *Biocatalysis in Organic Media*, Proceedings of an International Symposium held at Wageningen, The Netherlands, 7–10 December 1986.

SYNTHESIS OF POLYSACCHARIDES BEARING A LIPOPHILIC CHAIN FOR THE CHEMICAL MODIFICATION OF ENZYMES

M. WAKSELMAN and D. CABARET

CNRS-CERCOA, 2 rue Henry Dunant, 94320 Thiais (France)

SUMMARY

Amphiphilic reagents have been designed for the chemical modification of proteins. They possess a lipophilic alkyl chain, an hydrophilic part and a reactive functional group. Some reagents of this type in which the hydrophilic region and the reactive group are a reducing disaccharide have been synthesized. However the model reductive alkylation of N^{α} -Z-L-Lysine is a slow process. Therefore the synthesis of alkylated disaccharides containing an aldehyde function which is not involved in an hemiacetal formation has been undertaken.

INTRODUCTION

The use of enzymes as synthetic tools in organic synthesis is rapidly increasing (refs. 1,2). The main advantages of these biocatalysts are their selectivity, their stereospecificity, and mild conditions of reaction.

The problem of the stability and catalytic activity of enzymes in the presence of organic solvents (refs. 3,4) arises for two main reasons : a) dissolution in organic solvents of poorly water-soluble substrates, b) shift of the reaction equilibrium of some hydrolases towards the synthesis of esters or peptides by decreasing the thermodynamic activity of water (a_w ; refs. 5-7) i.e. replacing the bulk of water by an organic solvent. Many classifications of organic solvents have been proposed. Two of them are particularly relevant to our subject : distinction between water miscible and water immiscible (poorly miscible) solvents, and according to Ray, between solvents of class I (polyols ...), II (DMF ...) and III (alcools ...) (refs. 8,9). Numerous reviews have been devoted to enzyme denaturation and stabilization (refs. 10-13).

In aqueous solutions the hydrophobic interactions between non-polar aminoacid residues are generally considered to be the dominant factor involved in maintaining the native structure of proteins. Strong interactions of an enzyme with certain organic solvents can overcome the internal solvophobic forces, stabilize unfolded structures and favor denaturation. The mechanism of protein stabilization induced by polyhydric compounds (class I solvents which contain at least two or three centers capable of forming three-dimentional hydrogen-bonded network structures) has been studied. Polyols and sugars are preferentially excluded from the immediate environment of the globular protein which is preferentially hydrated (refs. 14,15). On the other hand many proteins bind carbohydrates, often in clefts on the protein surface. Formation of extensive networks of hydrogen bonds, sequestration of water molecules and numerous van der Waals contacts are observed (ref. 16). The maintenance of a layer of water around the enzyme seems essential for activity.

In the use of non-immobilized enzymes for chemical synthesis in the presence of organic solvents, four advances have recently been made :

1) Biphasic (polyphasic) systems (refs. 17,18)

In the aqueous phase, which contains the enzyme, the concentration of the poorly water miscible solvent is low. Substrates and products are transferred from one phase to the other. Therefore the inhibitory effects of high concentration of solvent, substrate and product are expected to be minimized. The reaction may also take place at the interface of the two phases.

2) Reverse micelles and microemulsions (refs. 19,20)

Water can be solubilized in the polar core of reverse micelles formed by the association of the polar heads of surfactants in an organic solvent. Stable water in oil microemulsions may be prepared. Then enzymes may be hosted in these water pools without loss of activity.

3) Solubilization of PEG-modified enzymes in organic solvents (refs.21, 22)

2,4-bis (o-methoxypolyethylene glycol)-6-chloro-s-triazine prepared from cyanuric chloride and a monomethylated polyethylene glycol reacts with the surface amino-groups of an enzyme to give a PEG-modified enzyme which is soluble and catalytically active in benzene or chlorinated solvents. Other means of introducing the PEG chains, such as acylation have been used. The reaction media are probably not completely dry.

4) Suspension in a nearly anhydrous non-polar solvent (refs. 23,24)

Hydrophobic solvents such as hydrocarbons do not remove the essential layers of water surrounding the enzyme. Moreover the biocatalyst remains in the conformation and "remembers" the pH corresponding to that of the latest aqueous solution from which it has been isolated. At the end of the synthesis the insoluble catalyst is easily recovered.

PROJECT

Our purpose differs from the ones above mentioned. We set out to synthesize well defined amphiphilic reagents of type I*. Conjugation of I with the exposed amino-groups of an enzyme will lead to a covalently modified enzyme II.

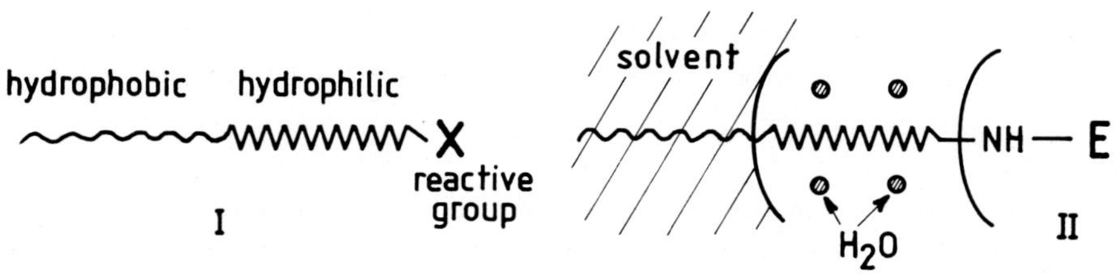

Fig.1

The system II-organic solvent containing a low proportion of water may have the organized "shell-like" structure shown on Fig. 1 : a few layers of water hydrogen-bonded to the added hydrophilic region could isolate the protein surface from the denaturating contact ot the water-immiscible solvent. Extension of the hydrophobic chain into the solvent should have a stabilizing effect. The presence of the amphiphilic chains will probably decrease the autoproteolysis of proteases. The modified enzyme will presumably be insoluble in the organic solvent and easily recoverable.

To begin with we chose a reducing polysaccharide as the hydrophilic and the reactive parts of the reagent I for the following reasons :

We have previously seen that polyols and sugars have a stabilizing effect on enzymes. Moreover some long chain alkyl glycoside detergents such as octyl- β-D-glucopyranoside or dodecyl-D-maltoside have been used for the solubilization of membrane proteins (refs. 25,26).

Synthetic glycoproteins formed by conjugation of carbohydrates with proteins often have a higher thermostability than the native proteins (refs. 27,28). Aldehydes and aldoses (more precisely the low amount of the hydroxy-aldehyde form in equilibrium with the cyclic hemiacetal structure) react reversibly and selectively with the amino-groups to give aldimines. These Schiff bases may be selectively reduced with sodium cyanoborohydride to give

---

*Reactive amphiphiles of type I may have other applications. For instance they can be used for protein immobilization on liposomes. They might also be useful for the design of covalent models of reverse micelles or the preparation of well defined mixed glycoconjugates as covalent membrane models.

stable alkylamines. Such reductive alkylation modification does not alter the net charge of the protein at neutral pH. The method has been successfully applied to the attachment of various polysaccharides to proteins (refs. 29-32).

Taking into account the slow rate of coupling of some alkylated derivatives of reducing disaccharides with N^{α}-Z-L-Lysine (<u>vide infra</u>) we also intend to prepare reagents of type I containing a free aldehyde function.

SYNTHESIS

We have already reported the preparation of two reducing disaccharides III and IV by osidic synthesis (ref. 33). The O-(2-decanamido-2-deoxy-β-D-glucopyranosyl)-(1 ⟶ 6)-D-galactopyranose (III) obtained by the oxazoline method has a long chain amido-substituent. The O-(6-O-octyl-β-D-galactopyranosyl)-(1 ⟶ 6)-D-galactopyranose (IV) prepared by a modified Koenigs-Knorr procedure possesses a long chain ether group (Fig. 2).

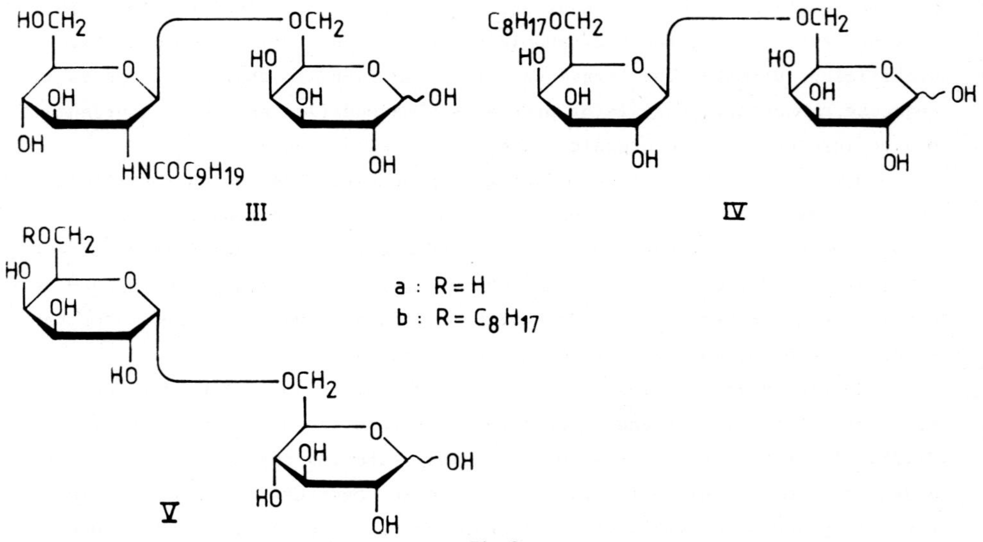

Fig. 2

A third alkylated reducing disaccharide, the O-(6-O-octyl- α -D-galactopyranosyl)-(1 ⟶ 6)-D-glucopyranose (Vb) has been obtained by functionalizing the O- α -D-galactopyranosyl-D-glucose (Va ; D(+)-melibiose). In the synthesis the primary hydroxyl function was temporarily protected by tritylation. Before the alkylation step, the secondary and hemiacetal alcoholic groups were masked by benzylation.

Then we attempted to prepare disaccharides having a free aldehyde function (ref. 34). Reductive ozonolysis (refs. 35,36) of the 1-β-allyl-6'-octyl-2,3,4,2',3',4'-hexaacetyl-melibiose (VIa) gave the acetylated aldehyde VIb (Fig. 3). Cleavage of the acetate groups gave a product, probably the hemiacetal VIIa, devoid of the characteristics of an aldehyde (IR, NMR). The direct reductive ozonolysis (O_3, Φ_3P) of the polyol VIc yielded the same compound. Increasing the chain length led to analogous results : the butenyl glycoside VId gave the hemiacetal VIIb.

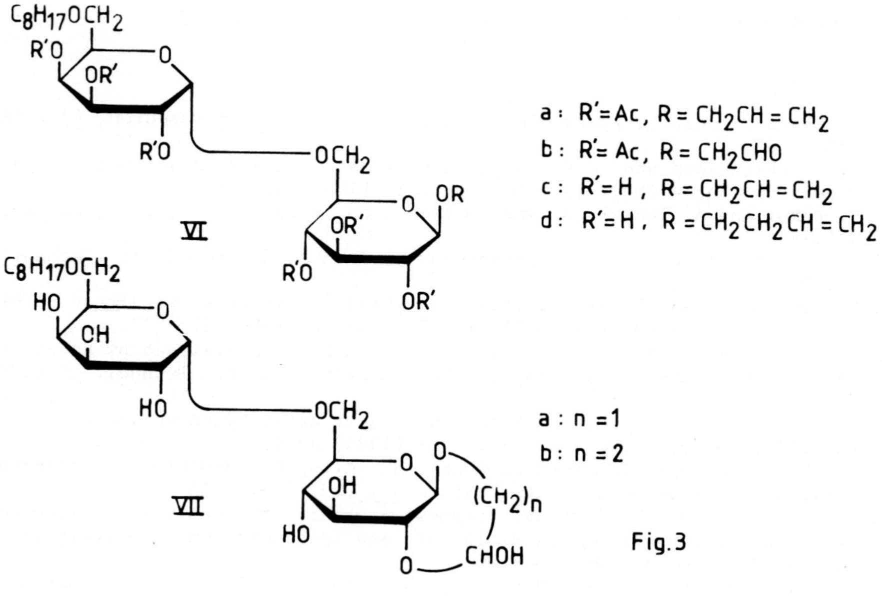

Fig.3

MODEL REACTION WITH N^{α}-Z-L-LYSINE

As a model of the amphiphilic reducing disaccharides III-V reaction with the ε-amino groups of the exposed lysine residues of a protein, the reductive alkylation of N^{α}-Z-L-Lysine has been studied in water with a ten fold excess of sodium cyanoborohydride (ref. 37). The long chain amido-disaccharide III which is poorly soluble is unsuitable for the conjugation.

The water soluble alkylated disaccharides IV and V reacted very slowly at 25°C. Products of the reaction were isolated and identified as the expected N^{α}-Z-N^{ε}- (6-O-octyl-β-D-galactopyranosyl-(1 ⟶ 6)-(1-deoxyhexitol-1-yl) -L-Lysines by high resolution mass spectroscopy. The reaction can be followed by HPLC (Lichrosorb NH_2, acetonitrile-water).

CONCLUSION

The model reaction reported above shows that the reductive condensation of the amphiphilic reducing disaccharides with the amino groups of a protein will probably be very slow. The reductive ozonolysis of allyl or butenyl β-D-glucopyranosides has not given the expected (ref. 35) disaccharides containing a free ω-aldehydoaglycon. Other methods of preparation of disaccharides, possessing both a long chain alkyl ether substituent and a free aldehyde function, are under study.

REFERENCES

1 J. Bryan Jones, Enzymes in organic synthesis, Tetrahedron 42 (1986) 3351-3403.

2 G.M. Whitesides and C.H. Wong, Enzymes as catalysts in synthetic organic chemistry, Angew Chem. Int. Ed. Engl. 24 (1985) 617-637.

3 S. Fukui and A. Tanaka, Enzymatic reactions in organic solvents, Endeavour 9 (1985) 10-17.

4 L.G. Butler, Enzymes in non-aqueous solvents, Enzyme Microb. Technol. 1 (1979) 253-259.

5 B. Hahn-Hägerdal, Water-activity : a possible external regulator in biotechnical processes, Enzyme Microb. Technol. 8 (1986) 322-327.

6 P.J. Halling, Effects of water on equilibria catalysed by hydrolytic enzymes in biphasic reaction systems, Enzyme Microb. Technol. 6 (1984) 513-516.

7 P. Monsan and D. Combes, Effect of water activity on enzyme action and stability, Annals N.Y. Acad. Sci. 434 (1984) 48-60.

8 A. Ray, Solvophobic interactions and micelle formation in structure forming nonaqueous solvents, Nature 231 (1971) 313-315.

9 K. Martinek and A.N. Semenov, Enzymes in organic synthesis : physicochemical means of increasing the yield of end products in biocatalysis, J. Applied Biochem. 3 (1981) 93-126.

10 V.V. Mozhaev and K. Martinek, Structure-stability relationships in proteins : new approaches to stabilizing enzymes, Enzyme Microb. Technol. 6 (1984) 50-59.

11 G.D. Kutuzova, N.N. Ugarova and I.V. Berezin, General Characteristics of the changes in the thermal stability of proteins and enzymes after the chemical modification of their functional groups, Russ. Chem. Revs 53 (1984) 1078-1100.

12 A.M. Klibanov, Stabilization of enzymes against thermal inactivation, Adv. Applied Microbiol. 20 (1983) 1-28.

13 R.D. Schmid, Stabilized soluble enzymes, Adv. Biochem. Eng. 12 (1979) 41-118.

14 T. Arakawa and S.N. Timasheff, Stabilization of protein structure by sugars, Biochemistry 21 (1982) 6536-6544.
15 K. Gekko and S.N. Timasheff, Mechanism of protein stabilization by glycerol : preferential hydration in glycerol-water mixtures, Biochemistry 20 (1980) 4667-4676.
16 F.A. Quiocho, Carbohydrate-binding proteins : tertiary structures and protein-sugars interactions, Ann. Rev. Biochem. 55 (1986) 287-315.
17 G. Carrea, Biocalysis in water-organic solvent two-phase systems, Trends Biotechnol. 2 (1984) 102-106.
18 M.D. Lilly and J.M. Woodley, Biocatalytic reactions involving water-insoluble organic compounds, in : J. Tramper, H.C. Van der Plas and P. Linko (Eds), Biocatalysis in organic Syntheses, Elsevier, Amsterdam, 1985, pp. 179-192.
19 P.L. Luisi and C. Laane, Solubilization of enzymes in apolar solvents via reverse micelles, Trends Biotechnol. (1986) 153-161.
20 K. Martinek, A.V. Levashov, N. Klyachko, Yu L. Khmelnitski and I.V. Berezin, Micellar enzymology, Eur. J. Biochem. 155 (1986) 453-468.
21 K. Tukahashi, A. Matsushima, Y. Saito and Y. Inada, Polyethylene glycol-modified hemin having peroxidase activity in organic solvents, Biochem. Biophys. Res. Comm. 138 (1986) 283-288.
22 K. Takahashi, A. Ajima, T. Yoshimoto, M. Okada, A. Matsushima, Y. Tamura and Y. Inada, Chemical reactions by polyethylene glycol modified enzymes in chlorinated hydrocarbons, J. Org. Chem. 50 (1985) 3414-3415.
23 A.M. Klibanov, Enzymes that work in organic solvents, Chemtech. (1986) 354-359.
24 A. Zaks and A.M. Klibanov, Enzyme-catalyzed processes in organic solvents, Proc. Natl. Acad. Sci USA 82 (1985) 3192-3196.
25 N.C. Robinson, J. Neumann and D. Wiginton, Influence of detergent polar and apolar structure upon the temperature dependence of beef heart cytochrome C oxidase activity, Biochemistry 24 (1985) 6298-6304.
26 P. Rosevear, T. Van Aken, J. Baxter and S. Ferguson-Miller, Alkyl glycoside detergents : a simpler synthesis and their effects on kinetic and physical properties of cytochrome C oxidase, Biochemistry 19 (1980) 4108-4115.
27 J.D. Aplin and J.C. Wriston, Preparation, properties and applications of carbohydrate conjugates of proteins and lipids, Crit. Rev. Biochem. 10 (1981) 259-306.
28 C.P. Stowell and Y.C. Lee, Neoglycoproteins. The preparation and application of synthetic glycoproteins, Adv. Carb. Chem. Biochem. 37 (1980) 225-282.
29 I. Yu Sakharov, N.I. Larionova, N.F. Kazanskaya and I.V. Berezin, Stabilization of proteins by modification with water-soluble polysaccharides, Enzyme Microb. Technol. 6 (1984) 27-30.
30 J.P. Lenders and R.R. Crichton, Thermal stabilization of amylolytic enzymes by covalent coupling to soluble polysaccharides, Biotech. Bioeng. 26 (1984) 1343-1351.
31 M.H. Remy and D. Thomas, Neoglycoenzymes : production and physicochemical properties, Enzyme Microb. Technol. 4 (1982) 381-384.
32 G.R. Gray, Antibodies to carbohydrates : preparation of antigens by coupling carbohydrates to proteins by reductive amination with cyanoborohydride, Methods Enzymol. 50C (1978) 155-160.
33 D. Cabaret, R. Kazandjan and M. Wakselman, Synthesis of reducing disaccharides bearing a lipophilic chain for the conjugation to proteins, Carb. Res. 149 (1986) 464-470.
34 R.T. Lee and Y.C. Lee, Preparation and some biochemical properties of neoglycoproteins produced by reductive amination of thioglycosides containing an ω-aldehydroaglycon, Biochemistry 19 (1980) 156-163.

35 L.D. Hall and K.R. Holme, Tailored-rheology : chitosan derivatives with branched pendant sugar chains, J. Chem. Soc. Chem. Comm. (1986) 217-219.
36 M.A. Bernstein and L.D. Hall, A general synthesis of model glycoproteins : coupling of alkenyl glycosides to proteins, using reductive ozonolysis followed by reductive amination with sodium cyanoborohydride, Carb. Res. 78 (1980) C1-C3.
37 D.J. Walton, E.R. Ison and W.A. Szarek, Synthesis of N-(1-deoxyhexitol-1-yl) amino acids, reference compounds for the nonenzymic glycosylation of proteins, Carb. Res. 128 (1984) 37-49.

C. Laane, J. Tramper and M.D. Lilly (Editors), *Biocatalysis in Organic Media*,
Proceedings of an International Symposium held at Wageningen,
The Netherlands, 7–10 December 1986.

DEAD MYCELIUM STABILIZED LIPOLYTIC ACTIVITY IN ORGANIC MEDIA : APPLICATION TO ESTER LINKAGE HYDROLYSIS AND SYNTHESIS IN A FIXED-BED REACTOR

C. GANCET and C. GUIGNARD
Lacq Research Centre, Société Nationale Elf-Aquitaine, P.O. Box 34, Lacq, 64170 Artix (France)

SUMMARY

A continuous process for triglycerides hydrolysis using a Rhizopus arrhizus mycelium has been developed at a laboratory scale. Medium used is t-butyl-methyl-ether/acetone, that allows solubilizing water as a co-substrate for fats concentrations up to 20% by weight. In these conditions the lipolytic activity of the mycelium has been retained during several months, with no visible decline. The hydrolytic reaction has been performed in a 2 to 4 steps segmented fixed-bed reactor, with adjustment of the water content at each step. The same system has been applied to ester linkage synthesis.

INTRODUCTION

Triglycerides enzymatic hydrolysis is an interesting alternative to the thermal scission, especially when low degrading conditions are needed for fatty acids manufacture. Within this field, two major companies designed catalytic systems, both using an immobilized thermostable lipase : NOVO with an enzyme from a special strain of Mucor miehei (ref.1,2), and NIPPON OIL & FATS with one from a Pseudomonas fluorescens (ref.3,4).

The process we built up uses a "dead" Rhizopus arrhizus mycelium bearing naturally a lipolytic activity that allows hydrolysis of triglycerides from several fats in organic solvent medium (ref.5.). Pioneering work of Bell and Patterson (ref.6-8) has been improved towards an efficient and economically viable continuous process, where key-points are :

a) Choice of a segmented fixed bed reactor allowing a high conversion rate of the substrate

b) Choice of a medium composition that eases the hydrolytic reaction, and gives good ageing properties for the enzymatic system.

Continuous synthesis of ester bonds was performed using the same enzyme/solvent system. These results have to be compared with those of Strobel (ref.9.) and Knox and Cliffe (ref.10.).

RESULTS

Reaction medium

The solvent used must have the following properties : good solubility of substrate, good compatibility with the enzyme, absence of reactivity, correct solubilization of water needed in the reaction. Ethers only are convenient and among them, tert-butyl-methyl ether(MTBE) has the best safety properties and potentially, the lowest cost. It appears that the optimum water level can only be reached by use of either a surfactant(sodium dioctyl sulfosuccinate) or a polar co-solvent like a ketone. Acetone was retained as being easier to recycle and thus more cost effective than DOSS.

Mycelium

Rhizopus arrhizus (ATCC 24563) mycelium cultivated through special conditions (ref.11) shows high bound lipase activity.Typically, after 5 days the biomass is harvested, and after washing to pH 6.5-7, freeze-dried and solvent extracted before milling (fig. 1.)

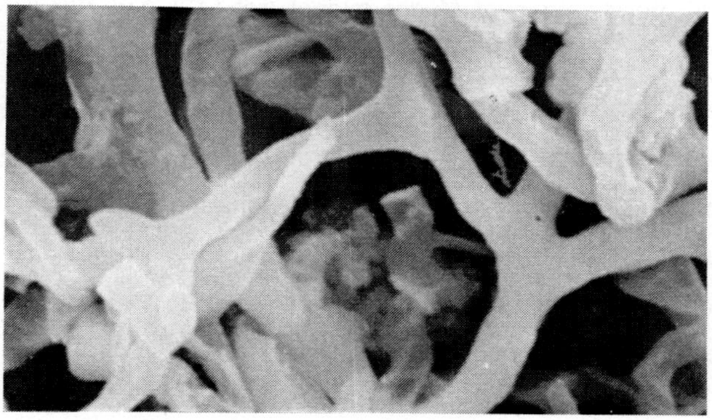

Fig.1. Rhizopus arrhizus mycelium (x 1,000)

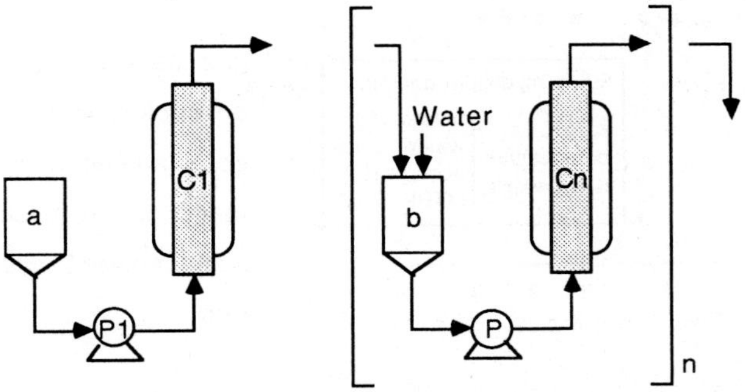

Fig.2. Hydrolysis segmented reactor : (a) fat solution , e.g. tallow 20, water 1.5, acetone 15, MTBE 63.5 (% by weight) ;(b) adjustment of water content.

Hydrolysis

Each segment of the fixed-bed reactor (fig.2.) contains a charge of R. arrhizus mycelium added with silica to ensure good flow properties. The substrate solution is injected through first segment, then water content is increased (water solubility gets higher as mono and diglycerides appear), and the reaction goes on through the other segments. Estimated contact time is 1 hour per segment and conversion rates are typically as follow (TABLE 1.) :

TABLE 1 : Hydrolysis of fat and oil

% Conversion rate at step n°	Cod liver oil	Evening primrose oil	Beef tallow
1	53	60	69,8
2	74	82,5	85
3	80	91	95
4	88	95	97

Total water added reaches about 5x stoichiometry ; e.g.in case of primrose oil hydrolysis, water added at each step was : 7.5, 11, 5.6 and 4.4 (g/ 100 g of triglycerides).

Ageing evaluation was tested through tallow hydrolysis : it appears that a constant yield can be maintained during several months(fig.3.). It seems that both the low water concentration and the protective effect of the fun-

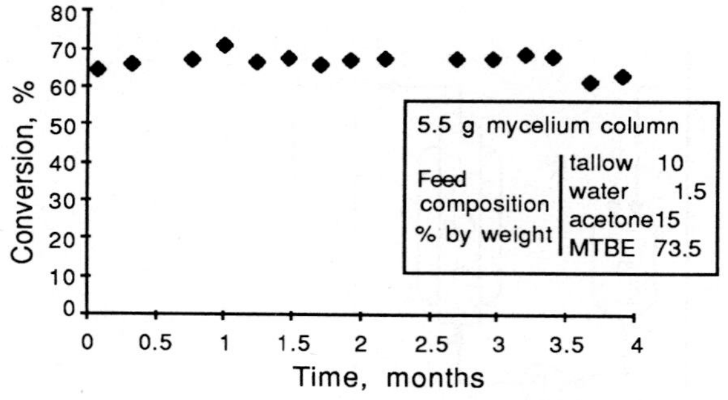

gus cell-wall could be implied in the good ageing resistance of the mycelium /solvent system.

Fig. 3. Influence of ageing on R.arrhizus mycelium lipolytic activity.

Unsaturated chains selectivity

No selectivity was observed during hydrolysis of tallow by R. arrhizus mycelium(fig.4.). A Geotrichum candidum(ATCC 34614) biomass was also tested and gives a good selectivity for conversion yields up to 20 %. Unfortunately, G. candidum activity decreases very quickly and cannot be used under these conditions.

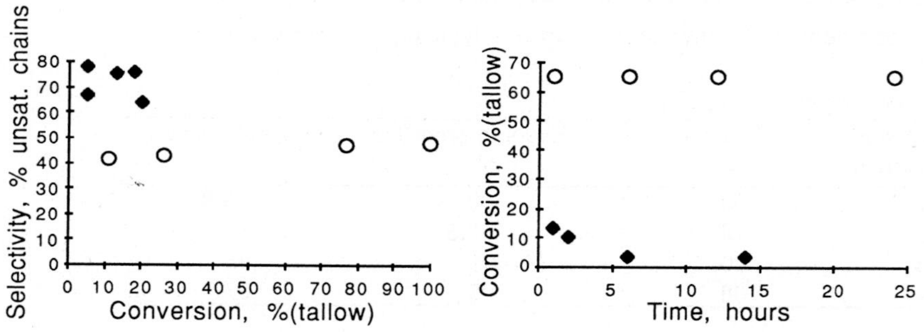

Fig. 4. Comparison of selectivity and stability between R. arrhizus (O) and G. candidum (◆) mycelial lipases.

Synthesis

The same R. arrhizus mycelium can also catalyse formation of ester linkage. Typically , a 0.5 M solution of the acid-alcohol couple in pure MTBE is passed through the reactor, and the ester is obtained continuously in high yield, even in two steps, as shown in table 2 :

TABLE 2 : n-Octanol esterification (initial water content : 10 ppm)

Acid		C9	C16	C18	C18:1
Yield	step 1	74	68	68	60
%	step 2	82	88	86	84

In this case, a column of molecular sieve is inserted between each segment in order to trap water produced during the reaction. Some batch experiments showed no marked influence of the alcohol chain length, but lower yields were obtained with short chain carboxylic acids. It was also shown that secondary alcohols give poor results and that tertiary alcohols do not react at all.

CONCLUSION

A mycelial lipase / organic solvent catalytic system has been designed that allows either continuous hydrolysis of fats or synthesis of aliphatic esters. It has to be considered that other reactions like amide, thio-ester and thio-amide bond synthesis can also be made ; for example, synthesis of pelargonic acid dodecylamide has been obained with a 40 % yield during a batch assay of 15 hours.

Synthesis of glycerides, transacylation, transesterification and interesterification of triglycerides appear also as a very promising field.

ACKNOWLEDGEMENTS

The authors wish to thank here M. Eyssautier, MRS(Satia), Baupte, for the cultivation of Rhizopus arrhizus.

They wish also to thank " La Revue Française des Corps Gras" for kind permission to reproduce figures 2 and 3, and table 2.

REFERENCES

1 P. Eigtved, An immobilized lipase preparation and use thereof, Eur. Pat. App., n°0140542, NOVO Industry A/S (1985).

2 P. Eigtved and T.T. Hansen, A new immobilized lipase for interesterification and ester synthesis, Proc. AOCS Conf., Cannes, France, November 3-8, 1985, NOVO Industry A/S, Bagsvaerd, Denmark.

3 Y. Kosugi and H. Suzuki, New parameters for simulating progress curves of the lipase reaction, J. Ferment. Technol., 61(3) (1983) 287-294.

4 Y. Kosugi, H. Susuki and A.Sato, Fermentation Research Institute, Jap. Pat. n°1092192, Japan Industrial Technology Association.

5 C. Gancet and C. Guignard, Process and device for oils and fats enzymatic hydrolysis, Eur. Pat. Appl., n° 0400915, SNEA-CECA (1986).

6 G. Bell, J.A. Blain, J.D.E. Patterson, C.E.L. Shaw and R. Todd, Ester and glyceride synthesis by Rhizopus arrhizus mycelia, Fems Microbiol. Lett., 3 (1978) 223-225.

7 J.D.E. Patterson, J.A. Blain, C.E.L. Shaw, R. Todd and G. Bell, Synthesis of glycerides and esters by fungal cell-bound enzymes in continuous reactor systems, Biotechnol. Lett., 1 (1979) 211-216.

8 G. Bell, J.R. Todd, J.A. Blain, J.D.E. Patterson and C.E.L. Shaw, Hydrolysis of triglycerides by solid phase lipolytic enzymes of Rhizopus arrhizus in continuous reactor systems, Biotech. Bioeng., 23 (1981) 1703-1719.

9 R. Strobel, Enzymatic esterification: an effective alternative to chemical synthesis, Biotechnol. News, 3(7) (1983) 5.

10 T.Knox and K.R. Cliffe, Synthesis of long-chain esters in a loop reactor system using a fungal cell bound enzyme, Process Biochem., 10 (1984) 188-192.

11 J.A. Blain, J.D.E. Patterson, C.E. Shaw and M. Waheed Akhtar, Study of bound phospholipase activities of fungal mycelia using an organic solvent system, Lipids, 11 (1976) 553-560.

C. Laane, J. Tramper and M.D. Lilly (Editors), *Biocatalysis in Organic Media*,
Proceedings of an International Symposium held at Wageningen,
The Netherlands, 7–10 December 1986.

THERMAL STABILITY OF IMMOBILIZED HORSERADISH PEROXIDASE (HRP) IN WATER-ORGANIC SOLVENT SYSTEMS

L. D'Angiuro*, S. Galliani*, and P. Cremonesi**

*Stazione Sperimentale per la Cellulosa, Carta e Fibre Tessili Vegetali ed Artificiali-Piazza Leonardo da Vinci 26 - 20133 Milano and

**Italfarmaco - Research Group - Via dei Lavoratori 54 - 20092 Cinisello Balsamo (MI) - Italy

SUMMARY

The results obtained studying the thermal stability of HRP immobilized into Sepharose graft-copolymers in water-dioxane and water-dimethylsulfoxide (DMSO) mixtures of different composition (from 0% to 50% of the organic solvent) are reported. Dioxane produces a remarkable deactivation of HRP. Better results have been obtained in DMSO solutions; in fact, at 60°C, no deactivation was shown with solutions up to 15% DMSO and, using this mixture, the residual activity was 45% after 240 min. The results suggest that HRP is stabilized against deactivation by immobilization.

INTRODUCTION

The need for the preparation of immobilized enzyme systems, suitable for biotechnological processes to be carried out at high temperature and in deactivating media, has been widely stressed (1,2). Different strategies have been employed to improve the stability of enzymatic macromolecule; among these, attention has focused on rigidity of the protein structure to prevent its unfolding and much success has been achieved by using the enzyme multipoints attachment methodology (3).

A new and very interesting approach to increase the enzymatic stability has recently been proposed by Klibanov et al. (4-7): several enzymes retain their catalytic activity if they work in organic solvent in the absence of water even at elevated temperature. The use of active proteins in such a medium has numerous advantages among which the high solubility of most substrates is that we are mainly interested to.

Since in the field of energy recovery from biomass most materials are in aqueous media - because they come from pretreatment with inorganics - we need biocatalyst able to work efficiently in aqueous organic solvent mixtures. The aim of this work is to begin to study the behavior in water-organic solvent mixtures of immobilized enzyme, obtained by graft copolymerization reaction, which showed good thermal stability properties when employed in water.

METHODS

Agarose (27-5497-XX) was purchased from Pharmacia Fine Chemicals. Hexhydro-1,3,5-triacryloyl-s-triazine (HTsT) was purchased from BASF. Horseradish peroxidase (HRP, E.C. 1.11.1.7) was purchased from Boehringer Mannheim. Other reagents and enzyme substrates were of analytical grade.

The immobilization of the HRP was carried out according to the methodologies previously described (8,9). Three different samples were prepared: A) 10% p-HTsT-agarose graft copolymer; B) 70% p-HTsT-agarose graft copolymer; C) 100% p-HTsT polymer. A given amount of dissolved agarose (55 mg or 15 mg in a final volume of 10 ml) was placed into a copolymerization vassel with 2 mg of lyophilized HRP and 5 mg, 35 mg or 50 mg of p-HTsT. The immobilization reaction was photochemically initiated at 20°C using $FeCl_3$, as catalyst (0.5 g/L).

For all the obtained samples after their lyophilization, the percentage of recovered solid material and composition (% of agarose, % of synthetic polymer) were measured. The enzyme content of active material was obtained by amino acid analysis after hydrolysis of the samples in 6M HCl for 24 h at 105°C under nitrogen.

The measure of enzymatic activity of solid samples was carried out following the same procedures as for the free enzyme (8). The dependence of free and immobilized HRP activity on substrate concentration was tested in the range of 0.04×10^{-4} and 8×10^{-4} M. The activity of solid materials was given as enzyme units per g of copolymer and it was respectively 450 U/g for the 10% p-HTsT-agarose graft copolymer, 570 U/g for the 70% p-HTsT-agarose graft copolymer and 200 U/g for the 100% p-HTsT polymer.

The kinetic parmaters (K'_m) of HRP samples were obtained from Lineweaver-Burk plots using the least square procedure.

The stability of free and immobilized enzyme in aqueous-organic solvent mixture, at different concentration of organic solvent and treating the samples at different temperatures, was determined at scheduled time.

RESULTS AND DISCUSSION

In previous paper we reported that the enzymes immobilized onto Sepharose or agarose graft copolymers had an improved thermal stability (8-10). This was related to the reduction of the enzyme flexibility which was obtained by its multipoints attachment to the support (10). To synthesize materials with the same properties independently of their composition, we had to control the immobilization reaction parameters (8-11).

Using HTsT as functional vinyl monomer and HRP as enzyme to prepare materials with a low polymer content (less than 10%), having the highest thermal stability, we carried out the immobilization reaction at 20°C (11). Materials with 50-60% polymer content, having the same thermal behavior, could only be prepared increasing the reaction tmperature to 35°C or using the polysaccharide matrix in the soluble form (9,10). In both cases, the immobilized HRP samples had similar kinetic behavior. The K'_m values were not two times higher than that of the free enzyme ($0.25x10^{-4}$ mol/L). No hindrance to substrate and products diffusion was observed because, in general, the enzyme should be located outside the support since the coupling to the solid occurs in the last stages of the graft copolymerization reaction.

In order to study the influence of the water-organic solvent mixtures on the stability of immobilized HRP graft copolymers, three preparations, having different synthetic polymer content (10%, 70%, 100%), were used (see experimental part). In Fig. 1 the residual activity of the considered samples as a function of treatment time at 60°C in aqueous solution, are compared. No significative differences are evidenced in agreement with their kinetic features which does not differentiate the three samples. However, the thermal stability of the enzyme resulted to be considerably increased by immobilization.

The effect of the concentration of dioxane-water mixtures on the sample containing 70% synthetic polymer is shown in Fig. 2. The thermal stability measures were carried out at 50°C because, at higher temperatures, there was a drammatic loss of activity (at 60°C the residual activity after 240 min with 5% dioxane was less than 10%). The activity of immobilized HRP decreases by increasing , from 0% to 15%, the solvent concentration in the mixture. From the comparison of the loss of activity of free and immobilized HRP it follows that the immobilization produced a remarkable stabilization against the deactivation

Fig. 1

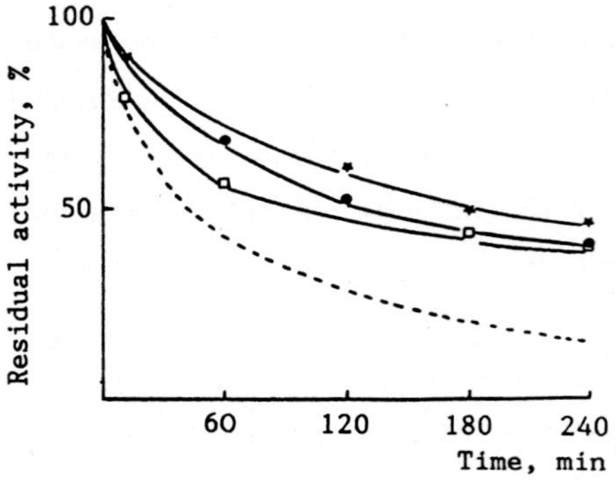

Fig. 2

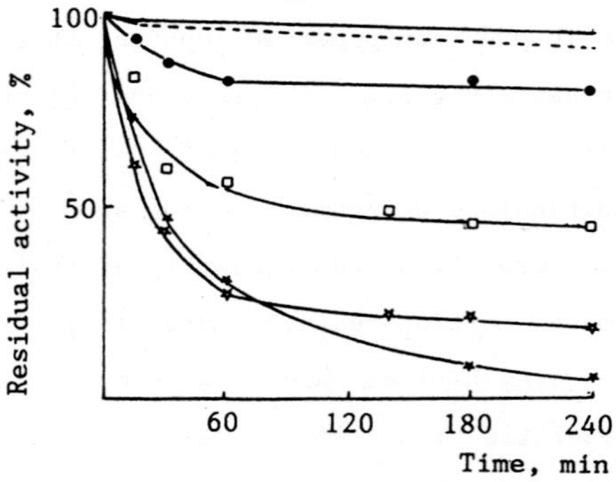

Fig. 1 - Thermal stability in 0.1M phosphate buffer (pH 7.0) at 60°C of HRP graft copolymers. (✶) 70% p-HTsT-agarose graft copolymer; (●) 10% p-HTsT-agarose graft copolymer; (□) 100% p-HTsT polymer; (---) free HRP.

Fig. 2 - Effect of dioxane concentration on the thermal stability at 50°C of the 70% p-HTsT-agarose HRP graft copolymer.
(——) 0%; (---) 0% on free HRP; (●) 5%; (□) 10%; (☆) 15%; (★) 5% on free HRP.

due to this harsh solvent. The measures carried out on the other two samples were not appreciably different from those of Fig. 2. It follows that the copolymer composition, as previously discussed, is uneffective on the enzyme behavior which seems dependent mainly on the kind and number of links between the enzyme and the support.

Better results were obtained in dimethylsulfoxide (DMSO) solution. In fact, the thermal stability measures were carried out at 60°C instead of 50°C used for dioxane solutions. The obtained results are shown in Fig. 3. It is interesting to note that no deactivation was evident with solutions up to 15% DMSO. From the Fig. 3 we observed a further evidence of the extent at which the HRP was stabilized by the attachment to graft copolymers. The free enzyme had a residual activity not higher than 5% of the initial, whereas the insoluble form was about 10 times more stable.

Increasing the DMSO concentration over 15% the loss of activity of immobilized HRP linearly increased as shown in Fig. 4. Unspectedly, at 50% DMSO concentration, the residual activity of the immobilized samples is not so different from that of the free enzyme tested in water solutions.

The influence of the temperature of the treatment on the immobilized enzyme stability in the considered water-organic solvent solutions, is summarized in Fig. 5. It is remarkable that the presence of DMSO concentration up to 15% does

Fig. 3

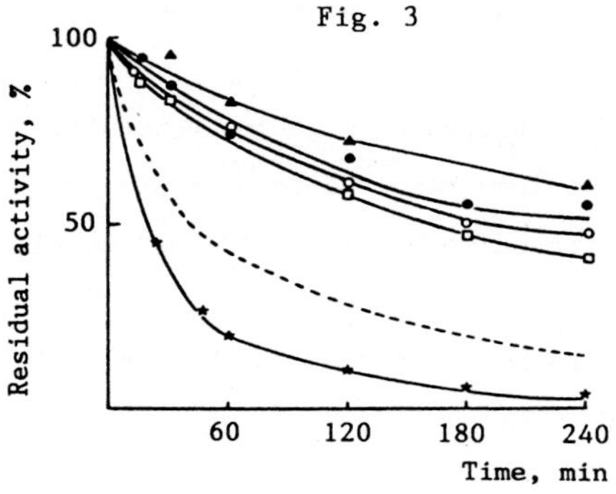

Fig. 4

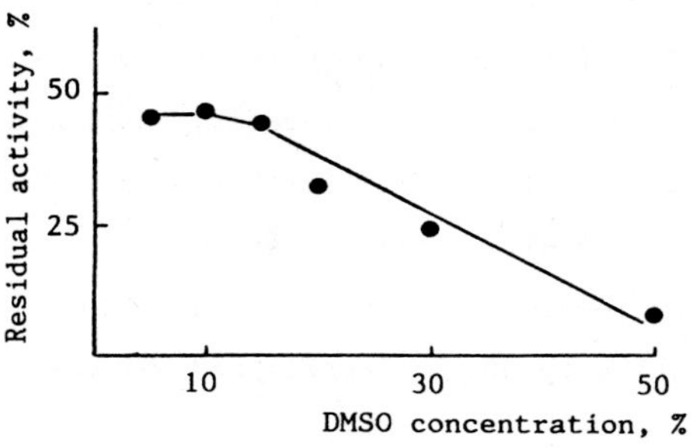

Fig. 3 - Effect of DMSO on the thermal stability at 60°C of the 70% p-HTsT-agarose HRP graft copolymer. (▲) 10%; (●) 5%; (○) 0%; (□) 15%; (---) 0% on free HRP; (★) 5% on free HRP.

Fig. 4 - Residual activity of the 70% p-HTsT-agarose HRP graft copolymer, at 60°C, as a function of DMSO concentration.

Fig. 5

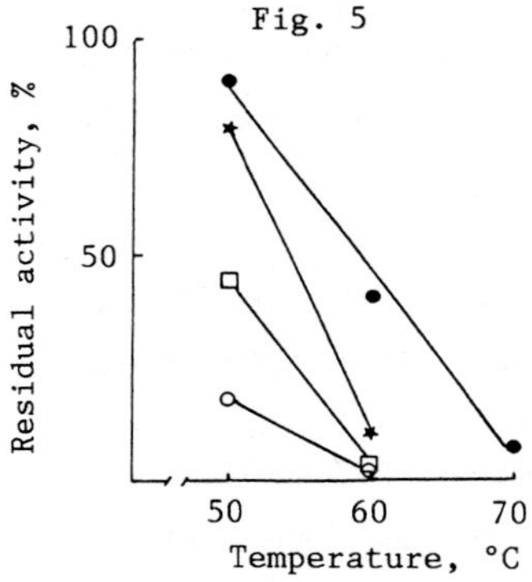

Fig. 5 - Residual activity of the 70% p-HTsT-agarose HRP graft copolymer as a function of temperature. (●) 0%, 5%, 10%, 15% DMSO; (★) 5% dioxane; (□) 10% dioxane; (○) 15% dioxane.

not act on the linear deactivation produced by the increase of the temperature of the treatment solution. This was not what occured with dioxane solution. Even at 5% concentration, the presence of the solvent modified the deactivation courses due to the temperature.

From the presented data concerning the behavior of immobilized HRP graft copolymers in the considered aqueous-organic solvent solutions the following consideration can be drawn:

1) The stability of the fixed enzyme is dependent mainly on the nature of the solvent. In our case, dioxane is much more destabilizing than DMSO.
2) The percentage of p-HTsT, the synthetic polymer present in the studied graft copolymers, does not influence the enzyme stability. We will verify if other

vinyl monomers can produced a positive effect against the enzyme deactivation induced by organic solvents.

3) The free HRP was very sensitive to the solvent used also at low concentration (less than 5%). Conversely, the immobilized form resulted considerably more stable. These data further support the proposed mechanism of thermal stabilization of immobilized enzymes.

4) As regards DMSO, the concentration (15%) which do not reduced the catalytic activity of the immobilized HRP was found.

REFERENCES

1 Martinek, K., Mozhaev, V., and Berezin, I., (1980) in Enzyme Engineering (Wingard L., Berezin I. and Klysov, Eds.), Plenum Press, New York.

2 Martinek, K., Klibanov, A.M., Goldmacher, V.S., and Berezin, I.V., (1977) Biochem. Biophys. Acta, 485, 1.

3 Martinek, K., Klibanov, A.M., Goldmacher, V.S., Tcherhvsheva, A.V., Mozhaev, V.V., and Berezin, I.V. (1977) Biochim. Biophys. Acta 485, 13.

4 Zaks, A., and Klibanov, A.M., (1985) Proc. Natl. Acad. Sci. USA, 82, 3192-3196.

5 Zaks, A., and Klibanov, A.M., (1984) Science, 224, 1249-1251.

6 Ahern, T.J., and Klibanov, A.M. (1985) Science, 228, 1280-1284.

7 Kazandjian, R.Z., Dordick, J.S., and Klibanov, A.M., (1986) Biotechnol. Bioeng., 28, No. 4.

8 D'Angiuro L., Cremonesi, P., (1982) J. Appl. Biochem., 25, 496-507.

9 D'Angiuro L., Galliani, S., Cremonesi, P., (1986) Biotech. and Appl. Biochem., in press.

10 D'Angiuro, L., de Lalla, C., Cremonesi, P., (1985) Biotechnol. Bioeng., 27, 1548-1553.

11 Cremonesi, P., D'Angiuro, L., (1983) Biotechnol. Bioeng., 25, 735-744.

C. Laane, J. Tramper and M.D. Lilly (Editors), *Biocatalysis in Organic Media*,
Proceedings of an International Symposium held at Wageningen,
The Netherlands, 7–10 December 1986.

PROPERTIES AND SPECIFICITY OF AN ALCOHOL DEHYDROGENASE FROM THERMOPHILIC ARCHAEBACTERIUM SULFOLOBUS SOLFATARICUS

R. RELLA[1], C.A. RAIA[1], A. TRINCONE[2], A. GAMBACORTA[2], M. DE ROSA[2] and M. ROSSI[3,4]

[1]Istituto Internazionale di Genetica e Biofisica, CNR, Via Marconi 10, 80125 Napoli (Italia)

[2]Istituto di Chimica di Molecole di Interesse Biologico, CNR, Via Toiano 6, 80072 Arco Felice, Napoli (Italia)

[3]Dipartimento di Chimica Organica e Biologica, Università, Via Mezzocannone 16, 80134 Napoli, (Italia)

[4]Istituto di Biochimica delle Proteine ed Enzimologia, CNR, Via Toiano 6, 80072 Arco Felice, Napoli (Italia)

SUMMARY

An NAD^+-dependent alcohol dehydrogenase has been obtained in homogeneous form from the archaebacterium Sulfolobus solfataricus grown at 87°C. The enzyme, a dimer, is thermophilic, thermostable, resistant to organic solvents and has a very broad substrate specificity. Reduction of 3-methyl-butan-2-one involves a hydride attack at the *re* face of the carbonyl to produce the corresponding S-alcohol.

INTRODUCTION

Thermophilic microorganisms in general and archaebacteria in particular offer undoubted advantages for microbial technology. Higher temperatures in microbial fermentation processes increase the solubility of many compounds, as well as the diffusion rate, reduce the viscosity of the medium and allow even the distillation of volatile products that could inhibit cell growth. Furthermore, immobilized cells can carry out biotransformation reactions also in mixed solvents.

Attention has been focused particularly on enzymes (1). In fact, whereas in conventional organisms enzymes are irreversibly inactivated by heat, the thermophilic archaebacterial enzymes have been found to be thermostable and capable of functioning at high temperatures, their activity being low at moderate temperatures, at which conventional enzymes are optimally active. They are not only stable to heat but, in general, show an enhanced stabili ty in the presence of common protein denaturants and organic solvents and this intrinsic stability is further enhanced by immobilization. In addition to other enzymes (2,3,4) we have recently found a novel NAD^+-linked alcohol dehydrogenase from the extreme thermophilic archaebacterium Sulfolobus solfataricus (SSADH). This enzyme, unlike the one isolated by Lamed and Zeikus from the eubacterium Thermoanaerobium brockii (TBADH), (5,6) was NAD^+ dependent, had a different molecular structure and stereospecificity. Like the other one, it had a very broad substrate specificity, was remarkably stable to temperature and exhibited tolerance toward organic solvents.

METHODS

Sulfolobus solfataricus cells (strain MT-4) were grown at 87°C, as previously described (2), and collected after 40 hours growth.

Alcohol dehydrogenase activity was assayed spectrophotometrically at 65°C by measuring the change of absorbance at 340 nm, in a reaction mixture, 1 ml, containing 25 mM barbital-HCl pH 8.0, 1 mM $ZnSO_4$, the coenzyme and 0.5-1.0 ug of enzyme. The concentrations of substrate and cofactors were, respectively, 1.0 mM NAD^+ and 10 mM benzylic alcohol for alcohol oxidation and 0.1 mM NADH and 1 mM anisaldehyde for aldehyde reduction.

One unit of enzyme was defined as the amount of enzyme catalyzing the transformation of 1 umole of substrate in 1 min at 65°C in the described conditions.

The detailed purification procedure will be published elsewhere.

Gram-scale production of chiral alcohol 3-methyl-butan-2-ol.

SSADH was immobilized on Eupergit C, according to the procedure suggested by Röhm Pharma. 4.4 gr of Eupergit C under N_2 atmosphere was mixed with 20 ml of purified SSADH, (4 mg), in 1 M potassium phosphate pH 7.5. After 24 hours at room temperature the resin was washed with 0.1 M potassium phosphate pH 7.5 and stored in the presence of sodium azide. No activity was found in the supernatant and the activity of the immobilized enzyme was estimated by suspending a known volume of the resin in the standard incubation mixture at 65°C. The estimated enzymatic activity was about 0.3 units per ml of gel. Gram-scale production of the chiral alcohol was performed with the coupled substrate approach, using propan-2-ol as oxidable co-substrate. In this experiment the reaction mixture (100 ml) contained 0.1 M 3-methylbutan-2-one, 1.3 M propan-2-ol, 2.5 uM NADH, 0.025 M Tris-HCl pH 7.0 and 15 ml of SSADH-Eupergit C. The reaction was carried out at 50°C for 24 hours under gentle shaking. The catalyst was removed by centrifugation and the chiral alcohol was extracted with ethyl ether from the $(NH_4)_2SO_4$ saturated solution. After drying over $MgSO_4$ the organic phase was concentrated under nitrogen flux and the final purification of the chiral alcohol was achieved by preparative HPLC on Microporasil column (n-hexane/ethyl acetate, 95/5 v/v, flow rate 4 ml/min). The purified product was characterized by its $[\alpha]_D^{20}$ value. Further information on the optical purity of alcohol was obtained by its derivatization with (R)-(+)-α-Methoxy-α-trifluoromethyl phenylacetic chloride (MTPA) at room temperature in anhydrous pyridine. The diastereoisomer was purified by TLC preparative silica gel plates, eluted with n-hexane/ethyl acetate, 85/15 (v/v) and was characterized by ^1H-NMR spectroscopy on 500 MHz Bruker spectrometer in C^2HCl_3.

RESULTS AND DISCUSSION

Enzyme purification

S.solfataricus cells contain an NAD^+-dependent alcohol dehydrogenase that was purified to homogeneity by means of three columns: a DEAE cellulose, a Matrex gel RedA and a Blue Dextran-Sepharose-4B. The crucial step was the specific elution of the enzyme from Blue Dextran-Sepharose-4B with 1 mM NAD^+ and 1 mM $ZnSO_4$ in 20 mM Tris-HCl pH 7.4, 5% glycerol.

The enzyme was shown to be homogeneous by different methods (7). The recovery was over 50% on centrifuged homogenate with a purification of about 280 times.

Subunit structure

The subunit structure of SSADH was studied by gel filtration, sucrose or glycerol gradient centrifugation and electrophoresis in sodium dodecylsulfate of the protein native and crosslinked with dimethyl suberimidate (DMS). With these methods (7) the enzyme was shown to be composed of two very similar or identical subunits of about 36,000 D, differently from TBADH, which is a tetramer (5).

Reaction requirements and enzyme substrate specificity

The purified SSADH enzyme, which can be defined as an NAD^+-linked alcohol aldehyde/ketone oxidoreductase, displayed activity toward aldehydes, a variety of primary and secondary alcohols as well as linear and cyclic ketones (Table).

TABLE

Substrate specificity of alcohol dehydrogenase from Sulfolobus solfataricus.

Substrate	Km (um)	Kcat (min^{-1})
Methanol	27,000	186
Ethanol	260	113
Propan-1-ol	19	242
Propan-2-ol	500	415
Butan-1-ol	55	118
Butan-2-ol	350	213
Pentan-2-ol	50	229
Pentan-3-ol	225	175
Benzylic alcohol	13	101
3-Methyl-cyclohexanol*	21	463
Anisaldehyde	2.5	59
Acetone	6,600	19
Butan-2-one	14,000	116
3-Methyl-Butan-2-one	7,500	73
Cyclopentanone	1,300	61
(±)-3-Methyl-cyclohexanone	250	130

*Mixture of diastereoisomers obtained by sodium borohydride reduction on enantiomeric 3-Methyl-cyclohexanone.

Zn^{++} ions at a concentration of 1 mM were required for maximal activity.

The apparent Kms observed for the different substrates were in the order of umolar concentrations except, for ketones like acetone, butanone and cyclopentanone. The apparent Kms calculated for coenzymes were also in the order of umolar concentration. With benzyl alcohol, as substrate, the apparent Km for NAD^+ was 100 uM, whereas the apparent Km for NADH, with anisaldehyde as substrate, was 3 uM.

The optimal pHs for benzylic alcohol oxidation and anisaldehyde reduction, in barbital buffer, were 8.5 and 7.5, respectively.

Thermophilicity, thermostability and effect of organic solvents.

The enzyme thermophilicity was quite extraordinary. In fact, the activity increased with temperature up to 95°C and the activation energy, calculated from the Arrhenius plot was 47.8 KJ. SSADH displayed a considerable thermostability not influenced by the presence of certain organic solvents. In the standard buffer the enzyme lost about 30% of its activity, after 24 hours at 50°C whereas in the presence of methanol, 2-propanol, ethyl acetate, at a concentration of 10%, no loss of activity was detected. In addition, ethyl acetate, at a concentration of 10% in the standard assay mixture, did not affect the enzyme activity.

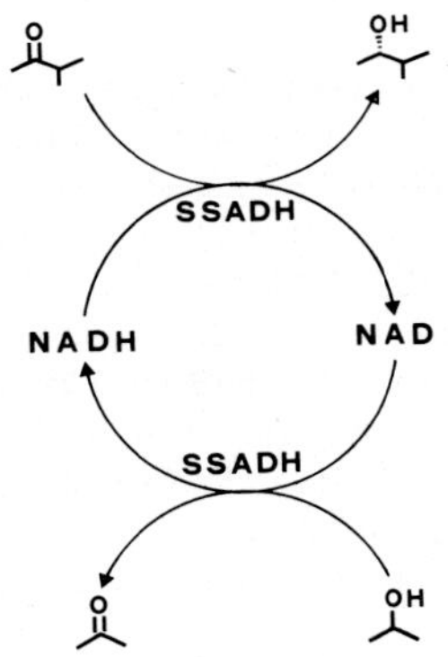

Scheme I. Representation of coenzyme recycling by the coupled-substrate methods (8).

Stereospecificity

The stereospecificity of SSADH was tested on 3-methyl-butan-2-one. The reduction was performed on gram scale with Eupergit C immobilized pure enzyme. NADH regeneration was achieved "in situ" with the coupled substrate approach (Scheme I), using propan-2-ol as oxidable co-substrate in 13-fold molar excess with respect to ketone. The number of NADH cycles obtained in the experiment was about 20,000 after 24 hours reaction, when 50% of the keton was converted into alcohol and 3,8% of propan-2-ol was oxidized to acetone. Reduction of keton gives rise to the (S)-(+)-3-methyl-butan-2-ol with 100% optical yields as judged by ^{1}H-NMR spectroscopy of the MTPA diastereoisomer derivative. δ 7.65-7.35 (aromatic protons); δ 5.0 (1H, five lines, C$\underline{H}$-O-CO-R); δ 3.6 (3H, singlet, OC$\underline{H}_3$); δ 1.28 (3H, doublet, C$\underline{H}_3$-CH-O-CO-R); δ 1.81 (1H, eigth lines, $(CH_3)_2$-C$\underline{H}$-); δ 0.83-0.86 (6H, two doublets, $(C\underline{H}_3)_2$-CH-).

It is therefore evident that SSADH catalyzes the hydride attack to re face of carbonyl of the 3-methyl-butan-2-one and in this respect differs from TBADH, which give rise to a preferential si attack on this ketone.

Work supported by ECC Biotechnology Action Programme Contract N°.0052 -I and by Progetto Strategico Biotecnologie CNR, Italy.

REFERENCES

1 A. Fontana, Thermophilic Enzymes and their Potential Use in Biotechnology, Dechema, Weinheim, 1984, pp. 221-232.
2 M. Rossi, R. Rella, M. Pensa, S. Bartolucci, M. De Rosa, A. Gambacorta, C.A. Raia and N. Dell'Aversano Orabona, System. Appl. Microbiol, 7 (1986) 337-341.
3 V. Buonocore, O. Sgambati, M. De Rosa, E. Esposito, A. Gambacorta, J. Appl. Biochem. 2 (1980) 390-397.
4 P. Giardina, M.G. De Biase, M. De Rosa, A. Gambacorta and V. Buonocore, Biochem. J. 239 (1986) 517-522.
5 R. Lamed and J.G. Zeikus, Biochem. J. 195 (1981) 183-190.
6 E. Keinan, K.K. Seth and R. Lamed, J. Am. Chem. Soc. 108 (1986) 3474-3480.
7 R. Rella, F.M. Pisani, C.A. Raia, C. Vaccaro, M. De Rosa, A. Gambacorta and M. Rossi, 17th FEBS Meeting, Berlin (West), 1986, Abstr. FRI 06.01.23.
8 J.B. Jones and J.F. Beck, Application of Biochemical System in Organic Chemistry, Techniques of Chemistry, Vol. X, 1976, pp. 254-319.

C. Laane, J. Tramper and M.D. Lilly (Editors), *Biocatalysis in Organic Media*, Proceedings of an International Symposium held at Wageningen, The Netherlands, 7–10 December 1986.

ENANTIOSELECTIVE HYDROLYSIS AND TRANSESTERIFICATION OF GLYCIDYL BUTYRATE BY LIPASE PREPARATIONS FROM PORCINE PANCREAS

M.Chr. Philippi, J.A. Jongejan and J.A. Duine

Delft University of Technology, Laboratory of Microbiology and Enzymology
Julianalaan 67, 2628 BC Delft, The Netherlands

SUMMARY

The submaximal enantioselectivity of crude pancreas lipase preparations for glycidyl butyrate is not due to the aspecific hydrolysis of contaminating enzymes but is inherent to the submaximal enantioselectivity of lipase. Preliminary results of glycidyl butyrate transesterification in 1-alkanols point to a somewhat lower enantioselectivity of the enzyme in these solvents. An analysis procedure was developed enabling detection and quantification of glycidol enantiomers by HPLC.

INTRODUCTION

Enantiomer pure glycidol (2,3-epoxy-1-propanol) is an attractive intermediate for the production of optically pure pharmaceuticals. Production of this chiral synthon can be performed via a biocatalytic route, as recently described by Ladner and Whitesides [1]. Starting with racemic glycidyl esters, a crude porcine pancreatic lipase preparation hydrolyzed the (+)-ester more or less preferentially and the (-)-ester could be subsequently isolated by extraction.

As it was expected that the crude lipase preparation could be contaminated by interfering enzymes (e.g. esterases) responsible for submaximal results, the effect of enzyme purification on enantioselectivity was studied. Transesterification with lipase has been described but so far the effect of the organic solvents on enantioselectivity is unknown. Preliminary results of transesterification with alcohols are presented. However, since the specific

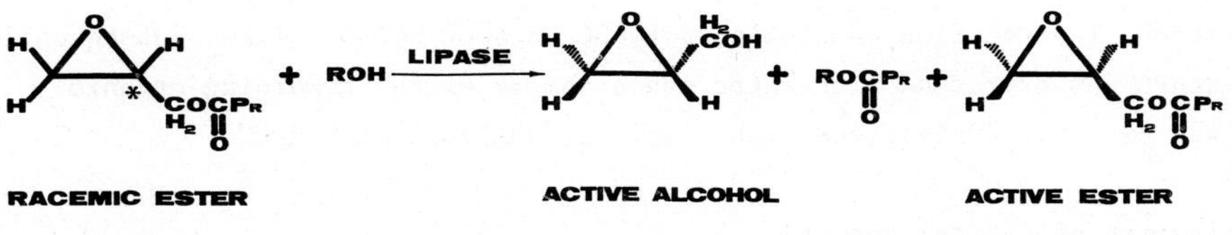

Fig. 1. Enantioselective transesterification of glycidyl butyrate with an alcohol

rotation of only a few glycidyl esters is known and measurement of optical rotation requires purification of the product, a search for an alternative method was made. Based on the method described by Gal [2] for related epoxides, a procedure was developed to separate and quantify the glycidyl enantiomers by HPLC.

MATERIALS AND METHODS

Materials

Lipase (E.C. 3.1.1.3.) from porcine pancreas (Sigma type II) was purchased from Sigma Chemical Co. Glycidyl butyrate was synthesized from epichlorohydrin and sodium butyrate, following the procedure of Aserin et al. [3]. The chiral reagent 2,3,4,6-tetra-O-acetyl-β-D-glucopyranosylisothiocyanate (TAGIT) was prepared according to the procedure of Nimura et al. [4].

Determination of lipase activity

Activity with tributyrin and glycidylbutyrate was measured using a pH-stat at pH 7.8. Activity with p-nitrophenylacetate was determined by following the increase in absorbance at 400 nm.

Partial purification of lipase

The first step of the procedure of Verger et al. [5], that is ion exchange chromatography on DEAE-Sepharose and elution by a NaCl gradient, was used. The combined and collected fractions were electrophorised on polyacrylamide gels (7.5%) and stained with Coomassie-Blue G250.

Hydrolysis of glycidylbutyrate

Hydrolysis experiments were performed according to the procedure by Ladner and Whitesides [1], using 1 gram glycidyl ester,10 ml water, and crude or purified lipase.

Transesterification of glycidylbutyrate

A mixture of 2 g of the ester, an 8-fold molar excess of a 1-alkanol and 150 ul toluene (as a marker for GC) was shaken with 200 mg crude lipase preparation. The reaction was followed by GC on a CP Sil 5 column. Subsequently the mixture was extracted with water and a sample of the remaining organic phase was directly derivatized with TAGIT for analysis with HPLC.

Determination of optical purity

Derivatization of the glycidyl esters in a sample was performed with n-butylamine and TAGIT (procedure A) as described by Gal [2]. The reaction time with the amine was terminated after 8 hours. Separations were performed on a Nova-pack C18 column (Waters Assoc.) while monitoring the eluate at 260 nm. Isocratic chromatography occurred with methanol/water (50/50 v/v) at a flow rate of 1.0 ml/min.

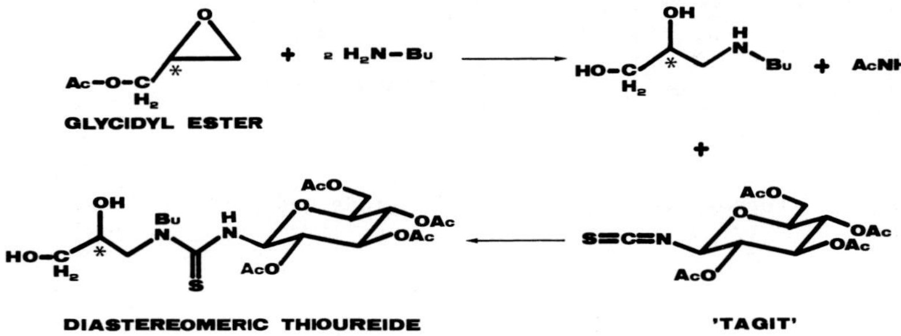

Fig. 2. Derivatization of a glycidyl ester with n-butylamine and TAGIT

RESULTS AND DISCUSSION

DEAE-Sepharose chromatography of crude pancreas lipase resulted in three peaks having tributyrin hydrolase activity (Fig. 3). The collected fractions of the three peaks appeared to be homogeneous, as revealed by polyacrylamide gel electrophoresis followed by staining with Coomassie-Blue. The three fractions could hydrolyze tributyrin, glycidylbutyrate as well as p-nitrophenylacetate, although the specific activities varied somewhat (Table 1).

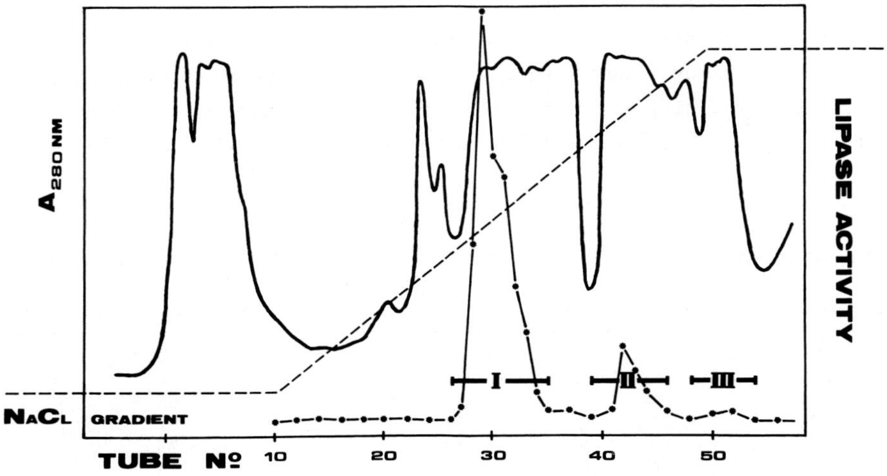

Fig. 3. DEAE-Sepharose chromatography of crude pancreas lipase with an NaCl gradient (10-500 mM NaCl)

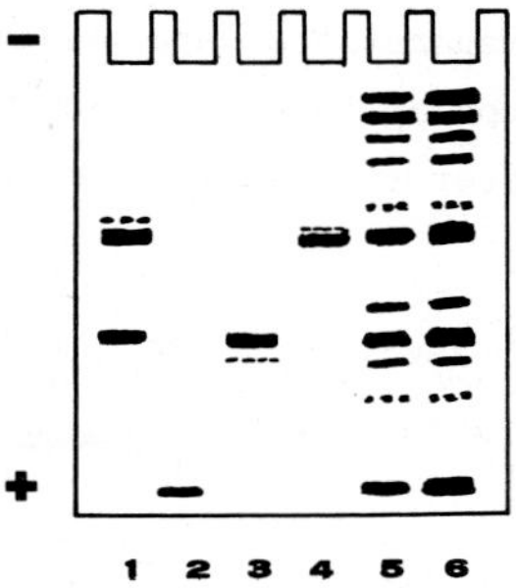

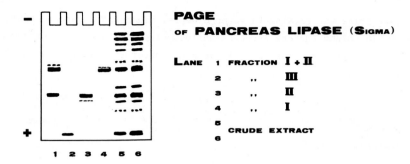

Fig. 3a. Gel electrophoresis at pH 8.3 of the crude and purified lipase preparations

TABLE 1

Relative substrate specificities for crude and purified lipase preparations

preparation	relative activity on: tributyrin	glycidylbutyrate	p-nitrophenylacetate
crude	5.7	1	44 x 10^-5
peak 1	8.0	1	31 x 10^-5
peak 2	4.3	1	87 x 10^-5
peak 3	3.4	1	31 x 10^-5

As expectable, the ee-values obtained for the remaining glycidyl ester increase when the extent of hydrolysis is higher (Table 2). Therefore, in order to compare the enantioselectivity of different preparations, ee-values should be compared for the same extent of conversion. This was accomplished by calculating the factor E according to Chen et al. [6]. (E representing the ratio of the specificity constants (V/K) of the enantiomers). Assuming a constant E value, ee-values are given for a conversion of 60% (Table 2). From

TABLE 2

Optical purities and enantiomeric ratios (E) for different lipase preparations

preparation	conversion(%)	ee(%)	E	ee at 60% conversion
crude preparation	53	44	3.4	53
,,	60	64	4.6	64
,,	60	86	9.7	86 *
peak 1	52	54	5.0	67
,,	58	70	6.1	73
peak 2	50	40	3.4	53

* In this experiment 15 g ester was used.

this it is clear that the submaximal enantioselectivity of the crude preparation is not due to aspecific, contaminating enzymes but to the inherent submaximal specificity of lipase.

Transesterification of glycidylbutyrate in different 1-alkanols was possible, although the ee-values calculated for 60% conversion were somewhat lower than for the hydrolytic procedure. However, it should be realized that these are preliminary results and a change in conditions could give higher values. Therefore further studies are necessary before conclusions on the enantiomer selectivity of glycidyl ester transesterification in organic solvents by lipase can be made.

TABLE 3
Optical purities and enantiomeric ratios (E) for transesterification in 1-alkanols

alcohol	conversion(%)	ee(%)	E	ee at 60% conversion
1-butanol	51	11	1.4	13
1-pentanol	71	26	1.5	19
1-heptanol	50	50	4.8	65
1-decanol	39	34	4.5	63

Figure 4 shows that adequate resolution and quantification of glycidol enantiomers were obtained by derivatization and HPLC. Since only separation of glycidol from its esters is required, this method is indispensable when only small samples are available or when the specific rotation of the esters is unknown.

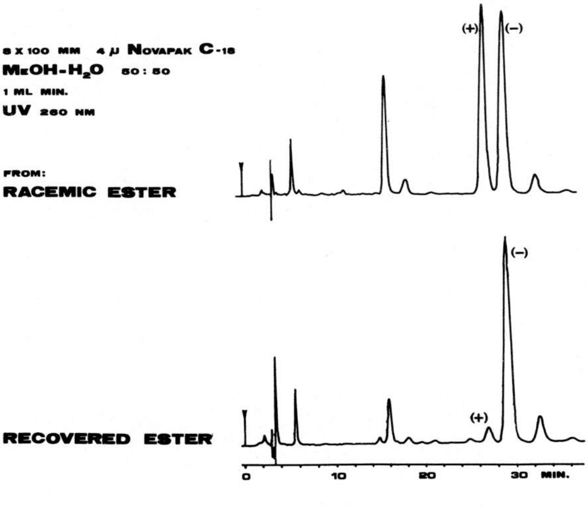

Fig. 4. Resolution of a racemic and an optically active glycidyl ester after derivatization with TAGIT

REFERENCES

1 W.T. Ladner and G.M. Whitesides ,J. Am. Chem. Soc. ,106 (1984) 7250-7251

2 J. Gal, J. Chromatogr. ,331 (1985) 349-357

3 A. Aserin, N. Garti and Y. Sasson, Ind. Eng. Chem. Res. Dev. 23 (1984) 452-454

4 N. Nimura, H. Ogura and T. Kinoshita ,J. Chromatogr. 202(3) (1980) 375-379

5 R. Verger, G.H. de Haas, L. Sarda and P. Desnuelle, Biochim. Biophys. Acta ,188 (1969) 272-282

6 C. Chen, Y. Fujimoto, G. Girdaukas and C.J. Sih ,J. Am. Chem. Soc. 104 (1982) 7294-7299

C. Laane, J. Tramper and M.D. Lilly (Editors), *Biocatalysis in Organic Media*, Proceedings of an International Symposium held at Wageningen, The Netherlands, 7–10 December 1986.

REVERSED MICELLAR EXTRACTION OF ENZYMES; EFFECT OF NONIONIC SURFACTANTS ON THE DISTRIBUTION AND EXTRACTION EFFICIENCY OF α-AMYLASE

M. DEKKER[1], J.W.A. BALTUSSEN[2], K. VAN 'T RIET[1], B.H. BIJSTERBOSCH[3], C. LAANE[2,4] AND R. HILHORST[2]

Agricultural University, De Dreijen 12, 6703 BC Wageningen (The Netherlands)
[1]Department of Food Science, Food- and Bioengineering group
[2]Department of Biochemistry
[3]Department of Physical- and Colloid Chemistry
[4]Present address Unilever Research Lab, PO Box 114, Vlaardingen (The Netherlands)

ABSTRACT

Enzymes can be concentrated by a continuous forward and back extraction with a reversed micellar phase. In this process the enzymes are transferred from one aqueous phase to another via a circulating reversed micellar phase. The effect of nonionic surfactants in addition to cationic surfactants in the apolar phase on the distribution behaviour and extraction efficiency of α-amylase has been investigated. Both the pH range in which solubilization occurs and the degree of solubilization increase by the addition of nonionic surfactants. By performing the continuous extraction in two mixer/settler units the enzyme could be concentrated up to a factor of 11.5 with a recovery of enzyme activity of 75%.

INTRODUCTION

The recovery of enzymes from a fermentation broth by conventional processes is laborious and usually accounts for a substantial part of the total cost of the final product. New separation processes, that are selective for the desired enzyme and that can be scaled up easily, are therefore desirable.

Liquid-liquid extraction with an apolar solvent containing reversed micelles has very promising features in this respect (refs. 1-5). Reversed micelles are aggregates of surfactant molecules in an apolar solvent surrounding an inner core of water. Polar compounds such as enzymes can be solubilized in an apolar solvent by reversed micelles. To enable the use of a reversed micellar phase for enzyme extraction, the transfer of the enzyme from an aqueous phase to the reversed micellar phase and *visa versa* must be possible (see Fig. 1).

Using the anionic surfactant Aerosol OT, Göklen and Hatton (ref. 5) have shown that differences in the distribution behaviour of low molecular weight proteins can be used to separate a mixture of three proteins.

Recently we have investigated the performance of a continuous forward and back extraction with a reversed micellar phase in two mixer/settler units (ref. 4). Using the surfactant trioctylmethylammonium chloride (TOMAC) in isooctane, the enzyme α-amylase could be concentrated by performing these two extractions at different pH values in the aqueous phases. During the forward extraction the pH in the aqueous phase favoured the transfer of the enzyme towards the reversed micellar phase, while during the back extraction the opposite transfer was favoured.

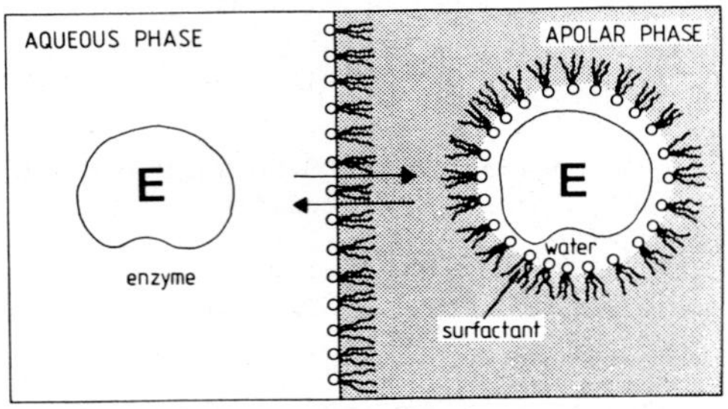

Fig. 1. Transfer of an enzyme between an aqueous phase and an apolar phase containing reversed micelles.

RESULTS

The distribution of α-amylase between a reversed micellar phase and an aqueous phase was studied in relation to the pH of the aqueous phase (see legend ~~Fig. 3~~ for the composition of the phases).

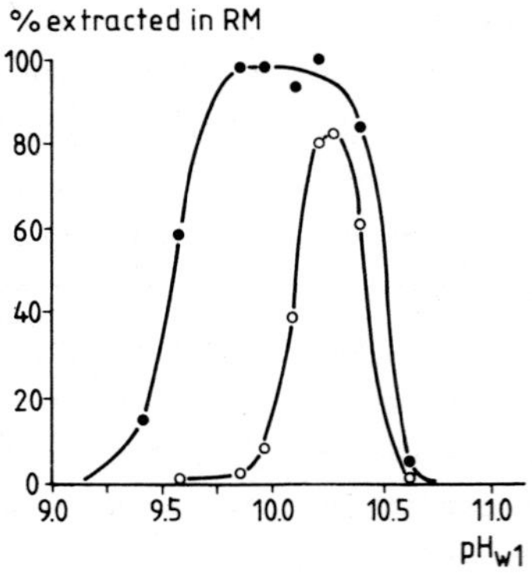

~~Fig. 3~~. Solubilization of α-amylase in a reversed micellar phase in relation to the pH in the aqueous phase (pH_{W1}).

Aqueous phase: 1 g/l α-amylase (50 U/mg) in 50 mM ethylene diamine buffer, adjusted to pH with HCl.

Reversed micellar phase: 0.4% (w/v) TOMAC, 0.1% (v/v) octanol in isooctane. (o): no extra surfactant added. (•): 2 mM Rewopal HV5 (nonyl phenolpentaethoxylate) added to the reversed micellar phase.

As can be seen from Fig. 2 the distribution of α-amylase is strongly pH dependent. The distribution behaviour is markedly influenced by the addition of a nonionic surfactant (rewopal HV5) to the reversed micellar phase. An increase both in the degree of solubilization of the enzyme and in the pH range, in which solubilization occurs, was observed.

These effects will possibly be caused by changes in the structure of the reversed micelles and in their adaptability due to the addition of the nonionic surfactant. More fundamental studies on this effect are necessary to understand this effect in more detail.

It should be noted that the pH value at which maximum solubilization occurs, depends on the type and concentration of the buffer used in the aqueous phase (e.g. 0.5 M tris buffer resulted in an optimum at pH 9.9 (ref. 4).

For an efficient extraction process, a high distribution coefficient of the enzyme is favourable, since the amount of equilibrium stages to reach the same concentration factor can be reduced. From the data in Fig. 2 it can be concluded that the addition of the nonionic surfactant will improve the extraction efficiency of the enzyme.

To test whether these results on the distribution behaviour can be used to improve a continuous forward and back extraction on a larger scale, two mixer/settler units were used for the two extractions (see Fig. 3).

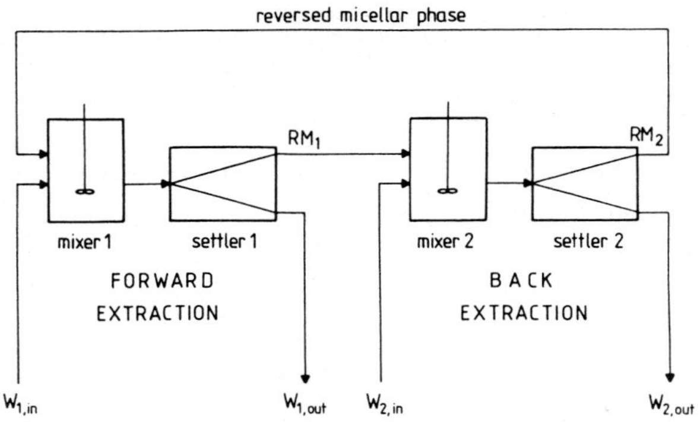

Fig. 3. Flowsheet of a combined forward and back extraction for two mixer/settler units with the reversed micellar phase circulating between the two extraction units (see ref. 4 for construction details).

In the first mixer/settler unit the pH in the aqueous phase favoured the transfer of the enzyme to the reversed micellar phase. The reversed micellar phase containing the enzyme was transferred continuously to the second extraction unit. In this unit the composition of the aqueous phase favoured the transfer of the enzyme to the aqueous phase. The empty reversed micellar phase was recirculated to the first extraction unit. The results of the continuous extractions with and without nonionic surfactant in the reversed micellar phase are given in Fig. 4.

The yield of active α-amylase in the second aqueous phase is considerably improved by the addition of the nonionic surfactant to the reversed micellar phase. A concentration factor of 11.5 was reached with a ratio of the flows of the first and second aqueous phase of 18. The total recovery of active enzyme during the extractions was 75%.

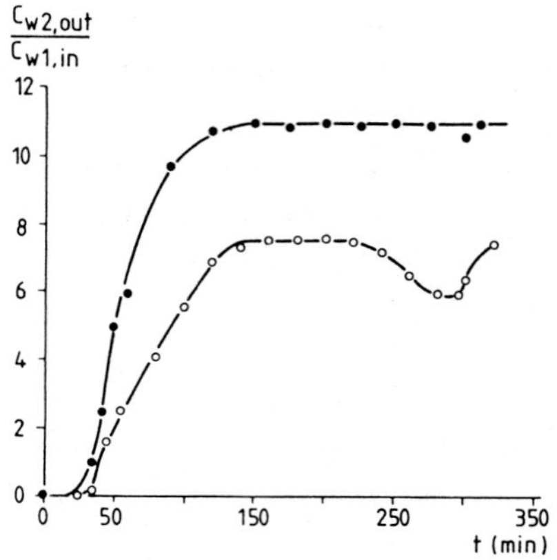

Fig. 4. Concentration of active α-amylase in the second aqueous phase relative to the concentration in the first aqueous phase entering the extraction unit. (o) Reversed micellar phase without nonionic surfactant added (data from ref. 4) (•) Reversed micellar phase with nonionic surfactant (Rewopal HV5) added.

The losses of surfactant (TOMAC) from the reversed micellar phase in the case of added nonionic surfactant was ca. 5% per circulation of the reversed micellar phase (60 min).

CONCLUSIONS

Enzyme concentration by continuous forward and back extraction using a reversed micellar phase is possible. The extraction efficiency can be increased by the addition of a nonionic surfactant to the reversed micellar phase. This addition resulted in a higher concentration factor of active enzyme, a higher total recovery of active enzyme and less surfactant losses from the reversed micellar phase. Because of the large increase in enzyme concentration that can be obtained by the reversed micellar extraction and the easy scaling up of the process, the extraction might be a valuable alternative to one or more conventional steps in the large scale recovery of industrial enzymes.

REFERENCES

1 P.L. Luisi, F.J. Bonner, A. Pellegrini, P. Wiget and R. Wolf, Helv. Chem. Acta, 62 (1979) 740.
2 K. Van 't Riet and M. Dekker, Proc. 3rd Eur. Congr. Biotechnol., München, Vol III, (1984) 541.
3 K.E. Göklen and T.A. Hatton, Biotechnol. Progr., 1 (1985) 69.
4 M. Dekker, K. Van 't Riet, S.R. Weijers, J.W.A. Baltussen, C. Laane and B.H. Bijsterbosch, (Bio-) Chem. Eng. J., 33 (1986) B27.
5 K.E. Göklen and T.A. Hatton, Proc. Int. Solv. Extr. Conf., München, Vol III, (1986) 587.

C. Laane, J. Tramper and M.D. Lilly (Editors), *Biocatalysis in Organic Media*, Proceedings of an International Symposium held at Wageningen, The Netherlands, 7–10 December 1986.

HALOPEROXIDASES IN REVERSED MICELLES: USE IN ORGANIC SYNTHESIS AND OPTIMISATION OF THE SYSTEM.

M.C.R. Franssen, J.G.J. Weijnen, J.P. Vincken, C. Laane[1] and H.C. van der Plas

Laboratory of Organic Chemistry, Agricultural University, De Dreyen 5, 6703 BC Wageningen (The Netherlands)

[1] Laboratory of Biochemistry, Agricultural University, De Dreyen 11, 6703 BC Wageningen (The Netherlands)

SUMMARY

The chloroperoxidase from the mold Caldariomyces fumago and the bromoperoxidase from the brown alga Ascophyllum nodosum were encapsulated into reversed micelles and used for the chlorination and bromination of monochlorodimedone and 1,3-dihydroxybenzene. The medium consisted of an aqueous buffer, a detergent (CTAB or CTAC), a cosurfactant (pentanol) and an organic solvent (octane). It appeared that the enzyme activity was highest at maximal water content of the micelles. The composition of the micellar interphase was optimal at 10-11% pentanol. The enzyme activity decreased slowly on incubation in reversed micelles alone and quickly when hydrogen peroxide was added to the system. The halogenation reactions in reversed micelles proceeded 1.7 - 2 times faster than in water.

INTRODUCTION

Haloperoxidases are enzymes which are capable of halogenating a large variety of organic compounds by means of hydrogen peroxide and chloride, bromide or iodide ions, depending on the enzyme under study (ref. 1). We have used the chloroperoxidase from the mold Caldariomyces fumago for the halogenation of barbituric acid (ref. 2,3). The bromoperoxidase from the brown alga Ascophyllum nodosum has been used for the bromination of phenol red (ref. 4). Although the yields of the reactions are very high, the use of the mold enzyme is confined to the conversion of water - soluble substrates, since the enzyme is rapidly denaturated by organic solvents (ref. 5). The algal bromoperoxidase is much more stable towards water - miscible organic solvents (ref. 4, 6) but up to now the enzyme has not been studied in water - immiscible solvents. Enzymes can be very active in such organic solvents when they are protected by encapsulation in reversed micelles. These transparent systems consist of tiny water droplets embedded in a water - immiscible organic solvent and stabilized by surfactants (see ref. 7 for a review). Apolar compounds like steroids were efficiently transformed in this way using cetyltrimethylammonium bromide (CTAB) as surfactant and octane as organic solvent (ref. 8). Hexanol was added as cosurfactant to regulate the polarity of the micellar interphase, because it was recognized that enzyme activity is maximal when the solubility of the substrate in the interphase is maximal (ref. 9).

The only haloperoxidase which has been studied in reversed micelles so far is horseradish peroxidase (ref. 10). We here present halogenation reactions in reversed micelles using the above - mentioned chloroperoxidase and bromoperoxidase. The micellar medium consists of buffer, pentanol, cetyltrimethylammonium bromide or chloride (CTAB resp. CTAC) and octane; the halide ion serves as counter ion for the ammonium ion and at the same time as substrate for the enzyme.

MATERIALS AND METHODS

Definition. $w_0 = \frac{[\text{water}]}{[\text{surfactant}]}$

Preparation of reversed micelles. For the chloroperoxidase (CPO) from *Caldariomyces fumago* (Sigma) the conditions were as follows. To a 0.2 M suspension of CTAX (X = Cl or Br) in octane containing 0.1 mM of the organic substrate were successively added suitable amounts of pentanol, 0.1 M potassium phosphate buffer pH 3.0 and 10 mM H_2O_2 in the same buffer. The suspension was mixed on a Vortex; only compositions which gave clear solutions within 5 sec. were used. A suitable amount of enzyme was then injected into the solution and the UV-absorbance at 279 nm was monitored. For a system with w_0 = 25 and 11% pentanol the medium was composed as follows: 2064 µl 0.1 mM organic substrate/0.2 M CTAX/octane; 250 µl pentanol; 123 µl buffer; 60 µl 10 mM H_2O_2 in buffer and 3 µl (150 ng) chloroperoxidase. The specific activity of the enzyme was 400 - 450 µmol MCD.mg protein^{-1}.min^{-1}.

For the bromoperoxidase (BPO) from *Ascophyllum nodosum* (a gift from mr. E. de Boer and dr R. Wever, University of Amsterdam) the conditions were as follows. To a 0.2 M suspension of CTAB containing 50 µM MCD in octane were successively added suitable amounts of pentanol, 0.1 M potassium phosphate buffer pH 6.5 and 100 mM H_2O_2 in the same buffer. The suspension was mixed till clear, enzyme was added and the activity recorded at 291 nm. For a system with w_0 = 25 and 11% of pentanol the medium was composed as follows: 2064 µl 50 µM organic substrate/0.2 M CTAB/octane; 250 µl pentanol; 163 µl buffer; 20 µl 100 mM H_2O_2 in buffer and 3 µl (5 µg) bromoperoxidase. The specific activity of the enzyme was 10 µmol MCD.mg protein^{-1}.min^{-1}.

The pseudo-ternary phase diagram was constructed as described in ref. 9.

RESULTS AND DISCUSSION

When buffer, CTAX, pentanol and octane are mixed in suitable concentrations a transparent solution is obtained. This reversed micellar solution is only obtained within a small area of the pseudo-ternary phase diagram that can be constructed by varying the concentrations of buffer, pentanol and octane/CTAX and keeping the ratio octane/CTAX constant (see Figure 1). When the chloroperoxidase from C. fumago was added to the micellar system supplied with hydrogen peroxide and monochlorodimedon (MCD, **1**), the organic substrate was rapidly converted into its bromo derivative **2a** (see Scheme). With CTAB as surfactant the bromide ion serves as counter ion for the ammonium ion and at the same time as substrate for the enzyme. The rate of this reaction was about 80% higher than in water (see Table), clearly showing the advantages of this system.

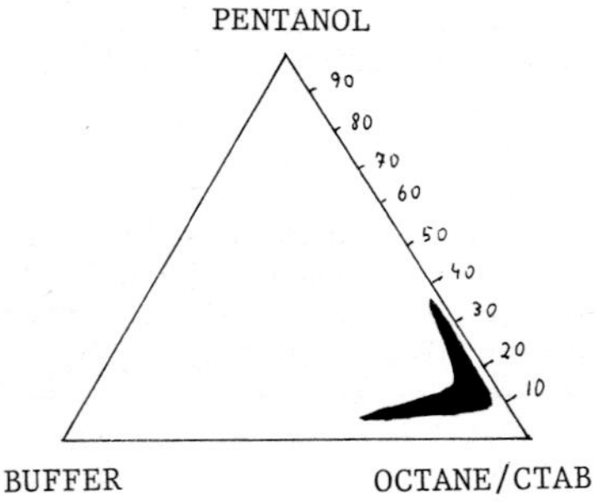

Fig.1. Pseudo-ternary phase diagram for the system CTAB/octane/pentanol/0.1 M H_3PO_4 pH 2.7. The black area reflects those compositions which gave clear solutions within 5 sec mixing time.

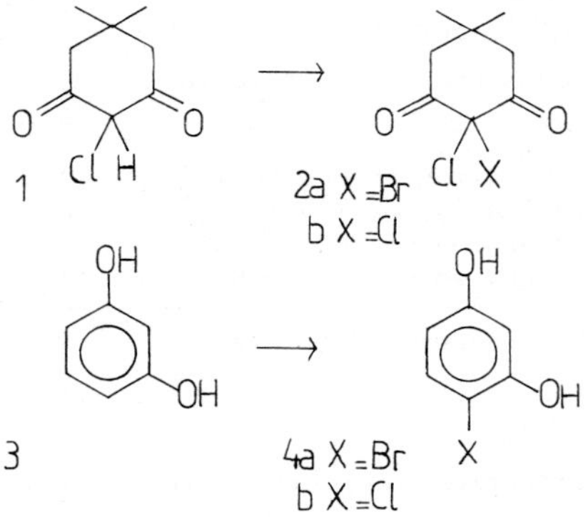

Scheme. Reactions catalyzed by CPO in reversed micelles.

To determine the optimal conditions for catalysis some parameters were varied and the effect on the enzyme activity recorded. The enzymatic bromination rate is maximal at maximal water content of the reversed micelles (w_0), (Figure 2a). That the reaction rate is maximal with high water content is probably due to the fact that in these solutions the bromide concentration is lower, thereby decreasing the inhibitory effect of the bromide ions to the enzyme at low pH (ref. 11). The enzyme activity decreases strongly with increasing pentanol content of the medium (see Figure 2b). Increasing the pentanol content of the medium makes the interphase more apolar and the continuous phase more polar, causing an extraction of the substrate (of intermediate polarity) into the continuous phase. This makes the substrate less accessible to the enzyme and a decreased activity is observed. The pH - profile of the

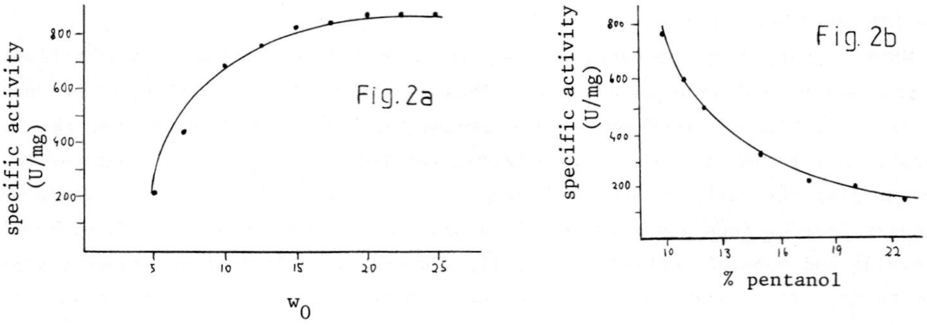

Fig. 2a. The effect of w_0 on the CPO-activity. The reversed micelles contain 0.2 M CTAB and 11% pentanol. Fig. 2b. The effect of the pentanol concentration on the CPO-activity. The reversed micelles contain 0.2 M CTAB; w_0 = 10.

TABLE

Specific activity (μmol substrate.mg^{-1}.min^{-1}.) for MCD and resorcinol in aqueous and micellar media.

chlorination of MCD in buffer	408	chlorination of resorcinol in buffer	196
chlorination of MCD in RM[1]	885	chlorination of resorcinol in RM	478
bromination of MCD in buffer	489	bromination of resorcinol in buffer	199
bromination of MCD in RM	872	bromination of resorcinol in RM	508

[1]RM = Reversed Micelles

bromination reaction in reversed micelles (w_0 = 25) is comparable to that in water.

Chloroperoxidase can also be used for the chlorination of apolar compounds in reversed micelles. When cetyltrimethylammonium chloride (CTAC) is used as surfactant the chloride ion serves as counter ion for the surfactant and at the same time as substrate for the enzyme. With this system MCD can be chlorinated giving **2b** (see Scheme) about twice as rapidly as in water (see Table).

Chloroperoxidase slowly loses its activity when kept in reversed micelles, as can be seen in Figure 3a. The rate of inactivation in reversed micelles without H_2O_2 and in water are comparable. The stability of the enzyme towards hydrogen peroxide is depicted in Figure 3b. With higher concentrations of H_2O_2 (0.4 mM overall concentration) 90 % of the enzyme activity is lost in 4 minutes when no organic substrate is added. With low concentrations (0.04 mM) 40% of the enzyme molecules are still active after 10 minutes of incubation.

In an extension of our work we were also interested in the conversion of an apolar substrate such as 1,3-dihydroxybenzene (resorcinol, **3**). **3** was chlorinated or brominated efficiently by chloroperoxidase in reversed micelles

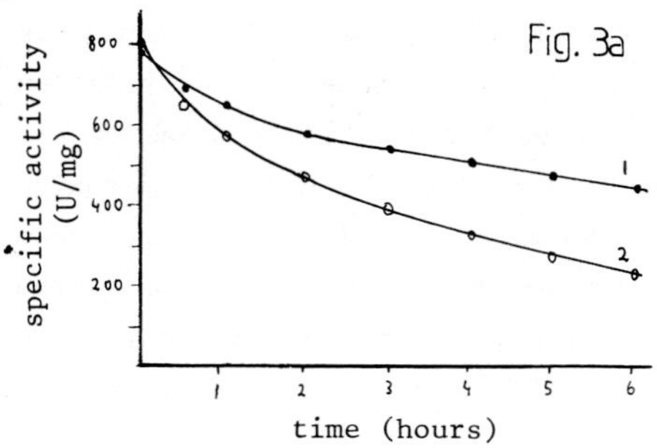

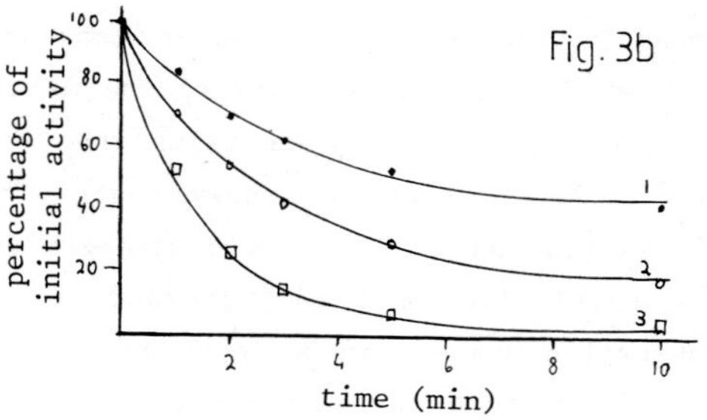

Fig. 3. Stability of CPO in reversed micelles. Fig. 3a without H_2O_2; trace 1: surfactant = CTAC, 11% pentanol, w_0 = 17.5; trace 2: surfactant = CTAB, 11% pentanol, w_0 = 25. Fig. 3b with H_2O_2; trace 1: 0.04 mM overall concentration; trace 2: 0.20 mM; trace 3: 0.40 mM. The vertical coordinate reflects the activity at t = t compared to t = 0. The surfactant is CTAB; 11% pentanol; w_0=25.

giving the 4-monohalo derivative **4** (see Scheme) at a higher rate than in water (see Table). Methoxybenzene, p-methoxyphenol and p-dimethoxybenzene were halogenated by the enzyme in water but not in reversed micelles; the reason for this is unknown.

The bromoperoxidase from Ascophyllum nodosum was also encapsulated in reversed micelles. The enzyme activity is found to increase twofold. Bromoperoxidase is not as vulnerable to hydrogen peroxide as chloroperoxidase, which can be seen in Figure 4a. Maximal activity is achieved at maximal water content of the micelles (Figure 4b).

We conclude that the enzymatic activity of haloperoxidases is increased when they are encapsulated into reversed micelles, improving the conversion of apolar compounds by these enzymes. Chloroperoxidase is extruded by C. fumago into its aqueous environment, which is reflected by the fact that the enzyme is

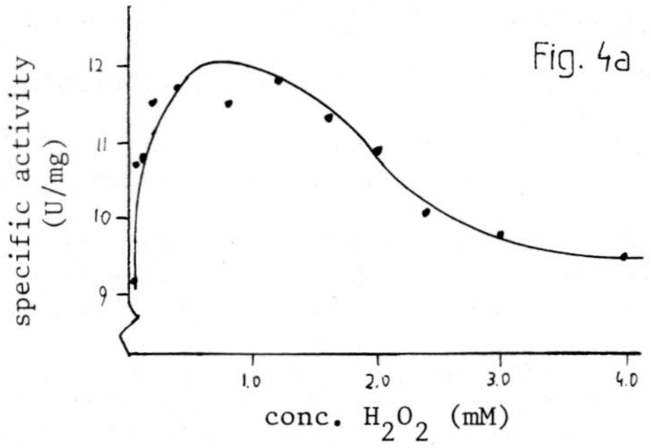

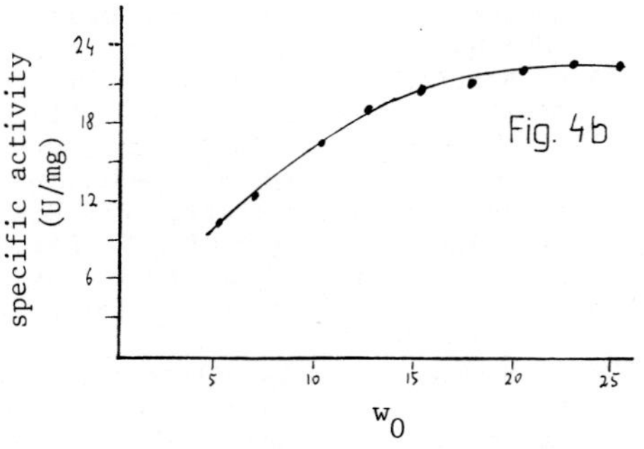

Fig. 4a. The effect of the H_2O_2-concentration on the BPO-activity. The reversed micelles contain 0.2 M CTAB, 11% pentanol and w_0 = 25. Fig. 4b The effect of w_0 on the BPO-activity. The reversed micelles contain 0.2 M CTAB and 11% pentanol.

most active at maximal water content of the micelles. Maximal activity is achieved when the polarity of the micellar interphase is tuned to the polarity of the organic substrate using the "logP - concept" described by Laane et al. (ref. 9). However, the enzyme activity is affected by the hydrogen peroxide concentration and the ionic strength inside the micelles. The chloroperoxidase in particular is very vulnerable to hydrogen peroxide, so we feel that the described system can be improved by using a H_2O_2 - generator inside the micelle and by using a nonionic detergent. Studies in this direction are being conducted at the moment.

ACKNOWLEDGEMENTS

We are highly indebted to the following persons: to mr. E. de Boer and dr. R. Wever (Lab. of Biochemistry, Univ. of Amsterdam) for the generous gift of bromoperoxidase and helpful discussions; to prof. dr. E.M. Meijer, dr. H.E. Schoemaker and mr. W. Boesten (DSM) for stimulating discussions. We gratefully acknowledge the financial support of Naamloze Vennootschap DSM, Heerlen, The Netherlands.

REFERENCES

1) S.L. Neidleman and J. Geigert, Biohalogenation: Principles, Basic Roles and Applications, Ellis Horwood Ltd., Chichester.
2) M.C.R. Franssen and H.C. van der Plas, A new enzymatic chlorination of barbituric acid and its 1-methyl and 1,3-dimethyl derivatives, Recl. Trav. Chim. Pays-Bas, **103** (1984) 99-100.
3) M.C.R. Franssen and H.C. van der Plas, The chlorination of barbituric acid and some of its derivatives by chloroperoxidase, Bioorg. Chem., in press
4) E. de Boer, H. Plat, M.G.M. Tromp, M.C.R. Franssen, H.C. van der Plas, E.M. Meijer, H.E. Schoemaker and R. Wever, Vanadium (V) containing bromoperoxidase, an example of an oxidoreductase with high operational stability in aqueous and organic media, Biotechnol. Bioeng., in press
5) C.L. Cooney and J. Hueter, Enzyme catalysis in the presence of nonaqueous solvents using chloroperoxidase, Biotechnol. Bioeng. **16** (1974) 1045-1053
6) R. Wever, H. Plat and E. de Boer, Isolation procedure and properties of the bromoperoxidase from the seaweed Ascophyllum nodosum, Biochem. Biophys. Acta **830** (1985) 181-186
7) K. Martinek, A.V. Levashov, N. Klyachko, Y.L. Khmelnitski and I.V. Berezin, Micellar enzymology, Eur. J. Biochem. **155** (1986) 453-468
8) R. Hilhorst, C. Laane and C. Veeger, Enzymatic conversion of apolar compounds in organic media using a NADH - regenerating system and dihydrogen as reductant, FEBS Lett. **159** (1983) 225-228
9) R. Hilhorst, R. Spruijt, C. Laane and C. Veeger, Rules for the regulation of enzyme activity in reversed micelles as illustrated by the conversion of apolar steroids by 20β - hydroxysteroid dehydrogenase, Eur. J. Biochem. **144** (1984) 459-466
10) K. Martinek, N.L. Klyachko, A.V. Levashov and I.V. Berezin, Micellar enzymology, Catalytic activity of peroxidase in a colloidal solution of water in an organic solvent, Dokl. Akad. Nauk SSSR **269** (1983) 491-493, Engl. Transl. 69-71
11) J.A. Thomas, D.R. Morris and L.P. Hager, Chloroperoxidase. VIII. Formation of peroxide and halogen complexes and their relation to the mechanism of the halogenation reaction, J. Biol. Chem. **245** (1970) 3135-3142

C. Laane, J. Tramper and M.D. Lilly (Editors), *Biocatalysis in Organic Media*, Proceedings of an International Symposium held at Wageningen, The Netherlands, 7–10 December 1986.

POTENTIAL OF ORGANIC SOLVENTS IN CULTIVATING MICRO-ORGANISMS ON TOXIC WATER-INSOLUBLE COMPOUNDS

J.M. REZESSY-SZABÓ, G.N.M. HUIJBERTS AND J.A.M. de BONT

Department of Microbiology, Agricultural University, Hesselink van Suchtelenweg 4, 6703 CT Wageningen (The Netherlands)

SUMMARY

The effects of seven organic solvents on nine different aerobic bacteria were tested. The reaction of growing cells and of immobilized cells to the organic solvents was similar. Solvents with a relatively high logP-value were not harmful to cells but it was also noticed that great differences exist amongst the various bacteria tested in their reaction to the solvents. Using a benzene-utilizing bacterium it was observed that dibutyl phtalate is a suitable solvent in circumventing benzene toxicity during growth of the organism. From the results it is expected that organic solvents will be of use in cultivating micro-organisms on other and more toxic water-insoluble compounds.

INTRODUCTION

The use of water-immiscible organic solvents in cultivating micro-organisms on substrates that are poorly soluble in water and/or are toxic to growing cells has been restricted to only very few instances. Innovating work by Schwartz and MacCoy (1) on growth of *Pseudomonas oleovorans* on octane in the presence of cyclohexane demonstrated the potential of organic solvents in growing bacteria in the presence of a toxic substance. The emphasis of their work was on the production of toxic epoxides by the growing cells from alkenes and such two-liquid phase biocatalytic reactions have mostly been studied using non-growing cells or using enzymes. Such reactions in the presence of an organic solvent may have many potential advantages (2) and two of these advantages are that a relatively high concentration of poorly water-soluble substrates may be obtained within a bioreactor system and that inhibition of the biocatalyst by toxic substrates may be prevented. The use of organic solvents however is restricted since most solvents will inactivate the biocatalyst and also because the partitioning of the substrate between organic phase and aqueous phase may be inadequate.

The influence of many water-immiscible solvents on retention of activity of immobilized *Mycobacterium* cells was determined by Brink and Tramper (3) and it was found that retention of activity is related to solvent properties like polarity and molecular size. A good correlation between remaining cell activity and solvent properties was also observed when applying another solvent parameter (4) *i.e.* logP. This parameter is defined as the logarithm of the partition coëfficient of a certain solvent in a standard octanol/water two-phase system

Based on the results obtained by Brink and Tramper (3) and in view of the limited information available on the use of water-immiscible solvents in cultivating micro-organisms on toxic compounds, we have undertaken a survey to define the possibilities and the limits of the use of organic solvents in cultivating bacteria. Effects of seven organic solvents on nine different aerobic bacteria were investigated and in the present paper it is shown that depending on the combination of organism, solvent and substrate used, it may be convenient to grow bacteria on toxic non-polar compounds in the presence of an organic solvent.

METHODS

Organisms

The bacteria used in the present investigation have been described previously (5,6,7) or were obtained from the culture collection of our department.

Cultivation

Organisms were grown in 5-l Erlenmeyer flasks containing 250 ml of a mineral medium (8) supplemented with glucose (2 g/l) and yeast extract (1 g/l) or in serum bottles (100 ml) containing 4.5 ml of the mineral medium supplemented with yeast extract (2 g/l).

Effects of solvents on cells

Effects of solvents on cells were tested in serum bottles (100 ml) with 0.5 ml organic solvent added to the aqueous phase (4.5 ml).

For growing cells was the organic phase added to the yeast extract medium prior to heat sterilization. Growth was recorded after three days of incubation at 30 C in a shaking water bath.

The growth of *Pseudomonas* 50 on benzene in the presence of dibutyl phtalate was also in serum bottles (4.5 ml mineral medium; 0.5 ml dibutyl phtalate) at 30 C in a shaking water bath and carbon dioxide formed was measured after 1 day.

Washed cell suspensions in 50 mM phosphate buffer or cells immobilized in alginate gel were obtained as described previously (5) and activity of these cells in phosphate buffer (50 mM, pH 7.2) in the presence of solvents was determined by measuring carbon dioxide evolution from glucose (2 g/l) added to the buffer solution. Cells (1-3 mg protein/bottle) were incubated while shaken at 30 C and during 4 hours.

Concentrations of carbon dioxide present in serum bottles were determined by analyzing head space samples using a gas chromatograph.

RESULTS

Effects of solvents on growing cells

Nine different aerobic bacteria that were able to grow on a yeast extract medium were tested for growth in this medium in the presence of 10 % (v/v) organic solvent. Solvents tested were dimethyl phtalate, diethylphtalate, dibutyl phtalate, diphenyl ether, perfluorohexane, hexane and hexadecane because these solvents have been found to retain activity of immobilized cells (3). The logP-value of these solvents ranged from 2.8 to 8.8. The respons of the organisms to the various solvents depended on both the solvent and the organism (Table 1). Pseudomonas 50 and Pseudomonas LW-4 on the one hand were not restricted in their growth on the yeast extract medium by any of the solvents tested, whereas the Bacillus sp. on the other hand only grew in the presence of hexadecane and failed to grow whenever another solvent was present. Other organisms responded to the organic solvents in an intermediate way. Nevertheless, it is obvious from these results that a general tendency exists for cells to be less susceptible for solvents with increasing logP-values.

TABLE 1

Effects of solvents on growing cells

Solvent	Dimethyl phtalate	Diethyl phtalate	Hexane	Diphenyl ether	Dibutyl phtalate	Perfluoro hexane	Hexadecane
LogP-value	2.3	3.3	3.5	4.3	5.5		8.8
Pseudomonas LW-4	+	+	+	+	+	+	+
Pseudomonas 50	+	+	+	+	+	+	+
Escherichia coli	-	+	+	+	+	+	+
Azotobacter sp.	-	-	-	+	+	+	-
Mycobacterium E3	-	+	-	-	-	+	+
Acinetobacter sp.	-	-	-	-	+	+	+
Xanthobacter Py2	-	-	-	-	+	+	-
Nocardia TB1	-	-	-	-	-	+	+
Bacillus sp.	-	-	-	-	-	-	+

Organisms were inocculated in the yeast extract medium in the presence of 10 % (v/v) organic solvent and growth (+) or no growth (-) was recorded after three days of incubation.

During all incubations were flasks shaken and it thus seems that cells from time to time would be in direct contact with the bulk organic solvent rather than with the very limited amount of solvent dissolved in the water-phase only. The Bacillus sp. was used to determine if this direct contact of an organism with the pure solvent is responsible for the delitereous effect of dibutyl phtalate on this organism. Cells were grown in a yeast extract medium not only while shaking with the dibutyl ester present (10 % v/v), but also in a stationary manner, with the solvent present as a layer at the bottom of the flask. No growth occurred in the non-shaken flask within three days, whereas a control flask, also incubated stationary but containing no solvent, showed good growth already after one day of incubation.

For the purpose of the present work organisms should not be able to metabolize the solvents used because a particular solvent would then not be suitable in cultivating the organism on a toxic compound. It was therefore investigated whether the organic solvents used served as carbon and energy source for the bacteria tested by incubating cells in the mineral medium supplemented with the respective solvents at 0.2 % (v/v). As expected did no organism grow on perfluorohexane and diphenyl ether was also not metabolized by the bacteria. Some strains, however, grew on hexadecane (TB1, E3 and LW-4), on hexane (TB1), on dimethyl phtalate (Py2), on diethyl phtalate (TB1) and on dibutyl phtalate (TB1, Py2).

Effects of solvents on non-growing and on immobilized cells

Four strains (Nocardia TB1, Pseudomonas LW-4, Mycobacterium E3 and the Bacillus sp.) were selected for further study of the effect of the seven organic solvents on the activity of both washed cell suspensions and on immobilized cells. Organisms were precultured on glucose and activity of the cells, either as non-growing cells in phosphate buffer or as immobilized cells entrapped in alginate gel, was recorded by measuring the amount of carbon dioxide released by such cells from glucose (Table 2). Activity of Pseudomonas LW-4 cells was not affected by the solvents with an exception for dimethyl phtalate. Results obtained for Mycobacterium E3, for Nocardia TB1 and for the Bacillus sp. were comparable. Activity of these cells was not influenced by solvents having a high logP-value whereas solvents with a lower logP-value

TABLE 2

Effects of solvents on non-growing and on immobilized cells

Solvent	None	Dimethyl phtalate	Diethyl phtalate	Hexane	Diphenyl ether	Dibutyl phtalate	Perfluoro hexane	Hexa-decane
Pseudomonas								
free	1400	400	1500	1400	1700	1300	1700	1400
immobilized	1400	600	1500	900	1500	1500	900	1400
Mycobacterium								
free	70	40	40	40	30	70	nd	70
immobilized	40	30	30	20	30	40	nd	40
Nocardia								
free	470	150	230	240	190	420	450	450
immobilized	60	20	30	30	40	50	60	60
Bacillus								
free	80	30	30	40	40	80	nd	60
immobilized	30	10	10	10	10	30	nd	30

Organisms were precultured on a glucose/yeast extract medium and the amount of carbon dioxide (μmoles/mg protein) formed from glucose by the free and immobilized cells was determined after 4 hours of incubation. nd = not determined.

inhibited glucose metabolism.

It was also investigated whether the bacillus cells gradually lost their activity or whether an immediate loss of activity occurred when the cells were confronted with solvents with a low logP-value. Cells were incubated in the presence of dimethyl phtalate and the time course of carbon dioxide evolution from glucose by washed cell suspensions and by immobilized cells was recorded. The results of these experiments indicate that the effect of the solvent was immediate since the amount of carbon dioxide released from glucose in the presence of dimethyl phtalate was linear with time for both the free and immobilized cells.

Growth of Pseudomonas 50 on benzene in the presence of dibutyl phtalate

From the above results it appears that several bacteria are not affected by certain solvents. It was the purpose of our work to determine whether solvents could possibly be of use in growing micro-organisms on toxic water-insoluble

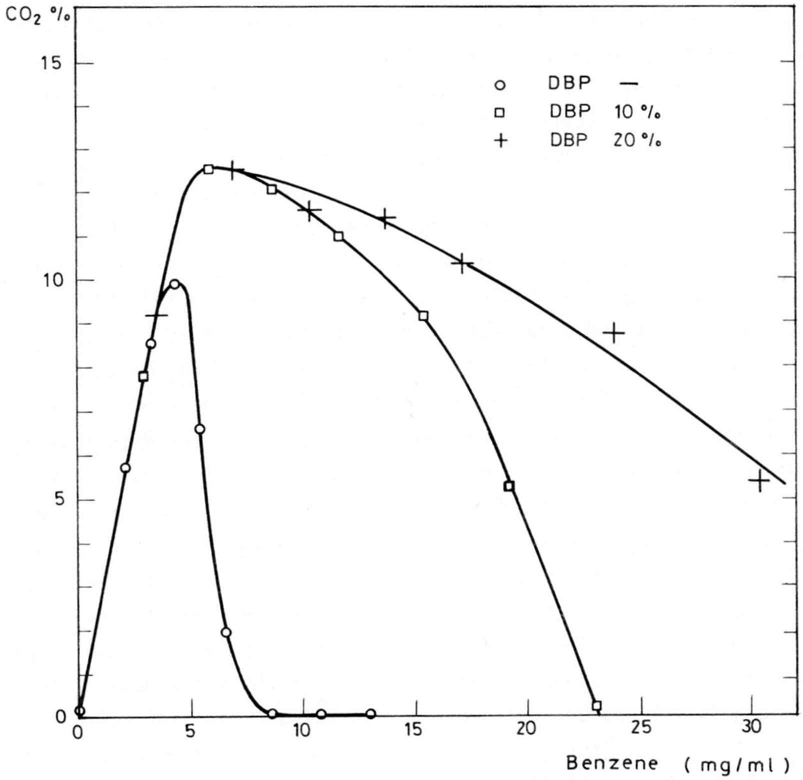

Fig. 1 Effect of dibutyl phtalate (DBP) on the toxicity of benzene for cells of *Pseudomonas* 50 growing on benzene.

compounds and consequently for further research we have chosen an organism that is able to grow on such a toxic compound. Pseudomonas 50 is able to grow on benzene and the organism was therefore cultivated in the mineral medium with varying amounts of benzene and also with varying amounts of dibutyl phtalate. As seen from the results presented in Fig. 1, was it possible to cultivate cells with much more benzene present in the incubation system whenever dibutyl phtalate was present. A higher dibutyl phtalate concentration (20 % v/v) was more effective in circumventing benzene toxicity than was the lower phtalate ester concentration used, showing the solvent served as a reservoir of benzene for the cells in the aqueous phase. In subsequent experiments it was demonstrated that it is also possible to cultivate Pseudomonas 50 on benzene in chemostat culture in the presence of dibutyl phtalate ester. These results will be presented elsewhere in more detail.

DISCUSSION

The use of organic solvents in product formation by either non-growing free cells or by immobilized cells has been amply documented and the present work shows the potential of organic solvents to serve as a reservoir of toxic water-insoluble compounds for cells growing on such compounds in an otherwise aqueous medium. In the present work were solvents tested that previously had been found satisfactory in retaining activity of immobilized cells (3). Interestingly, it was observed that growing cells reacted similarly to these solvents: solvents with a high logP-value were less harmful to both growing and immobilized cells. At the same time, it has to be realized however, that great diferrences exist amongst the bacteria in their reaction to solvents. The two Pseudomonas strains tested were less susceptible to damage by the solvents than the other strains, while the Bacillus strain was effected very much by the solvents. The activity of the non-growing washed cell suspension and of the immobilized cells of the Bacillus, of the Nocardia and of the Mycobacterium was not fully repressed whereas growth of these organisms did not occur in the presence of most solvents. This observation may be ascribed to the rather short incubation periods of the non-growing cells as compared with the growing cells.

From the results it also appears that a direct contact between the bacterial cell and the solvent is not a prerequisite for damaging effects because immobilized cells showed a reaction to solvents that was comparable to the reaction of the washed cell suspensions. It is supposed that the entrapment of cells in the alginate gel will prevent a direct contact between cell and solvent but will still allow an interaction of the cells with the low concentration of solvent dissolved in the water phase. Results obtained with the Bacillus sp. grown in the presence of methyl phtalate either while shaking or while stationary also suggest that an interaction between the cell and the dissolved solvent will cause the observed effects of the solvent on the metabolic activity of the cell. We presently can offer no explanation for differences in reactions to solvents exhibited by the various bacteria (Tables 1 and 2) but it is likely that events at the cytoplasmic membrane are important (9).

The most important observation of the present work is that most organisms can grow in the presence of some selected solvents. Especially dibutyl phtalate is a useful solvent because it is not toxic to most bacteria, because most bacteria do not grow on it, and because it is cheap and easy to handle. The potential of dibutyl phtalate as a solvent that is of use in growing bacteria on toxic water-insoluble compounds was assessed in our prelimnary experiments in which the benzene-utilizing Pseudomonas 50 was studied. Growth of this organism on benzene was halted at approxiomately 0.5 % benzene present in the incubation system whenever no dibutyl phtalate was present. This benzene level could be increased to 3% present in the incubation mixture when 20 % dibutyl phtalate was included. Benzene is not a very toxic substance to bacteria, but in prelimnary experiments we observed that other and more toxic compounds as for instance halogenated aromatics, or aliphatic aldehydes and epoxides, could also be given to growing cells in the presence of a suitable organic solvent and we therefore conclude that organic solvents may be of use in growing micro-organisms on toxic water-insoluble compounds.

REFERENCES

1. R.D. Schwartz and C.J. McCoy, Epoxidation of 1,7-octadiene by Pseudomonas oleovorans: Fermentation in the presence of cyclohexane, Appl. Environ. Microbiol., 31 (1977) 47-49.
2. M.D. Lilly and J.M. Woodley, Biocatalytic reactions involving water-insoluble organic compounds, in: J. Tramper, H.C. van der Plas and P. Linko (Eds.), Biocatalysts in Organic Syntheses, Elsevier, Amsterdam, 1985 pp 179-192.
3. L.E.S. Brink and J. Tramper, Optimization of organic solvent in multiphase biocatalysis, Biotechnol. Bioeng., 27 (1985) 1258-1269.
4. C.Laane, S. Boeren and K. Vos, On optimizing organic solvents in multi-liquid phase biocatalysis, Trends in Biotechnology, 3 (1985) 251-252.
5. A.Q.H. Habets-Crützen, L.E.S. Brink, C._. van Ginkel, J.A.M. de Bont and J. Tramper, Production of epoxides from gaseous alkenes by resting-cell suspensions and immobilized cells of alkene-utilizing bacteria, Appl. Microbiol. Biotechnol., 20 (1984) 245-250.
6. W.J.J. van den Tweel, J.P. Smits and J.A.M. de Bont, Microbial metabolism of D- and L-phenylglycine by Pseudomonas putida LW-4, Arch. Microbiol., 144 (1986) 169-174.
7. C.G. van Ginkel, H.G.J. Welten and J.A.M. de Bont, Epoxidation of alkenes by alkene-grown Xanthobacter spp., Appl. Microbiol. Biotechnol., 24 (1986) 334-337.

8. W.M. Wiegant and J.A.M. de Bont, A new route for ethylene glycol metabolism in *Mycobacterium* E44, J. Gen. Microbiol., 120 (1980) 325-331.
9. M. de Smet, J. Kingma, H. Wijnberg and B. Witholt, *Pseudomonas oleovorans* as a tool in bioconversions of hydrocarbons: growth, morphology and conversion characteristics in different two-phase systems, Enzyme Microb. Technol., 5 (1983) 352-360.

C. Laane, J. Tramper and M.D. Lilly (Editors), *Biocatalysis in Organic Media*, Proceedings of an International Symposium held at Wageningen, The Netherlands, 7-10 December 1986.

VIABILITY AND ACTIVITY OF FLAVOBACTERIUM DEHYDROGENANS IN ORGANIC SOLVENT/CULTURE TWO-LIQUID-PHASE-SYSTEMS

Sjef Boeren, Colja Laane* and Riet Hilhorst
Department of Biochemistry, Agricultural University, De Dreijen 11, 6703 BC Wageningen, The Netherlands
* Unilever Research Laboratorium, P.O.Box 114, 3130 AC Vlaardingen, The Netherlands

INTRODUCTION

A large number of steroids are produced on industrial scale by microbial conversions (refs. 1, 2). The reaction medium consists mainly of water although the steroid solubility in water is rather low even in the presence of solubilizing agents like Tween or Span. The low substrate solubility may limit the conversion rate and the low product solubility decreases the yield due to entrapment of substrate molecules in the crystalizing product. This co-crystalization can be avoided by the use of water-immiscible organic solvents as a reactant reservoir to keep the product as well as the substrate in a soluble form till the conversion is complete.

When the enzyme responsible for the desired conversion has to be induced it is essential that the micro-organisms remain viable in the presence of the organic solvent since proteins are rapidly degraded in dead cells. The quantity of proteins in a living cell is maintained by a contemporary process of synthesis and degradation. Hydrophilic organis solvents are known to kill micro-organisms but little is known about the toxic effect of more hydrophobic solvents.

This paper describes the effect of homologous series of hydrophobic organic solvents on viability and steroid conversion rate of Flavobacterium dehydrogenans in two-liquid-phase-systems.

The results show: 1) that the conversion rate in these two-liquid-phase systems can be significantly higher than in aqueous media if the growth stage of the micro organism at the moment of substrate and organic solvent addition is chosen properly,
2) that viability is high even in alkane-substituted solvents with a hydrophobicity between logP=2 and logP=4 if the hydrophobicity of the substituent is low enough.

MATERIALS AND METHODS

Materials

Flavobacterium dehydrogenans ATCC 13930, androstenolone-acetate (AAc), dehydro-epiandrosterone, 4-androstene-3,17-dione and yeast-extract (Gistex) were kindly supplied by Organon Int. B.V., Oss, The Netherlands. All other chemicals were obtained from commercial sources.

Cultivation
Cells were cultivated on a rotary shaker (200 rpm) at 30 ^{o}C in 2 L Erlenmeyer flasks containing 200 or 100 mL medium of the following composition (per liter): yeast extract, 15g; $(NH_4)_2SO_4$, 1g; KH_2PO_4, 1g; glucose, 5g; final pH 6.8.

The cultures were shaken up to the optical densities mentioned in the result section. The optical density (OD) was measured at 650 nm, in a 1mL glass cuvette with a Zeiss PMQ_2 spectrometer.

Reaction conditions and determination of the weight percentage 4-AD
At the specified cell density three times 10 mL not AAc induced culture was transferred to three different 100 mL Erlenmeyer flasks. To the first Erlenmeyer only AAc (the amount is specified in the result section) was added; to the second flask AAc together with Tween 80 (0,1 mL per liter culture) and to the third AAc and 10 mL organic solvent. These cultures were incubated (150rpm) for 5 hours. The reactions were stopped and the steroids were dissolved by adding 40 mL ethanol to the respective reaction media. The weight percentage of product to total steroid was measured by high-performance liquid chromatography (HPLC) as described before (ref. 3).

Determination of viability
10 mL organic solvent and 100 mg AAc were added to 10 mL culture in a 100 mL Erlenmeyer flask. After 4 hours incubation on a rotary shaker (150 rpm., 30 ^{o}C) 50μL aliquots from the water phase were transferred to 10 mL 1% (w/v) soya-peptone + 0,9% (w/v) NaCl. As a reference, a culture was taken to which 100 mg AAc but no organic solvent was added. The soya-peptone tubes were mildly shaken at 30 ^{o}C for one day after which the OD at 650 nm was measured. The relative cell viability can be calculated by dividing the OD_{650} as measured for the sample soya-peptone culture by the OD_{650} measured for the reference soya-peptone culture.

Determination of steroid solubility
A suspension of AAc or 4-AD in alkane was vigorously shaken for 2 hours at 21 ^{o}C. After centrifugation (6 minutes, 7200g) the concentration of the steroid was determined by HPLC by relating the determined peak area with the peak areas of a known steroid concentration series.

Calculation of solvent hydrophobicity
LogP was used as a parameter for solvent hydrophobicity. P denotes the partition coefficient of a solute in a water-octanol two-phase system:

$$\text{partition coefficient } P = \frac{[\text{solute}]\ \text{octanol}}{[\text{solute}]\ \text{water}}$$

Statistical calculations (carried out by Rekker (ref.4)) have shown that hydrophobic fragmental constants can be assigned to functional groups which enables the calculation of logP values of small molecules (ref. 5). Calculated values are in good agreement with experimental data.

RESULTS AND DISCUSSION

Model system

Figure 1 shows the reactions catalyzed by Flavobacterium dehydrogenans. The first reaction step, the deacetylation of androstenolone-acetate (AAc) into dehydro-epiandrosterone, is catalyzed by an esterase which has to be induced by AAc. The induction time is about 3 hours. The deacetylated product (dehydro-epiandrosterone) is rapidly converted into 4-androstene-3,17-dione (4-AD) by the combined action of an oxidase and an isomerase and completely excreted into the medium.

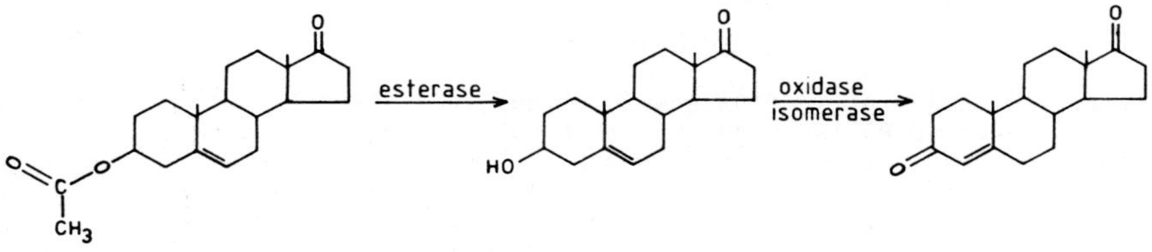

Figure 1: The conversion catalyzed by Flavobacterium dehydrogenans.

4-AD production in different media

Figure 2 shows that at all cell densities below OD_{650} = 6.6 the addition of a surfactant (0.1 mL Tween 80 /L culture) (o) resulted in a two fold enhancement of the 4-AD production rate as compared to the aqueous medium alone (Δ). This beneficial effect of Tween 80 abolished completely at higher densities.
In the octane/culture two-liquid-phase system (□) the steroid conversion rate at optical densities between 4.0 and 5.3 was comparable to that of the Tween containing medium. However at higher cell densities, up to an OD of 6.3, the activity in the octane/culture two-liquid-phase system increased about 5-fold after which it dropped to rates comparable to those obtained in the other media. The average maximum occured at densities between 6.0 and 6.6 which coincides with cells growing at the end of the exponential growth phase. In general this growth phase involves marked changes in cell metabolism, morphology as well as membrane structure and function (ref.4). In our case, the effect of the organic solvent could be explained by viability changes of the cells at different densities during growth (ref.3) which can be seen as an overall effect of the changes mentioned above.

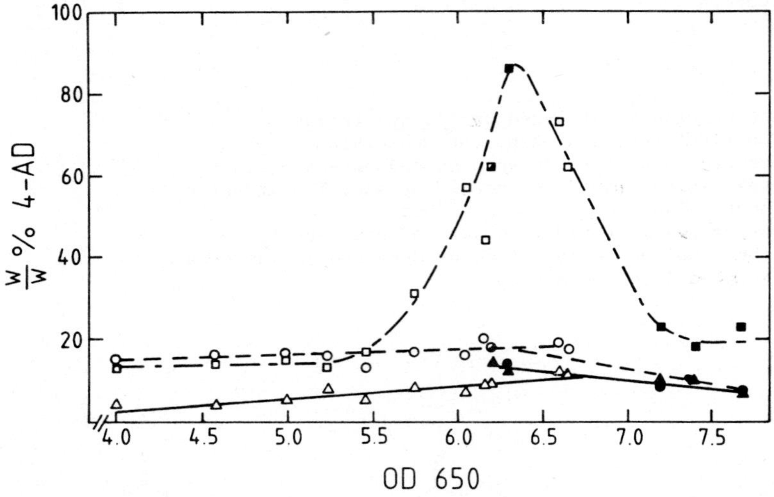

Figure 2: 4-AD production at varying growth stages in different media: (Δ,▲) aqueous medium; (O,●) Tween 80 (0,1 mL/L culture) medium and (□,■) octane/culture (1:1 v/v). Open symbols show 2 g/L substrate, dark symbols show 7 g/L substrate with respect to water. All additions were performed at t=0 hours to not AAc induced cultures. At the OD indicated in this figure the amount of 4-AD produced was determined after 5 hours of incubation at 21 °C.

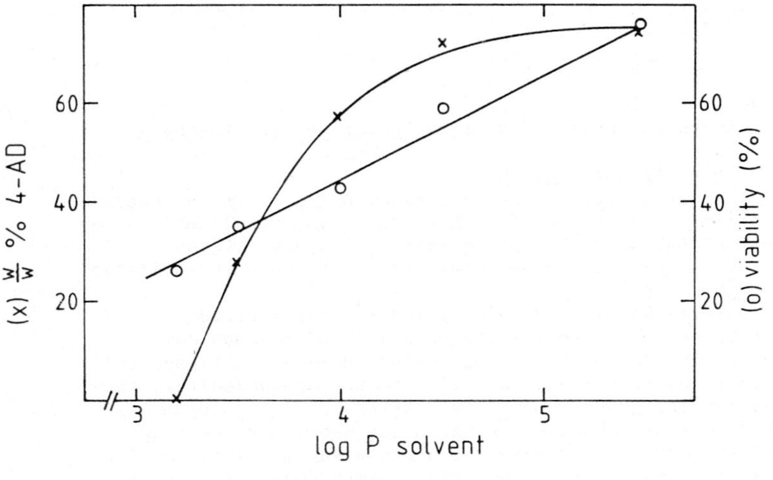

Figure 3: Viability and activity in alkane/culture two-liquid-phase systems (v/v=1:1). Viability was measured after 4 hours of exposure to alkane, and the amount of 4-AD produced was determined after 5 hours of incubation (overall concentration of substrate 5 g/L, incubation temperature 30 °C).

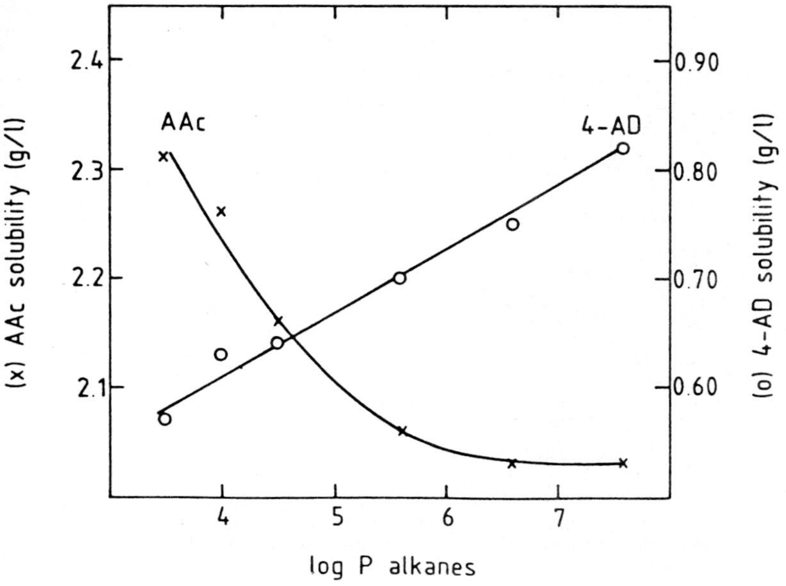

Figure 4: Solubilities of AAc and 4-AD in a series of alkanes.

Viability in organic solvent/culture two-liquid-phase systems

The effect of homologous series of hydrophobic organic solvents on viability and activity of cells growing in the late exponential phase (OD_{650}=6.3) was investigated in more detail. Figure 3 shows that an increase in solvent hydrophobicity (cyclohexane, logP=3.2 to decane, logP=5.6) results in an almost linear increase in viability. The AAc conversion rate increases with increasing solvent hydrophobicity up to logP=4.5 (octane) where it levels off. This stabilization is not a result of viability limitation but is probably due to a decreased substrate solubility in the more hydrophobic solvents as illustrated in figure 4.

From the results presented so far it can be concluded that viability and solubility determine the steroid conversion rate to a major extend. Introducing small substituents on or into the alkyl backbone of the organic solvent may increase viability and/or substrate solubility.

The influence of different substituents on viability was investigated using substituted alkanes (viz. 1-aminoalkanes, alcanones, alkylethers, 1-alcanols, alcanals, 1-chloroalkanes, 1-bromoalkanes, 1-alkenes, dialkylphthalates and alkylbenzenes). The alkylamines spontaniously mixed with the culture thereby making viability measurements impossible. For the other solvents a sigmoidal curve for viability versus organic solvent hydrophobicity was observed (figure 5).

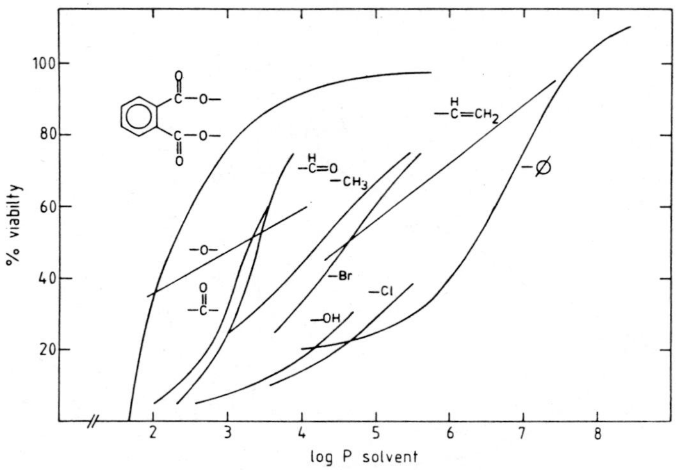

Figure 5: Viability of F. dehydrogenans in organic solvent/culture two-liquid-phase systems. For the conditions see the legend to figure 3.

Compared to the corresponding alkanes hydrophilic substituents (C=O, -O-, -CH=O) made the viability increase and maximum viability was reached within a smaller $\log P_{solvent}$ region (with an exception for the alcanols whose behaviour was more like that of the hydrophobic substituents). Hydrophobic substituents (-Cl, -Br, $-CH=CH_2$, ϕ-[COO-]$_2$, -ϕ) were found to cause the opposite behaviour as illustrated by the decrease in slope of the curves.

Since the hydrophobic fragmental constants can be used to assign hydrophobicity values to substituents it is possible to combine $\log P_{substituent}$, $\log P_{solvent}$ and viability into one figure. On doing so, figure 6 was obtained which shows that with increasing substituent hydrophobicity (from alcanones to benzenes) the solvent hydrophobicity has to increase disproportionally more to achieve high viability. Hence, for carrying out microbial conversions in two-liquid-phase systems, solvents with substituents of low hydrophobicity are preferred since within a broad region of solvent hydrophobicity no significant loss in viability was observed.

The sigmoidal viability versus organic solvent hydrophobicity curve can theoretically be explained by assuming that transport of solvent to the cell interiour occurs through aqueous pores in the cell membrane. This kind of transport was shown to give a sigmoidal curve for the (non-ionizable) compound permeability versus the compound hydrophobicity (ref.7).

The lethal effect of organic solvents with a low hydrophobicity can be due to distortion of the cell metabolism or the cell membrane and its function. In general little is known about the effect of hydrophobic solvents on membrane structure and function but the results of Ingram (ref. 8) are in favour of the latter possibility. He has shown that ethanol caused a dose dependant increase in the leakage of magnesium and nucleotides of Zymomonas mobilis cells.

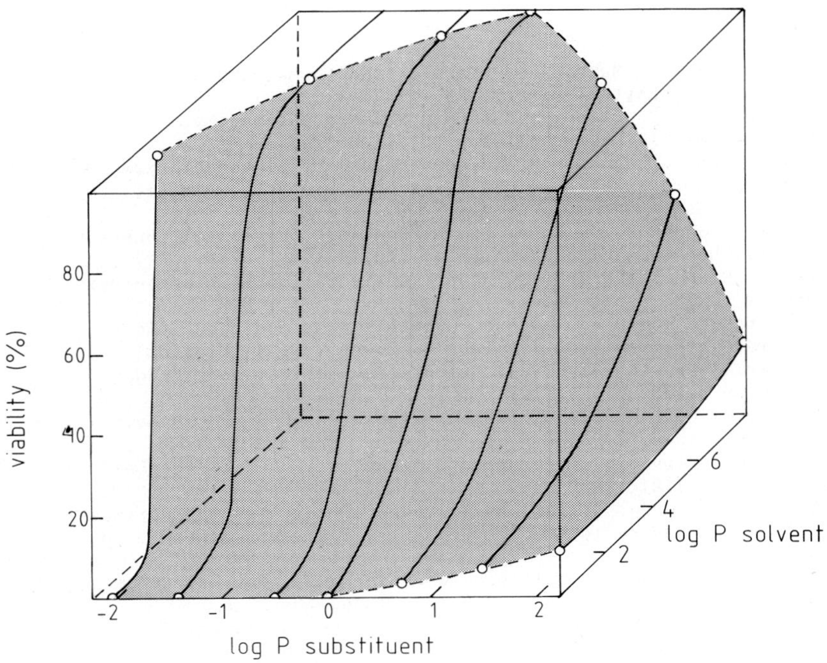

Figure 6: The viability versus solvent hydrophobicity curve as a function of substituent hydrophobicity.

We would like to point out that a high viability does not automatically imply a high activity. Although a hydrophobic solvent may limit the distorting effect of organic solvent on enzyme production, activity and stability (ref.9), in some cases it will be necessary to use less hydrophobic solvents e.g. when the product(s) or substrates(s) have a low hydrophobicity (logP<1.5). Here, cell viability may be increased by shielding of the biocatalyst by use of a membrane (ref.10) or immobilization in a hydrophilic matrix (ref. 11) to avoid direct contact of the cells with the bulk organic phase.

In summary the following conclusions can be reached:
1) Organic solvent/culture two-liquid-phase systems are more suitable for microbial conversions of hydrophobic substances than aqueous media with or without solubilizing agents.
2) The growth stage of the micro-organism at the moment of solvent addition must be chosen properly since changes in cell metabolism, morphology and/or cell membrane structure and function occuring during growth may effect viability.
3) Bacterial viability and steroid solubility determine the steroid conversion rate to a major extend.
4)High viabilities can be reached over a broad region of solvent hydrophobicity when the solvent used is of the alkyl-X type and X is a hydrophilic substituent.
5)Basic understanding of organic solvent effects on whole cells is required.

REFERENCES

1 S.B. Mahato and A. Mukherjee, Steroid transformations by microorganisms, Phytochemistry, 23 (1984) 2131

2 S.B. Mahato and S. Banerjee, Steroid transformations by microorganisms-II, Phytochemistry, 24 (1985) 1403

3 S. Boeren and C. Laane, Steroid conversions by Flavobacterium dehydrogenans in two-liquid-phase systems, Biotechn. Bioeng., 29 (1987) in press

4 R.F. Rekker and H.M. de Kort, The hydrophobic fragmental constant, Eur.J.Med.Chem.Chim. Therapeutica 14 (1979) 479

5 C. Laane, S. Boeren and R. Hilhorst, Optimization of biocatalysis in organic media, this volume

6 R.H. Pritchard and D.W. Tempest, Growth: Cells and populations, in: J. Madelstam, K. McQuillen and I. Dawes (eds.), Biochemistry of Bacterial Growth, Blackwell Scientific Publications, London, 1982, pp.99 - 123.

7 F.H.N. de Haan, The role of drug transport in QSAR, PhD thesis, University Leiden, (1985)

8 L.O. Ingram, Microbial tolerance to alcohols: Role of the cell membrane, TIB 4(2) (1986) 40

9 C. Laane, S. Boeren, K. Vos and C. Veeger, Rules for the optimization of biocatalysis in organic solvents, Biotechn. Bioeng., in press

10 T. Cho and M.L. Shuler, Multimembrane bioreactor for extractive fermentation, Biotechn.Progress 2 (1) (1986) 53

11 J.M.C. Duarte and M.D. Lilly, The use of free and immobilized cells in the presence of organic solvents: the oxidation of cholesterol by Nocardia rhodochrous, in H.H. Weetall and G.P. Royer (eds.), Enzyme Engineering, Plenum Press, New York, Vol. 5, 1980, pp. 363 - 367

C. Laane, J. Tramper and M.D. Lilly (Editors), *Biocatalysis in Organic Media*,
Proceedings of an International Symposium held at Wageningen,
The Netherlands, 7–10 December 1986.

THE LIQUID-IMPELLED LOOP REACTOR*: A NEW TYPE OF DENSITY-DIFFERENCE-MIXED BIOREACTOR

J. Tramper, I. Wolters and P. Verlaan,
Agricultural University, Department of Food Science,
Food and Bioengineering Group, De Dreyen 12,
6703 BC Wageningen, The Netherlands.

*Dutch Patent Application OA.86.03105 Ned.

SUMMARY

Biotechnology is currently a rapidly expanding field of interdisciplinary research. This appears amongst others from the development of a number of new type of bioreactors. The traditional stirred tank is not any longer a priori the standard bioreactor. Especially the airlift loop reactor, as result of several attractive features, is a good example of the coming bioreactor. The characteristic property of this reactor is the density-difference-driven mixing, making mechanical stirrers redundant. Another current breakthrough in biotechnology is catalysis in media, which are significantly less polar than aqueous solutions. Especially the use of water-immiscible organic solvents has recently been given increasing attention as a possible approach to diminishing the fundamental restrictions often inherent to biotechnological processes. Some of these restrictions are the low solubility of (gaseous) substrates and/or products in water, substrate and/or product inhibition or hydrolysis, and laborious product recovery. In this paper a novel type of bioreactor is described. The principle of mixing is based on the density difference between an organic solvent and water. The advantages of the airlift loop reactor and of the introduction of an organic phase in biocatalysis are thus combined.

INTRODUCTION

The liquid-impelled loop reactor is a development which logically results from two research projects which are executed in our laboratory. This is on the one hand a fundamental study of the airlift loop reactor (ref. 1-3) and on the other hand a study concerning the design of an organic-liquid-phase/immobilized-cell reactor (ref. 4-8).

The airlift loop reactor

An airlift loop reactor is a reactor which essentially consists of two vertical parts which have an open connection at the bottom and the top (Figure 1).

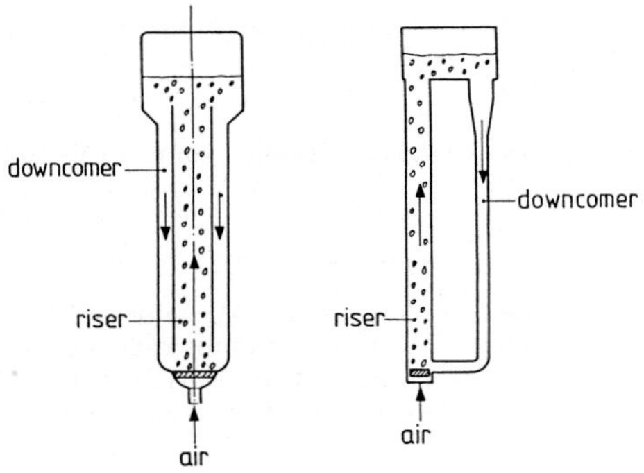

Figure 1.
Airlift reactors.
Left : with internal loop.
Right: with external loop.

Continuous injection of air at the bottom of one of the two vertical parts, the so-called riser, creates a density difference with the other vertical part, the so-called downcomer. Due to this density difference, a liquid flow from the bottom to the top exists in the riser and from the top to the bottom in the downcomer, thus resulting in the circulation of the continuous liquid phase. The study of this type of reactors was at the time initiated in view of its application as bioreactor for conversions with immobilized biocatalysts and for growth of animal (ref. 9) and plant cells (ref. 1). The reason for this was the fragility of many of these biological systems and the absence of mechanically-moved parts in the airlift loop reactor. Other advantages of this loop reactor in comparison to more conventional bioreactors like the standard fermentor are the relatively simple construction and with that a low fault sensitivity, an adequate and adjustable gas phase disengagement at the top, a large specific interfacial contact area at a low energy-input, a controllable heat-exchange, a unique combination of controlled flow and good mixing properties, and an easy access for measurement and control devices, in particular if it concerns an airlift loop reactor with an external loop (Figure 1). Furthermore, it should be noted that the airlift loop reactor can be easily operated under sterile conditions as result of its simple construction.

Biocatalysis in organic media

One of the main, current breakthroughs in biotechnology is biocatalysis in more or less apolar and hydrophobic, organic media. The use of biocatalysts in these apparently anomalous environments is, however, not as amazing as it might seem at first thought. Many enzymes, for instance, are associated in their natural state with more or less non-polar, cellular elements, especially membranes. Therefore, a strictly aqueous, highly polar microenvironment could often even be more hostile to the biocatalyst than a less polar medium and could thus result in a reduced activity, specificity and/or stability. From this point of view, the at times remarkable reports on bioconversions in non-aqueous solvent systems, e.g. by Klibanov's group (ref. 10), who carried out succesfully enzymic reactions in nearly anhydrous solvents at sometimes elevated temperatures, can be more easily understood. Substitution of part or all of the free water by a water-immiscible organic solvent can in principle largely solve many of the problems restricting the number and scope of biotechnological applications. Indeed there could be many reasons to justify the use of organic media instead of aqueous solutions (see TABLE I).

TABLE I. Reasons for the use of organic media in biocatalysis

* Possibility of high concentrations of poorly water-soluble substrates/products.
* Shift of reaction equilibria as a result of the altered partitioning of substrates and products between the phases of interest.
* Shift of reaction equilibria (with water as one of the products) as a result of a substantially reduced water activity (ref. 11).
* Reduction of substrate and/or product inhibition.
* Prevention of hydrolysis of substrates/products.
* Facilitated recovery of products and biocatalyst, even when the latter is not immobilized.
* Less risk of microbial contamination.
* Stabilization of the biocatalyst.

Naturally, it is not likely that all the mentioned reasons are relevant for one particular bioconversion and of course there can be disadvantages involved too. These are primarily biocatalyst denaturation and/or inhibition by the organic solvent and the increasing complexity of the reaction

system.
The degree of miscibility of an organic solvent with water will be highly decisive for the attractiveniss of the solvent system. Water/organic solvent two-phase systems (with the biocatalyst in the aqueous phase) are generally believed to be more promising than water/water-miscible organic solvent systems. The latter do not show the advantages of reduced substrate/product inhibition and facilitated product/biocatalyst recovery. Also, the chance of inhibition or denaturation by the solvent is markedly smaller in biphasic systems as a result of the low concentration of the water-immiscible solvent in the aqueous biocatalyst phase. A potential danger for the biocatalyst in a solvent/water biphasic system is denaturation at the liquid/liquid interface. However, if required, this problem can be overcome by immobilizing the biocatalyst in a solid support. Immobilization has other, well-proven benefits as well, like facilitated product/biocatalyst separation, continuous processing and, sometimes, increased stability.

THE LIQUID-IMPELLED LOOP REACTOR

The principle of mixing in the new developed liquid-impelled loop reactor is the same as applied in the airlift loop reactor. In the liquid-impelled loop reactor the pressure difference between the two vertical parts is created by injection of a water-immiscible solvent with a density different than that of water. The attractive features of the airlift loop reactor and the advantages of biocatalysis in organic media can thus be combined in such a loop reactor. Since loop reactors can be equiped with an internal or external loop and since the water-immiscible solvent can have a density smaller or larger than that of water, 10 ways of operating a liquid-impelled loop reactor are possible:

1. A reactor with an internal loop, water as the continuous phase, and an organic solvent as dispersed phase with a density smaller than that of water (Figure 2a).
2. A reactor with an external loop, water as the continuous phase, and an organic solvent as dispersed phase with a density smaller than that of water (Figure 2b).

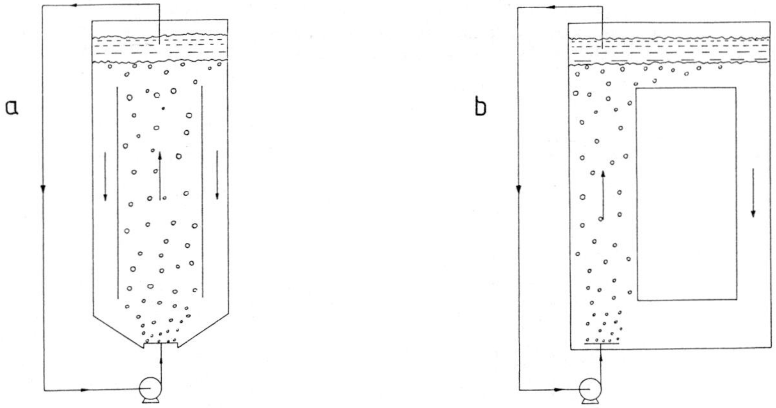

Figure 2. Liquid-impelled loop reactors with the dispersed phase having the lowest density.

3. A reactor with an internal loop, water as the dispersed phase, and an organic solvent as the continuous phase with a density larger than that

of water (Figure 2a applies here as well).

4. A reactor with an external loop, water as the dispersed phase, and an organic solvent as the continuous phase with a density larger than that of water (Figure 2b applies here as well).
5. A reactor with an internal loop, water as the continuous phase, and an organic solvent as the dispersed phase with a density larger than that of water (Figure 3a).
6. A reactor with an external loop, water as the continuous phase, and an organic solvent as the dispersed phase with a density larger than that of water (Figure 3b).

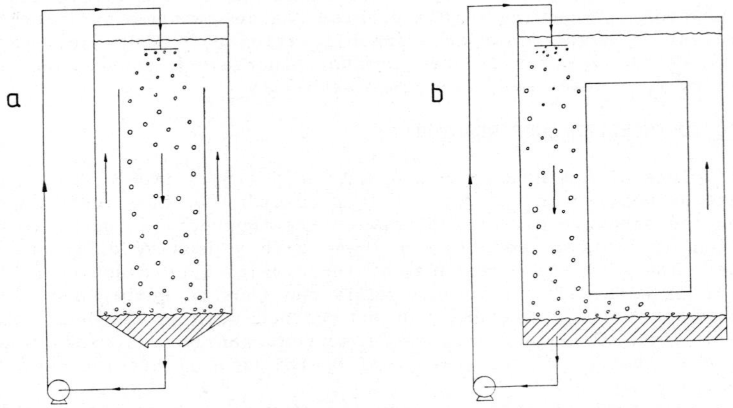

Figure 3. Liquid-impelled loop reactors with the dispersed phase having the highest density.

7. A reactor with an internal loop, water as the dispersed phase, and an organic solvent as the continuous phase with a density smaller than that of water (Figure 3a applies here as well).
8. A reactor with an external loop, water as the dispersed phase, and an organic solvent as the continuous phase with a density smaller than that of water (Figure 3b applies here as well).

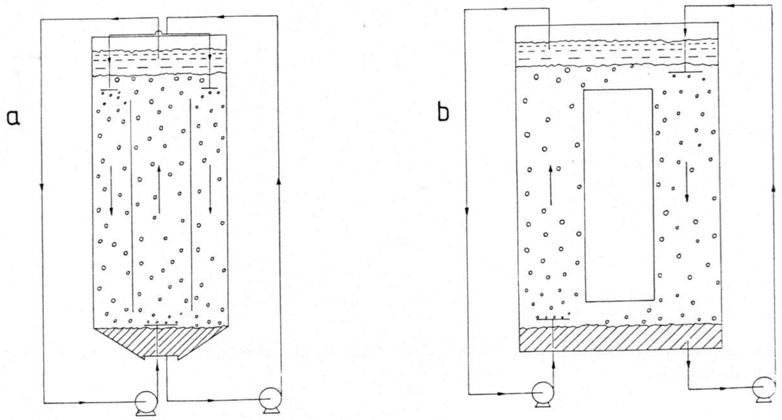

Figure 4. Dual-liquid-impelled loop reactors.

The last two, more special configurations are presented in Figure 4. These what we have called dual-liquid-impelled-loop reactors make use of two organic solvents which are immiscible with water, one with a density smaller and one with a density larger than that of water.

DISCUSSION.

In the liquid-impelled loop reactor the attractive feactures of the airlift loop reactor can be combined with those of biocatalysis in organic media. So far we have tested the various possibilities in available loop reactors in the volume range of 1-2 dm^3. None of these reactors has been optimized for this purpose, but especially the reactors with an external loop performed well, because phase separation was easier controlled. Due to dispersion problems the systems with water as dispersed phase worked less well than those with water as continuous phase. Improvement of the injector, however, very likely can solve this problem. Petroleum ether with a density of about 0.65 $kg.dm^{-3}$ has mostly been used as water-immiscible organic solvent but, as Table II shows, a large variety of solvents is in principle attractive, also with respect to retention of biocatalytic activity. The systems have also been tested in the presence of alginate beads with a diameter of about 3 mm. In all cases the liquid flow could easily suspend and circulate the beads with ciculation times of the order of 3 seconds. The liquid-impelled loop reactors are thus also suitable for bioconversions with immobilized biocatalysts.

Tabel II. Properties of organic solvents and retention of activity of immobilized Mycobacterium cells in two-liquid-phase systems (ref. 5).

Solvent	Solvent number	Source	Purity (%)	Molecular weight (g/mol)	Specific gravity (20°C) (g/cm^3)	Solubility in water (20°C) (% w)	Solubility parameter $[(cal/cm^3)^{0.5}]$	Activity retention (%)
Saturated aliphatic hydrocarbons								
n-Hexane	1	Merck	97	86.18	0.660	0.00095	7.3	39
n-Hexadecane	2	Baker	>99	226.45	0.7733	—	8.0	85
Aromatic hydrocarbons								
Toluene	3	Merck	99.5	92.15	0.8669	0.052	8.9	~0
Ethylbenzene	4	Merck	99	106.17	0.867	0.0152	8.8	29
Unsaturated hydrocarbons								
Styrene	5	Merck	99.5	104.15	0.906	0.031	9.3	45
Aliphatic alcohols								
1-Butanol	6	Merck	98	74.12	0.8098	9	11.4	~0
1-Hexanol	7	Merck	98	102.18	0.814	0.594	10.7	6
Aliphatic ethers								
Diisopropyl ether	8	Merck	98	102.18	0.723	1.197	7.1	56
Diisopentyl ether	9	Merck	97	158.29	0.777	0.020	7.2	77
Aromatic ethers								
Methyl phenyl ether	10	Merck	99	108.13	0.993	1.04	9.5	5
Diphenyl ether	11	Fluka	>98	170.21	1.0748	0.39	9.6	~95
Aliphatic aldehydes								
Hexanal	12	Merck	98	100.16	0.815	0.5	9.2	~0
Aliphatic ketones								
Methyl isobutyl ketone	13	Merck	99	100.16	0.798	1.693	9.4	~0
Methyl octyl ketone	14	Fluka	>95	156.27	0.825	—	7.8	53
Esters of saturated aliphatic monocarboxylic acids								
Butyl acetate	15	Merck	99	116.16	0.883	0.43	8.5	3
Ethyl trichloro acetate	16	Pfaltz B.	—	191.44	1.3836	—	8.9	16
Esters of unsaturated aliphatic monocarboxylic acids								
Butyl acrylate	17	Merck	99	128.17	0.898	—	8.4	~0
Esters of aromatic monocarboxylic acids								
Butyl benzoate	18	Merck	99	178.23	1.004	—	8.4	14
Esters of dicarboxylic acids								
Diethyl maleate	19	Merck	97	172.18	1.065	1.4	8.7	1
Dibutyl adipate	20	Fluka	>99	258.36	0.962	—	8.9	7
Dimethyl phthalate	21	Merck	99	194.19	1.188	0.43	10.7	19
Diethyl phthalate	22	Merck	99	222.24	1.117	—	10.0	74
Dibutyl phthalate	23	Merck	99	278.35	1.045	0.040	9.3	76
Dioctyl phthalate	24	L&I	99	390.54	0.985	—	7.9	108
Didecyl phthalate	25	Fluka	>95	446.68	0.966	—	7.2	101
Di-(2-methoxyethyl) phthalate	26	Fluka	>98	282.30	1.171	—	10.2	22
Diallyl phthalate	27	Fluka	~97	246.27	1.119	—	10.1	43
Esters of polybasic acids								
Tri-*n*-butyl phosphate	28	Fluka	99	266.32	0.976	0.039	8.6	12
Tri-2-tolyl phosphate	29	BDH	—	368.37	~1.16	0.000050	8.4	55
Aliphatic chlorinated hydrocarbons								
Chloroform	30	Merck	99	119.38	1.4832	0.815	9.3	~0
Tetrachloro methane	31	Merck	99	153.82	1.5940	0.077	8.6	~22
1,1,2,2-Tetrachloro ethane	32	Merck	98	167.85	1.594	0.287	9.7	10
Fluorinated hydrocarbons								
FC-40	33	3M	—	~650	~1.87	—	5.9	94
FC-70	34	3M	—	~820	~1.94	—	5.9	94

CONCLUSION

The liquid-impelled loop reactor is an attractive, new type of density-difference-mixed bioreactor for biocatalysis in organic media. In fact, promising applications as (continuous) bioreactor with free or immobilized biocatalysts, as (continuous) fermentor, but also as (continuous) extractor can be foreseen.

REFERENCES

1. P. Verlaan, A.C. Hulst, J. Tramper, K. van ´ t Riet and K.Ch.A.M. Luyben, Immobilization of plant cells and some aspects of the application in an airlift loop reactor, Proc. 3rd Europ. Congress on Biotechnol., vol. 1, München, GDR, 10-14 sept. 1984, pp. 151-157.
2. P. Verlaan, J. Tramper, K. van ´ t Riet and K.Ch.A.M. Luyben, Hydrodynamics and axial dispersion in an airlift-loop bioreactor with two and three-phase flow, Proc. Int. Conf. Fluid Dynamics, Cambridge, England, 15-17 april, 1986, pp. 93-107.
3. P. Verlaan, J. Tramper, K. van´ t Riet and K.Ch.A.M. Luyben, A hydrodynamic model for an airlift-loop bioreactor with external loop, Chem. Eng. J. 33 (1986) B43-B53.
4. L.E.S. Brink, J. Tramper, K. van ´ t Riet and K.Ch.A.M. Luyben. Automation of an experimental system for the microbial epoxidation of propene and 1-butane, Anal. Chim. Acta 163 (1984) 207-217.
5. L.E.S. Brink and J. Tramper, Optimazation of organic solvent in multiphase biocatalysis, Biotechnol. Bioeng. 27 (1985) 1258-1269.
6. L.E.S. Brink and J. Tramper, Optimazation and modelling of the microbial epoxidation of propene in an organic-liquid-phase/immobilized-cell systems, Modelling and Control of biotechnological Processes (A. Johnson, ed.) Pergamon Press, Oxford 1985, pp. 111-117.
7. L.E.S. Brink and J. Tramper, Facilitated mass transfer in a packed-bed immobilized-cell reactor by using an organic solvent as a substrate reservoir, J. Chem. Tech. Biotechnol. (1987) in press.
8. L.E.S. Brink, Design of an organic-liquid-phase/immobilized-cell reactor for the microbial epoxidation of propene, 1986, Ph.D.-thesis.
9. J. Tramper and J. Vlak, Some engineering and economic aspects of continuous cultivation of insect cells for the production of Baculoviruses, Ann. N.Y. Acad. Sci. 469 (1986) 279-288.
10. A.M. Klibanov, this book.
11. P.J. Halling, this book.

C. Laane, J. Tramper and M.D. Lilly (Editors), *Biocatalysis in Organic Media*,
Proceedings of an International Symposium held at Wageningen,
The Netherlands, 7–10 December 1986.

ALGAL VANADIUM(V)-BROMOPEROXIDASE, A HALOGENATING ENZYME RETAINING FULL ACTIVITY IN APOLAR SOLVENT SYSTEMS.

E. DE BOER, H. PLAT and R. WEVER

Laboratory of Biochemistry and Biotechnological Centre, University of Amsterdam, P.O. Box 20151, 1000 HD Amsterdam (The Netherlands)

SUMMARY

Vanadium(V)-containing bromoperoxidase from the marine brown seaweed Ascophyllum nodosum displays exceptional stability and retains full activity in strongly apolar media, such as 47% butanol/20% ethanol/33% water. Compared with haem-containing fungal chloroperoxidase, algal bromoperoxidase was more than an order of magnitude more resistant to its enzymic product, hypobromous acid. Since this enzyme - in contrast to fungal chloroperoxidase - does not oxidise organic substrates, and only catalyses the oxidation of bromide ions, this stability enhances considerably the applicability of this vanadium-enzyme in organic synthesis.

INTRODUCTION

Terrestrial as well as marine organisms are known to synthesize a wide variety of halogenated compounds. The iodinated mammalian thyroid hormones and antibiotics, such as chloramphenicol and chlortetracycline, are well-known examples (1). Haloperoxidases are involved in the biosynthesis of these halometabolites and the following general equation holds for the formation of carbon-halogen bonds catalysed by peroxidases:

$$AH + X^- + H_2O_2 + H^+ \xrightarrow{\text{haloperoxidase}} AX + 2H_2O$$

where: AH = nucleophilic reagent; X^- = chloride, bromide or iodide, and AX = halometabolite. Up to now a number of haloperoxidases have been purified from different sources, for instance chloroperoxidase from the mould Caldariomyces fumago (2), and several bromoperoxidases from marine algae (3). Haloperoxidases catalyse the halogenation of a broad range of substrates and the potential commercial applications for this enzymic process have recently been reviewed by Neidleman and Geigert (4). For commercial use of these enzymes, efficient utilisation of reactants and an extended operational stability are ultimate requirements. Unfortunately, these criteria are not met by fungal chloroperoxidase, the best studied haloperoxidase in organic synthesis. Therefore, we purified a bromoperoxidase from the marine brown alga Ascophyllum nodosum (5). This bromoperoxidase displays great stability and was shown to be resistant towards organic solvent systems, such as 60% acetone, ethanol or methanol. Moreover, this enzyme retained full operational activity under turnover conditions for three weeks at room temperature (6). In the present paper experiments on the activity of algal bromoperoxidase in homogeneous mixtures of even more strongly apolar solvents are described. Furthermore, the effect of exposure of bromoperoxidase to its reaction product, hypobromous acid, was studied. Differences in reaction pathway between fungal chloroperoxidase and algal bromoperoxidase are discussed in the light of potential commercial use of these halogenating enzymes.

METHODS

Bromoperoxidase from A. nodosum and chloroperoxidase from the mould C. fumago were purified as described before (2,5). Brominating activity was measured by the conversion of 2-chlorodimedone into 2-bromo-2-chlorodimedone ($\Delta\varepsilon$ = 19.9 $mM^{-1} \cdot cm^{-1}$ at 290 nm, pH 6.5) by bromoperoxidase and chlorinating activity by the conversion of 2-chlorodimedone into 2,2-dichlorodimedone ($\Delta\varepsilon$ = 12.2 $mM^{-1} \cdot cm^{-1}$ at 278 nm, pH 2.75) by chloroperoxidase, according to Refs. 5 and 7. Protein content was measured by the method of Lowry et al. (8) with bovine serum albumin as a standard. The concentration of hypobromous acid was measured as described by Kanofsky (9).

RESULTS

Bromoperoxidase purified from the brown alga A. nodosum was shown to be quite different from other haloperoxidases from both terrestrial and marine origin. Normally a haem molecule is the prosthetic group present at the active site of haloperoxidases. Algal bromoperoxidase, however, contains a vanadium(V) ion as a prosthetic group (10). Furthermore, bromoperoxidase displayed a marked thermal and chemical stability, and loss of brominating activity was not observed when the enzyme was incubated for several hours at 50 °C. In addition, the initial rate of bromination did not decrease when enzymic activity was measured in the presence of methanol, ethanol or propanol upto 30% (5).

The effect of even more strongly apolar media on the initial rate of bromination of 2-chlorodimedone by bromoperoxidase is shown in Table 1. It is clear that in homogeneous mixtures of butanol and ethanol, up to 47% and 20%, respectively, bromoperoxidase retained its original activity observed in pure aqueous media. It was not possible to increase the concentration of organic solvents further, since phase separation occurred.

TABLE 1
Brominating activity of algal bromoperoxidase in homogeneous mixtures of butanol/ethanol/water. Enzymic activity was measured as described in Methods. The reaction mixture contained 25 mM sodium acetate (pH 5.5), 50 mM potassium bromide, 2 mM hydrogen peroxide, 50 μM 2-chlorodimedone (MCD), 0.4 μg/ml of bromoperoxidase and the amounts of butanol, ethanol and water indicated.

butanol (%)[a]	ethanol (%)	water (%)	rate (μM MCD.min^{-1})
-	-	100	70
7	60	33	73
14	53	33	69
20	47	33	68
27	40	33	70
34	33	33	71
40	27	33	69
47	20	33	70

[a]Percentages expressed as vol/vol.

A major obstacle to extended operational lifetime of haloperoxidases is inactivation by enzymically produced hypohalous acid. For instance, fungal chloroperoxidase was reported to show hardly any resistance towards low concentrations of hypochlorous acid (0-50 μM, 2 min of exposure; Ref. 4). Also when exposed to small amounts of hypobromous acid (30-40 μM), chloroperoxidase was inactivated within 2 min of incubation at room temperature (Fig. 1). In contrast to chloroperoxidase algal bromoperoxidase was more than an order of magnitude more resistant towards hypobromous acid.

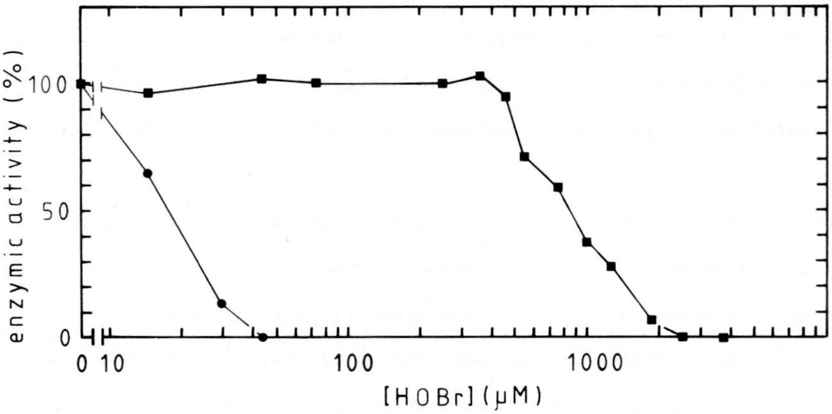

Fig. 1. Resistance of fungal chloroperoxidase and algal bromoperoxidase to hypobromous acid. Incubation of chloroperoxidase (5 μg/ml) and bromoperoxidase (0.3 mg/ml) was in 100 mM potassium phosphate (pH 7.0), 100 mM potassium sulphate and the indicated amount of hypobromous acid for 2 min at room temperature. Enzymic activity was measured as described in METHODS. Halogenating activity before incubation with HOBr was set to 100% for both enzymes. ●——●, chloroperoxidase; ■——■, bromoperoxidase.

DISCUSSION

Haloperoxidases possess a broad range of substrates and a multitude of halogenation reactions in vitro have been described in the literature (4). A great number of these reactions were done with fungal chloroperoxidase. Also bromoperoxidase from A. nodosum was shown to catalyse the bromination of several substances, e.g. 2-chlorodimedone (5), phenol red (6) and several barbituric acid derivatives (11) were found to be suitable substrates. Unlike other haloperoxidases (including fungal chloroperoxidase), which comprise the haem moiety protoporphyrin IX as a prosthetic group, algal bromoperoxidase contains a vanadium(V) ion at the active site (10). Therefore, it is conceivable that the mechanism of bromination by this vanadium(V)-containing enzyme differs from that of the haemoenzymes (12). In this regard it should be noted that fungal chloroperoxidase displays multiple pathways in the halogenation process. Competing reactions reported are e.g. decomposition of hydrogen peroxide and peroxidation/oxidation of the organic substrate (both favoured at low halide concentrations, cf. Ref. 4). Furthermore, chloroperoxidase is easily inactivated by an excess of hydrogen peroxide and is attacked by its own reaction products, hypochlorous or hypobromous acid. Since bromoperoxidase only catalyses the bromination of organic substrates (decomposition of hydrogen peroxide and peroxidation of the organic substrate were not observed), the thermal and chemical stability in aqueous as well as in apolar media enhances potential commercial applications of this vanadium(V)-containing halogenating enzyme.

ACKNOWLEDGEMENTS

We gratefully acknowledge the support by DSM N.V., Geleen, The Netherlands. This work is part of the research programme of the Netherlands Foundation for Chemical Research (S.O.N.) and was made possible by financial support from the Netherlands Technology Foundation (S.T.W.).

REFERENCES

1. J.F. Siuda and J.F. DeBernardis, Naturally occuring halogenated organic compounds, Loydia 36(2) (1973) 107-143.
2. D.R. Morris and L.P. Hager, Chloroperoxidase. I. Isolation and properties of the crystalline glycoprotein, J. Biol. Chem. 241(8) (1966) 1763-1768.
3. D.G. Baden and M.D. Corbett, Bromoperoxidases from Penicillus capitatus, Penicillus lamourouxii and Rhipocephalus phoenix, Biochem. J. 187(1) (1980) 205-211.
4. S.L. Neidleman an J. Geigert, Biohalogenation: principles, basic roles and applications, Ellis Horwood Limited, Chicester, England, 1986.
5. R. Wever, H. Plat and E. de Boer, Isolation and some properties of the bromoperoxidase from the seaweed Ascophyllum nodosum, Biochim. Biophys. Acta 830(1) (1985) 181-186.
6. E. de Boer, H. Plat, M.G.M. Tromp, M.C.R. Franssen, H.C. van der Plas, E.M. Meijer, H.E. Schoemaker and R. Wever, Vanadium(V)-containing bromoperoxidase, an example of an oxido-reductase with high operational stability in aqueous and organic media, (Biotechnol. Bioeng.), in press.
7. L.P. Hager, D.R. Morris, F.S. Brown and H. Eberwein, Chloroperoxidase. II. Utilisation of halogen anions. J. Biol. Chem. 241(8) (1966) 1769-1777.
8. O.H. Lowry, H.J. Rosebrough, A.L. Farr and R.J. Randall, Protein measurement with the folin phenol reagent, J. Biol. Chem. 193(1) (1951) 265-275.

9. J.R. Kanofsky, Singlet oxygen production by chloroperoxidase-hydrogen peroxide-halide systems, J. Biol. Chem. 259(9) (1984) 5596-5600.
10. E. de Boer, Y. van Kooyk, M.G.M. Tromp, H. Plat and R. Wever, Bromoperoxidase from Ascophyllum nodosum: a novel class of enzymes containing vanadium as a prosthetic group?, Biochim. Biophys. Acta 869(1) (1986) 48-53.
11. E. de Boer, M.C.R. Franssen, H.C. van der Plas, H. Plat and R. Wever, Bromoperoxidase from marine brown algae: a unique halogenating biocatalyst, Abstracts of the symposium on Biocatalysts in Organic Syntheses, Noordwijkerhout, The Netherlands, April 14-17, 1985, p.19.
12. E. de Boer, M.G.M. Tromp, H. Plat, G.E. Krenn and R. Wever, Vanadium(V) as an essential element for haloperoxidase activity in marine brown algae: purification and characterization of a vanadium(V)-containing bromoperoxidase from Laminaria saccharina, Biochim. Biophys. Acta 872(1) (1986) 104-115.

C. Laane, J. Tramper and M.D. Lilly (Editors), *Biocatalysis in Organic Media*,
Proceedings of an International Symposium held at Wageningen,
The Netherlands, 7–10 December 1986.

PRODUCTION OF L-TRYPTOPHAN IN A TWO-LIQUID-PHASE SYSTEM

M.H. Ribeiro[1], J.M.S. Cabral[2] and M.M.R. Fonseca[2]
[1]Faculdade de Farmácia, 1699 Lisboa Codex, Portugal
[2]Lab. Eng. Bioquímica, IST, 1096 Lisboa Codex, Portugal

SUMMARY

L-tryptophan was produced batchwise from indole and L-serine by free and k-carrageenan immobilized cells of an original ATCC E. coli strain with tryptophan-synthase activity in single liquid-phase and aqueous/organic phase systems. The organic solvent plays the role of a reservoir of indole. Amongst the solvents tested cyclohexane and n-hexane were the solvents which conferred the best performance.

The highest volumetric productivity (0,05 g $l^{-1} h^{-1}$) was obtained using immobilized cells in a two-liquid-phase system.

It was found that the volumetric productivity can be enhanced by increasing both the concentration of indole in the organic phase and the biomass hold-up per gel particle.

INTRODUCTION

Many important foodstuffs (e.g. cereal grains) lack L-tryptophan, an amino acid essential for mammals.

The annual production of L- and DL-tryptophan was estimated to be of the order of 1000 ton in 1983 (ref. 1). Most of the production is used by the food and pharmaceutical industries (ref. 2). The high selling price of tryptophan precludes its use by the fodder industry. It has been estimated that at a sufficiently low price the world demand for animal feeding could reach 11000 tons per year (ref. 1).

The production of L-tryptophan by microbial processes ("de novo" synthesis without precursor and bioconversions from precursors) was fully reviewed by Nyeste et al. (ref. 2). These authors refer a market research survey dated from 1977 where a bioconversion from indole is declared to be at least twice more economic than the synthetic method.

The bioconversion from indole poses several problems e.g. low solubility of indole in water and tryptophan synthase inhibition by indole. Attempts to circumvent these problems include (i) the use of organic solvents, non-ionic detergents and adsorbents as reservoirs of indole (ref. 3) and (ii) fed-batch operation with respect to indole (ref. 4).

In this work the presence of an organic phase during the bioconversion was explored. No microbial genetic transformation was attempted in order to obtain mutants e.g. with high L-tryptophan-synthase activity and tryptophanase negative. Instead, an E. coli strain ATCC 27553 was used. Hence, the productivities attained have only a relative significance, the comparison between the performances of single and two-phase systems being the focus of the work.

Whole cells collected at the end of the exponential phase were used (either free or gel-entrapped) for the production of L-tryptophan from indole and L-serine in the presence of the co-factor pyridoxal-5'-phosphate.

A range of solvents was assayed as to their potential value to become the second phase in the system. From these, cyclohexane and n-hexane were selected as the best performers.

The use of whole-cells in bioconversions processes is by itself a reason that justifies their immobilization. Furthermore, in aqueous/organic systems the immobilization matrix might play a protective role against solvent toxicity. Nevertheless, the performance of freely suspended cells, both in the presence and absence of solvent, was evaluated and compared with that of immobilized cells.

The ability of non-growing cells to maintain adequate production levels for extended periods of time is a pre-requisite for the economics of secondary metabolite production and bioconversion processes (cuts growth and immobilization costs).

In this context, the stability of the intracellular tryptophan-synthase was followed through the measurement of initial production rates for over two weeks. The effect of the presence of the organic phase upon the average productivity was studied.

MATERIALS AND METHODS

Microorganism

Escherichia coli B 1t-7A ATCC 27553 maintained on nutrient agar slopes at + 4 C.

Materials

K-carrageenan (Gelcarin CIC) was a gift from FMC, USA. Nutrient agar was from Difco. All other reagents were analytical grade from Merck.

Medium compositions and environmental conditions

Growth medium. Cells of E. coli were cultivated at 37 C and 200 rpm in a growth medium containing in g/l: glucose, 5 or 12; K_2HPO_4, 7; KH_2PO_4, 3; $(NH_4)_2SO_4$, 1.5; $MgSO_4.7H_2O$, 0.1; $CaCl_2.2H_2O$, 0.01; $FeSO_4.7H_2O$, 0.005; L-tryptophan, 0.01. This medium was inoculated with 1% liquid inoculum.

Production medium. (i) In single liquid-phase systems the productivity medium contained in g/l: L-serine, 2; pyridoxal-5'-phosphate, 0.01; indole, 2. (ii) In two liquid-phase systems three volumetric aqueous/organic ratios were tested - 2:1, 3:1 and 4:1. The aqueous phase contained in g/l: L-serine, 2(type A system) or 6 (type B system); pyridoxal-5'-phosphate, 0.01. In each system the mass of indole added to the organic phase was equal to the mass of L-serine added to the aqueous phase.

Production experiments were carried out in shake flasks incubated at 37 C and 200 rpm.

Distribution coefficient for indole and L-tryptophan.

Measurements of the distribution coefficient between a solvent and the aqueous phase (phosphate buffer, pH=8.0) were carried out at 37 C after shaking for 24 h at 200 rpm.

Analytical methods

Free cell concentration was estimated from optical density at 600 nm, using appropriate dilutions.

Glucose was assayed using the Somogyi-Nelson method (ref. 5).

Indole and L-tryptophan were assayed by a modified Erlich method (ref. 6) and by the method of Spies and Chambers (ref. 7) respectively.

Cell immobilization

Cell entrapment in k-carrageenan was carried out as described elsewhere (ref. 8).

Batch operational stability test

A one per day initial rate experiment was performed during a period of 17 days in order to generate stability data. The initial rate was taken as the productivity obtained at the end of the first hour. After 24 h the biomass (free or immobilized) was separated from the medium and washed (with a 9 g/l NaCl solution) before starting the next batch with fresh medium.

RESULTS

The maximum production of L-tryptophan is attained using cells collected 8 h after inoculation with an overnight grown inoculum (Fig. 1). Thus, the cells used in the production experiments were separated from the growth medium always before the stationary phase was reached.

In the absence of an indole reservoir, the optimum indole concentration in the production stage was found to be 2 g/l (Fig. 2), both for free and immobilized cells.

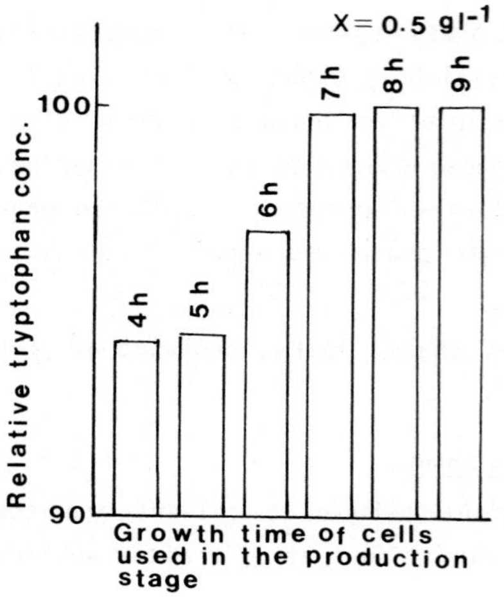

Fig. 1. Effect of growth time on the relative production of L-tryptophan

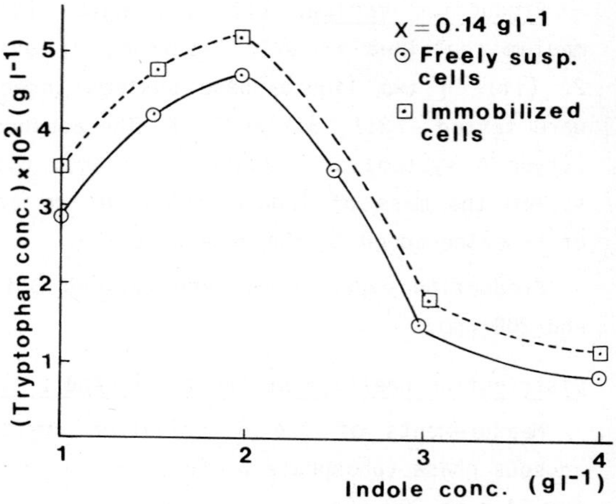

Fig. 2. Effect of indole concentration on the L-tryptophan produced after 20 h in single-phase systems.

Fig. 3 compares the relative productivities obtained with freely suspended cells in type A systems (see Materials and Methods) after 24 h of incubation, in the absence and in the presence of a variety of solvents. The results obtained with an aqueous/organic ratio of 3:1 (not shown for simplicity) were intermediate between those obtained with 2:1 and 4:1 ratios. Cyclohexane and n-hexane were the solvents selected.

The amounts of L-tryptophan present in the organic phase were not taken into account considering both (i) the low distribution coefficient (Table 1) and (ii) the relatively high aqueous : organic ratios used.

The results in Table 2 show that immobilized cells perform better than freely suspended cells as far as yield and volumetric productivities are concerned. The volumetric productivity can be further enhanced (Table 3) through the use of higher indole concentrations in the solvent phase together with increased biomass hold-ups per gel particle. The latter gave rise to a marked decrease in the specific productivity.

From the batch operational stability tests using immobilized cells (Fig. 4) no decrease of enzyme activity could be detected in the case of two liquid-phase systems. However, when immobilized biomass was used in single-phase systems, the enzyme activity was constant during the initial 5 days and then decreased at a rate of 0.087 day^{-1}. With freely suspended cells a complete deactivation occurred on the 12th day. These results reflect the stability conferred to the system by the use of an organic phase together with immobilized biomass.

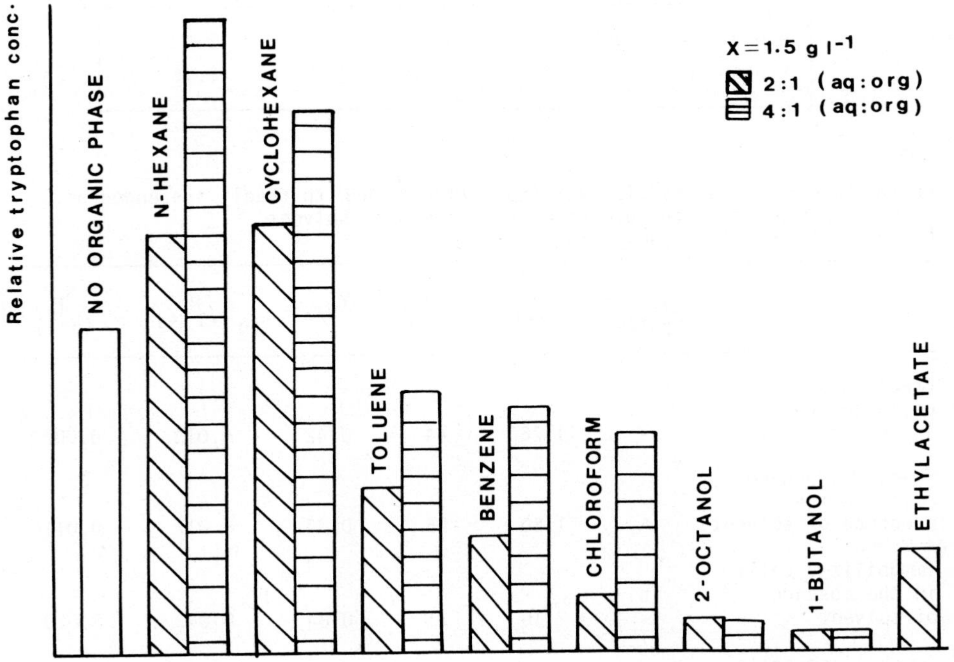

Fig. 3. Effect of the solvent on the production of L-tryptophan after 24 h by freely suspended cells in a single-phase (100%) and in two liquid-phase (type A) systems.

DISCUSSION

From an engineering point of view the use of adequate organic solvents seems an encouraging approach to improve the productivity of processes of bioconversion of indole to L-tryptophan. The results on solvent biocompatibility associated with this bioconversion agree with the scale for the optimization of biocatalysis in organic solvents proposed by Laane et al. (ref. 9) except for the case of ethyl-acetate. Experimenting other solvents selected according to Laane's scale is certainly a worthy exercise.

It is expected that the application of these conclusions coupled with the use of appropriate mutants might result in increased tryptophan yields and volumetric productivities.

TABLE 1

Distribution coefficients between the solvent and the aqueous phase

Compound \ Solvent	Cyclohexane	n-Hexane
Indole	3.5	2.5
Tryptophan	0.15	0.16

TABLE 2

Comparison of productivities attained after 48 h with freely suspended and immobilized cells in the presence and absence of solvent [4 (aq.) : 1 (cyclohexane)].

	Type of system	X $(g\,l^{-1})$	X_m $(g\,l^{-1})$	$Y_{P/S}$	r_P $(g\,l^{-1}\,h^{-1})$	R_P (h^{-1})
Freely suspended cells in the absence of solvent	-	1.76	44	0.42	0.013	0.0073
Freely suspended cells in the presence of solvent	A	1.50	66	0.35	0.017	0.0114
Immobilized cells in the absence of solvent	-	2.16	71	0.84	0.043	0.0200
Immobilized cells in the presence of solvent	A	2.16	50	0.94	0.034	0.0160

TABLE 3

Comparison of productivities attained after 24 h with immobilized cells in the presence and absence of solvent [4 (aq) : 1 (cyclohexane)].

	Type of system	X $(g\,l^{-1})$	X_m $(g\,l^{-1})$	$Y_{P/S}$	r_P $(g\,l^{-1}\,h^{-1})$	R_P (h^{-1})
Absence of solvent	-	2.16	8.64	0.45	0.032	0.015
	-	11.05	22.10	0.35	0.031	0.003
Presence of solvent	A	2.16	8.64	0.47	0.031	0.014
	B	11.05	22.10	0.18	0.050	0.005

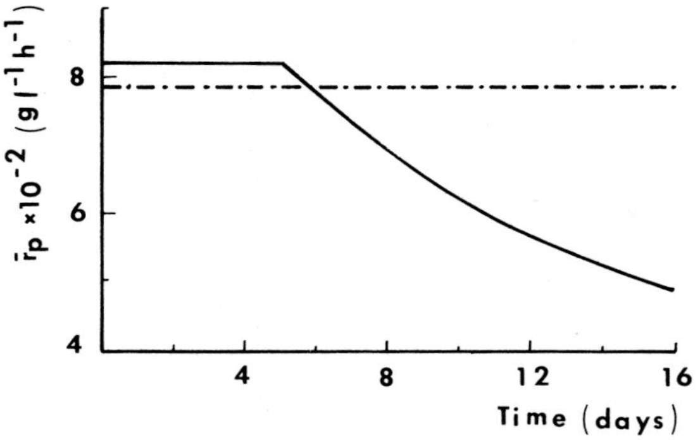

Fig. 4 Average productivity ($\bar{r}_p$) vs. time of operation obtained using 0.029 g (dry weight) of immobilized cells in a single-phase (——) and a type A [4(aq)::1(cyclohexane)] system (-·-·-).

NOMENCLATURE

r_p volumetric productivity (mass of tryptophan produced per hour per litre of aqueous phase)

R_p Specific productivity (mass of tryptophan produced per unit biomass per hour)

X Biomass concentration (cell dry weight per volume of aqueous phase)

X_m Cell dry weight per volume of immobilization matrix

$Y_{P/S}$ Product yield (n_o of moles of triptophan produced per mole of indole consumed)

x Indole conversion

REFERENCES

1 A. Toromanoff, Chimie Fine, Biofutur, 49 (Sept. 1986) 39-46.

2 L. Nyeste, M. Pécs, B. Sevella and J. Holló, Production of L-tryptophan by microbial processes, in A. Fiechter (Ed.), Adv. in Biochem. Eng., vol.26, Springer-Verlag, Berlin, 1983, pp 175-202.

3 W. Bang, S. Lang, H. Sahm and F. Wagner, Production of L-tryptophan by E. coli cells, Biotechnol. and Bioeng. 25 (1983) 999-1011.

4 J. Klein, personal communication.

5 G.R. Noggle, The identification and the quantitative determination of carbohydrates, in: W. Pigman (Ed.), The Carbohydrates: Chemistry, Biochemistry, Physiology, Ac. Press, N.Y., 1957, pp 602-624.

6 C. Yanofsky, Tryptophan synthase from Neurospora, Methods in Enzymology, vol 2, Ac. Press, 1955, pp 233-238.

7 J.R. Spies and D.C. Chambers, Chemical determination of tryptophan, Anal. Chem. 20 (1948) 30-39.

8 M. Wada, J. Kato and I. Chibata, Continuous Production of Ethanol Using Immobilized Growing Yeast Cells, Eur. J. Appl. Microbiol. Biotechnol. 10 (1980) 275-287.

9 C. Laane, S. Boeren, K. Vos and C. Veeger, Rules for the Optimization of Biocatalysis in Organic Solvents, Biotechnol. and Bioeng. (in press).

C. Laane, J. Tramper and M.D. Lilly (Editors), *Biocatalysis in Organic Media*, Proceedings of an International Symposium held at Wageningen, The Netherlands, 7–10 December 1986.

EFFECT OF WATER-MISCIBLE ORGANIC SOLVENTS ON THE CATALYTIC ACTIVITY OF PENICILLIN ACYLASE FROM KLUYVERA CITROPHILA.

J.M. GUISAN, G. ALVARO and R.M. BLANCO.

Instituto de Catâlisis. C.S.I.C. Serrano 119. 28006-Madrid. Spain.

SUMMARY

Water-miscible organic solvents exert very important effects on the catalytic activity of soluble and insolubilized/stabilized Penicillin G acylase(PA) from Kluyvera citrophila. Apolar solvents decrease the enzymatic activity dramatically even in very small concentrations, e.g. : residual activity of soluble PA is only 10% in 2.5% dioxane-water. The effects of organic solvents are reversible and depend on solvent and sustrate concentration as well as on ionic strength. Insolubilized/stabilized PA derivatives are slightly more resistant to the inhibitory effects than soluble PA. Adsorption of solvents on the active site of the enzyme seems to be the major component of this inhibition. Synthesis of Penicillin G with PA in glycerol-water mixtures is controlled by "product inhibition" instead of thermodynamic effects.

INTRODUCTION

From a thermodynamic point of view, it seems relatively easy to achieve very important yields in the synthetic reactions catalysed by Penicillin G acylases (e.g. acylation of 6-aminopenicillanic acid (6APA) with phenylacetic acid (PAA)) by using water-organic solvent mixtures with very low water content at slight acidic pH. However, the dimeric structure of bacterial Penicillin G acylases and the hydrophobic nature of their sustrates suggest important effects of high concentrations of organic solvents on activity and/or stability of these enzymes.

We have prepared insolubilized/stabilized derivatives (ISPA) of Penicillin G acylase from K. citrophila by multiple-point covalent attachment to aldehyde-agarose gels (1). The best ISPA retained a great percentage of catalytic activity (65%) and was more than 1000 fold more stable than soluble PA (irreversible thermal inactivation at pH 5.0, 55°C).

We present in this paper a study on the effects of water-miscible organic solvents on the catalytic activity of soluble PA and these ISPA derivatives. Preliminary tests were made by studying the hydrolysis of a synthetic sustrate 6-nitro-3-(phenylacetamido) benzoic acid (NIPAB), which was hydrolyzed more than 95% even in 75% glycerol-water, pH 5.0 . We have tested the role of different variables which might control the inhibitory effects of the organic cosolvents : i.- polarity of the solvent (dioxane, acetone, ethanol,glycerol and so on), ii.- solvent concentration, iii.- sustrate concentration and iv.- ionic strength. The reversibility of the solvent-enzyme interactions was also studied. Finally, syntheses of Penicillin G (Pen G) from 6APA and PAA, catalysed by ISPA in different glycerol-water mixtures, were performed. In this case, the effects of the solvent on the maximum synthetic yields were studied.

MATERIALS AND METHODS

Materials

The support Sepharose 6B-CL was obtained from Pharmacia (Uppsala, Sweden). Semipurified extracts of Penicillin G acylase (E.C. 3.5.1.11.) from K. citrophila , Pen G, PAA and 6APA were generously donated by Antibioticos S.A. (Madrid. Spain). The synthetic sustrate NIPAB was obtained from Sigma Chem. Co. (St. Louis. MO). Organic solvents and all other reagents used were of analytical quality.

Preparation of ISPA derivatives

Activation of agarose gels (etherification with glycidol and further oxidation with periodate) and preparation of insolubilized/stabilized agarose-acylase derivatives (ISPA) were carried out as previously described (2,1). Two identical ISPA, differing only in enzyme concentration, were prepared: i.- "dilute" ISPA, containing 2 I.U. of acylase per ml. of gel, used in experiments of NIPAB hydrolysis (there were not diffusional problems and so the effect of organic solvents on the intrinsic activity of the enzyme could be tested). ii.- "concentrate" ISPA, containing 200 I.U. per ml., used in the synthesis of Pen G in glycerol-water mixtures.

Hydrolysis of NIPAB

Activity of soluble PA and ISPA was measured by following the increase of absorbance at 405 nm. which accompanies the release of 3-amino-6-nitrobenzoic acid (3). Assays were performed in a 2 cm. pathlength cell with magnetic stirrer.

Synthesis/Hydrolysis of Penicillin G

The reactor was a jacked column containing 15 ml. of "concentrate" ISPA. In every case, the column was previously equilibrated by passing 50 ml. of reaction mixture. A small volume of mixture was next continuously recirculated through the column until the concentration of Pen G became constant, (volume of mixture / volume of column = 1.5, flow = 22 ml./h.). Analysis of reaction mixtures was performed using reverse-phase HPLC. Pen G, 6APA and PAA were separated on a Spherisorb S5 Octyl column by isocratic elution with methanol-0.067 M PO_4H_2K (35:65), pH 4.7 at 1 ml./min.. The eluted substances were monitored spectrophotometrically at 210 nm.

RESULTS

Effect of water-miscible organic solvents on NIPAB hydrolysis with PA

(i) Polarity of the organic solvent. A comparative study of the effect of organic solvents with different polarity on the activity of soluble PA and ISPA is shown in Table 1.

TABLE 1

Hydrolysis of NIPAB in 2.5% organic solvent-water mixtures. Assay: 7 ml. of 200 μM NIPAB in 50 mM phosphate buffer, containing 2.5% of organic solvent, pH 7.5, 25°C. Amount of enzyme in the reactor : 0.3 I.U.

	Relative Activity, %						
	No solv.	Glycerol	PEG-600	EG	Ethanol	Acetone	Dioxane
Soluble PA	100	96	92	88	60	43	12
ISPA	100	100	100	95	75	60	24

In the former Table we can observe that, as the polarity of the solvent decreases, its deleterious effect increases and this effect becomes dramatic in dioxane, specially with soluble PA, though a very small concentration of solvent was used. The effects of the solvents are not so intense for ISPA as for soluble PA in all the cases.

(ii) Sustrate concentration. The deleterious effect of the solvents clearly depends on sustrate concentration. Residual activity of ISPA in 2.5% dioxane is 80% with 3000 μM NIPAB, 50% with 1200 μM, 24% with 200 μM and 10.5% with 30 μM (Experimental conditions, except NIPAB concentration, as given in Table 1).

(iii) Ionic strength. The effect of the solvents on the activity of ISPA or soluble PA becomes more intense as the ionic strength increases. For instance, residual activity of ISPA in 2.5% dioxane decreases from 24% in 50 mM phosphate down to 3% in 50 mM phosphate 0.6 M ClNa. (Experimental conditions as given in Table 1). The same increase in ionic strength does not affect the activity of ISPA when solvents were not used (100% of activity at low and high ionic strength).

(iv) Reversibility of dioxane-ISPA interactions. ISPA derivatives preserve 100% of hydrolytic activity after they are incubated in 2.5% dioxane- 50 mM phosphate pH 7.5, 25 °C (40 ml. of solution / ml. of derivative) during 30 min. and washed again with 1 liter of distilled water.

(v) Effect of glycerol concentration. As it has been shown in Table 1, glycerol exerts the mildest effect on PA activity. We have studied the effect of glycerol concentration on PA activity at pH 5.0 (in these conditions more than 95% of NIPAB was hydrolyzed even in 75% glycerol).

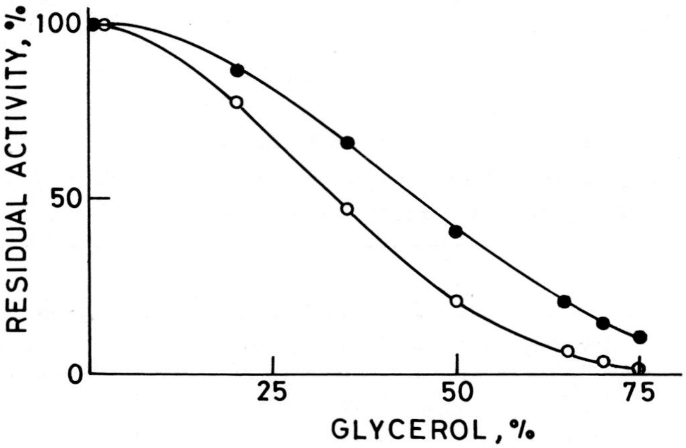

Fig. 1. Hydrolysis of NIPAB in glycerol-water mixtures. Assay: 7 ml. of 100 μM NIPAB in glycerol-water, 20 mM acetate buffer pH 5.0, 25 °C. Amount of enzyme in the reactor: 0.3 I.U. o: reaction catalysed by soluble PA, ●: reaction catalysed by ISPA.

We observe in the figure that the deleterious effect of glycerol increases considerably as its concentration does. In the same way we have observed in Table 1, the effect of the organic solvent is not so intense for ISPA as for soluble PA, this difference increasing as the effect of the solvent becomes more dramatic. From results of fig.1, we devised that, by using "concentrate" ISPA, we would be able to study the reaction of acylation of 6APA in a broad spectrum of glycerol-water mixtures.

Hydrolysis / Synthesis of Penicillin G in glycerol-water mixtures

We have chosen the synthesis of Pen G from 6APA and PAA as an example of acylation. We have studied the maximum synthetic yields of Pen G and compared them with the residual concentrations of Pen G obtained after its hydrolysis in the same experimental conditions. We observe in Table 2 that the yields of Pen G obtained in the synthesis are lower than those obtained in the hydrolysis and this difference increases as glycerol concentration does. Thermodynamically , identical final yields of Pen G must be expected.

TABLE 2

Final Pen G concentrations, expressed as % of the maximum possible ones, obtained either in the hydrolysis of 20 mM Pen G or in the acylation of 20 mM 6APA with 20 mM PAA. Both reactions were catalysed by "concentrate" ISPA (see Methods) in different glycerol-20 mM acetate mixtures, pH 5.0, 15 °C.

	Yield of Penicillin G, %			
	No glycerol	20% glycerol	50% glycerol	75% glycerol
Hydrolysis	40	55	68	80
Synthesis	30	33	26	7.5

These results cannot be explained by enzyme inactivation because: the residual activity of ISPA even in 75% glycerol is not negligible, the "concentrate" ISPA is very active and the stability of ISPA in all these conditions is very high (100% of activity is preserved after several days of storage in all these conditions)(4).

DISCUSSION

As it has been shown in Fig. 1, the deleterious effect of the organic solvents is not so intense for ISPA as for soluble PA. This difference increases when the effects of the solvents become more dramatic, e.g. in 2.5% dioxane or 75% glycerol (in this case, the residual activity is 10% for ISPA and only 1% for soluble PA).It seems that, specially in these drastic conditions, the deleterious effect of the solvents is partially due to conformational changes of the enzyme structure, which have to be very restricted in the case of ISPA. We have already proved the resistance of ISPA to conformational changes induced by heat (1).

The main component of the inhibition by organic solvents , which affects ISPA as well as Soluble PA, was mainly studied with ISPA. This inhibition is completely reversible and increases when the polarity of the solvent decreases and when solvent concentration and ionic strength increase. All these results suggest that concrete adsorption of solvent molecules on hydrophobic areas of the enzyme must occur. The clear dependence of solvent inhibition on the sustrate concentration, which can be adjusted to competitive inhibition (Lineweaver-Burk plots are not shown), indicates that hydrophobic areas of the enzyme, on which solvent adsorbs, must remain in the active site of PA. This explanation seems logical if we consider that PA is specific for acyl donors containing a hydrophobic residue (PAA and analogues), so sustrate recognition must imply hydrophobic acyl donor-enzyme interactions. Thus, the hydrophobic areas in the active site of PA must be accessible to the low molecular weight solvents used.

The pattern of Pen G yields obtained in the hydrolysis of Pen G in glycerol -water mixtures is qualitatively the one expected from thermodynamic effects: decreasing of hydrolytic yields as the concentration of organic solvent increases (Table 2). However, the pattern of synthetic yields in the reverse reaction results contrary to the thermodynamic predictions: there is a decrease of synthetic yields when the concentration of solvent increases. It is evident that enzymatic properties cannot change the thermodynamic equilibrium concentrations, but it is possible that certain enzymatic properties , mainly inhibition by product, can hinder equilibrium concentrations to be reached. To this point, the real maximum synthetic yields become controlled by "product inhibition". In this case, we can consider that adsorption of PAA on the active site of the enzyme is more intensively affected by one of the solvent than the adsorption of Pen G. The last one is a more complex multi-interaction, which can even involve electrostatic interactions in the binding site for the nucleophile moiety of the antibiotic (5), which are favoured in these less polar media. Thus, an increase in the difference between PenG and PAA affinities for the active site of PA, due to the organic solvent, can be responsible of the dramatic increase of product inhibition in this synthetic reaction.

We have made a more complete study of this extremely complex synthetic reaction. We have been able to achieve real synthetic yields of more than 90% in the best conditions (4).

REFERENCES

1 J.M. Guisán, R.M. Blanco and G. Alvaro, Stabilization of trypsin and penicillin G acylase by multipoint covalent attachment to aldehyde-agarose gels, Proc. 3rd. Portuguese Meeting of Biotechnology, Lisbon, October 6 - 9, 1986, In press.

2 J.M. Guisán, Aldehyde-agarose gels as supports for insolubilization/stabilization of enzymes, Submitted for publication.

3 C. Kutzbach and E. Rauenbusch, Preparation and General Properties of Crystalline Penicillin Acylase from Escherichia Coli ATCC 11105, Hoppe-Seyler´s Z. Physiol. Chem. Bd 354, S (1974), 45 - 53.

4 J.M. Guisán, R.M. Blanco and G. Alvaro, Synthesis of Penicillin G from 6APA and PAA in water-organic solvent mixtures, In preparation.

5 V. Kasche, U. Haufler and R. Zollner, Kinetic Studies on the Mechanism of the Penicillin Amidase-Catalysed Synthesis of Ampicillin and Benzylpenicillin, Hoppe-Seyler´s Z. Physiol. Chem. Bd 365, S (1984), 1435 - 1443.

C. Laane, J. Tramper and M.D. Lilly (Editors), *Biocatalysis in Organic Media*, Proceedings of an International Symposium held at Wageningen, The Netherlands, 7–10 December 1986.

ACTIVITY OF STAPHYLOCOCCAL NUCLEASE IN WATER-ORGANIC MEDIA

A. ALCANTARA[1], A. BALLESTEROS[2], M.A. HERAS[3], J.M. MARINAS[1], J.M.S. MONTERO[1] and J.V. SINISTERRA[1]

[1]Dept. de Química Orgánica, Universidad de Córdoba, 14004 Córdoba (Spain)

[2]Instituto de Catálisis, CSIC, 28006 Madrid (Spain)

[3]Depto. de Química Física, Universidad de Córdoba, 14004 Córdoba (Spain)

SUMMARY

The DNase activity of immobilized (on tosylated agarose) and native staphylococcal nuclease in water/water-miscible organic solvents (acetonitrile, dimethylformamide, dimethyl sulfoxide, methanol, tetrahydrofurane) has been studied. In the presence of the organic solvents the decrease of enzyme activity in the insoluble form was found to be lower than in the soluble one. Similarly, the immobilized enzyme is more resistant to the deleterious effects of the solvents.

INTRODUCTION

Enzymes are traditionally used in aqueous media, where most of their reactants and products are soluble. However, much wider applications of biocatalysts are being demanded (e.g., transformation of hydrophobic molecules), specially with the development of biotechnology. Thus, it has become absolutely necessary to introduce organic solvents in the reaction mixture to improve the water solubility of the organic compounds. Today, there is rather little information available on catalysis by enzymes in water/water-miscible organic solvents (refs. 1, 2) and also in water/organic solvents two-phase systems (refs. 3,4).

Micrococcal endonuclease is a well-studied (ref. 5) extracellular phosphodiesterase from Staphylococcus aureus which hydrolyzes either DNA or RNA to produce 3'-mononucleotides and dinucleotides, requiring Ca^{2+} for activity. In our laboratory, we have been interested in studying the behaviour in aqueous medium of this enzyme insolubilized on activated (by cyanogen bromide) agarose (refs. 6, 7). Now we are extending our knowledge to the study of the behaviour in water/organic media, of nuclease insolubilized on agarose activated with tosyl chloride. To select the solvents to be used, two of their properties are considered: dielectric constant and solubility in water (ref. 8). In this paper, we present our first data of nuclease hydrolytic activity -on DNA- in the presence of several water-miscible organic solvents (acetonitrile, DMF*, DMSO, methanol, THF).

METHODS

Agarose (BiO-Gel A-150 m, 100-200 mesh, containing 1 % (w/v) agarose, from Bio-Rad Laboratories) was activated with p-toluene-sulfonyl chloride following our modifications (ref. 9) of the method developed by Mosbach and collaborators (ref. 10). The number of tosyl groups in the activated gel per 10 nm^2 of surface was 7.

Tosylated agarose in 0.1 M $NaHCO_3/Na_2CO_3$ buffer, pH 9.0, was mixed with nuclease (from Boehringer Manheim) solution and left at 25°C with gentle stirring, as described earlier (ref. 11). The insoluble enzyme derivative -containing 10.3 μg enzyme per ml of gel- was stored until use in 20 mM citrate buffer, pH 6.4, 10 mM $CaCl_2$ and 0.1 % bovine serum albumin.

Initial activity of the soluble and insoluble enzymes towards salmon testes DNA (Type III, from Sigma) was measured by following graphically the increase in A_{260} at 30°C and pH 8.8 (refs. 5, 11) in a Cary 219 spectrophotometer equipped with a magnetic stirrer.

RESULTS AND DISCUSSION

Prior to the enzymic studies, the influence of the presence of organic solvents in the UV absorption of heat-denatured DNA was determined. Table 1 shows that the organic media produce a hyperchromic effect concomitant with a bathochromic shift in the wavelength of maximum absorption. Increments in absorptivity from 18 to 35 % are observed when the proportion of the organic solvent increases from 2 to 5 %. Hydrolytic activity towards DNA was, therefore, measured at the λ_{max} determined for each individual water-organic mixture.

Figure 1 depicts the effect of the amount of enzyme (soluble or insolubilized) added to the spectrophotometer cuvette, on enzyme activity. It is clearly shown that using in the assay an amount of enzyme below 20 ng, we are in the initial linear part of the graph. [From the initial slope of the curves, the specific activity of soluble and insolubilized enzyme can be evaluated: this

*Abreviations: DMF, N,N-dimethylformamide; DMSO, dimethyl sulfoxide; THF, tetrahydrofuran.

TABLE 1

Effect of organic solvents on absorptivity and λ_{max} of heat-denatured DNA

% (organic solvent)	λ_{max} (nm)	Extinction coefficient (arbitrary units)
0*	259.5	100
2 (DMF)	260	118
5 (DMF)	262	128
2 (DMSO)	260	119
5 (DMSO)	260.5	127
2 (THF)	259	125
5 (Acetonitrile)	260	134
5 (Methanol)	261	135

*0.1 M Tris buffer, pH 8.8.

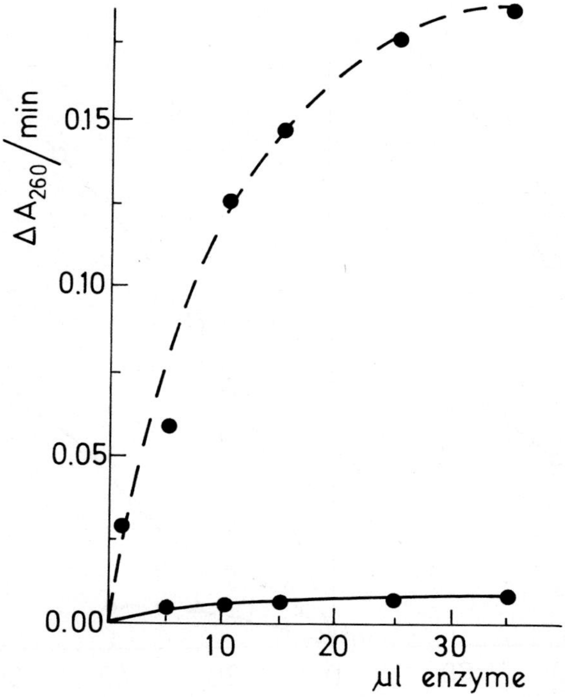

Fig. 1. Rate of hydrolysis of heat-denatured DNA *versus* amount of enzyme in the assay. DNA concentration: 60 µg/ml. Broken line: soluble enzyme added (from a 32.6 mg/ml solution). Solid line: immobilized enzyme derivative (10.3 µg enzyme/ml gel).

insoluble enzyme derivative posesses a specific activity which is 43 % of that corresponding to the native enzyme; this value is higher than that found in derivatives previously prepared with agarose activated by other method -by CNBr- (ref. 7)]. Henceforth (except in Figure 3), the amount of biocatalyst in the assays was always maintained in this region of linearity between activity and amount of derivative.

The hydrolytic rate of soluble nuclease in several organic-water media as a function of substrate concentration was determined. Some of the results obtained with DMSO and DMF appear in Fig. 2. Although the reaction rate at the lowest substrate concentration in H_2O - solvent (98:2) mixtures (Fig. 2a) differs little from that typical in water, it decreases substantially as the DNA concentrations increases up to 65 μg/ml. DMF produced a larger effect than DMSO. When the DMF proportion is rised to 5% (Fig. 2b), the enzymatic rate measured is less than 10% of the control. Several factors could be responsible

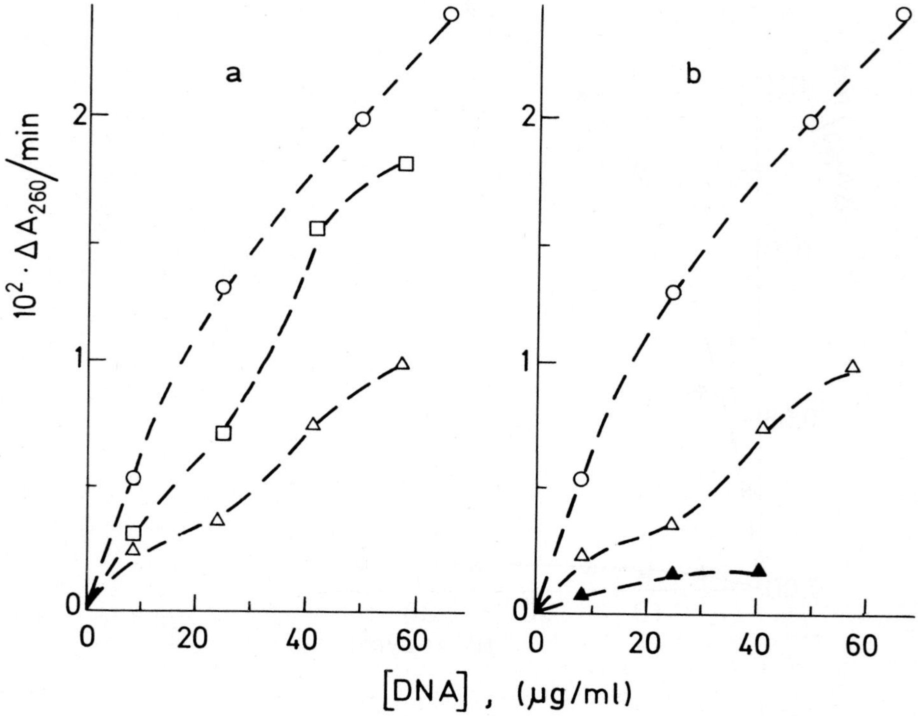

Fig. 2.- DNase activity of soluble nuclease as a function of substrate concentration. O, 0% organic solvent; □, 2% DMSO; △, 2% DMF; ▲, 5% DMF.

for this effect. On the one hand, if the dielectric constant (ε) is taken into account (ref. 12), we found that ε for DMSO and DMF have a value of 45 and 36.7, respectively, against 78.5 for water. On the other hand, DMSO and DMF, very common solvents to the protein chemist, act as proton acceptors (but not as proton donors), therefore they can break hydrogen bonds in proteins. Their very different structure could also be important in explaining the results obtained: the amide group might interact with the popypeptide chain of the enzyme or even be partially hydrolyzed; this would not be possible in the case of DMSO.

No activity of the nuclease was detected with THF –dielectric constant = 7.4–, even at 2 % concentration (results not shown). It is known that Ca^{++} binds to several acidic groups of the protein in a pocket next to the active site (ref.5). Similarly, ether molecules have strong affinity for alkaline-earth cations. Therefore, the binding of THF molecules to the calcium atom could explain the inactivation of the enzyme.

The dependence of the activity of insolubilized nuclease versus amount of enzyme sample, in the presence of 2 and 5% DMF, is presented in Fig. 3. We can see that there is very little effect of 2% DMF as opposed to the soluble enzyme (cf. Fig. 2): the specific activity is over 80% of the value in water. In general, much more resistance to decrements in specific activity, in presence

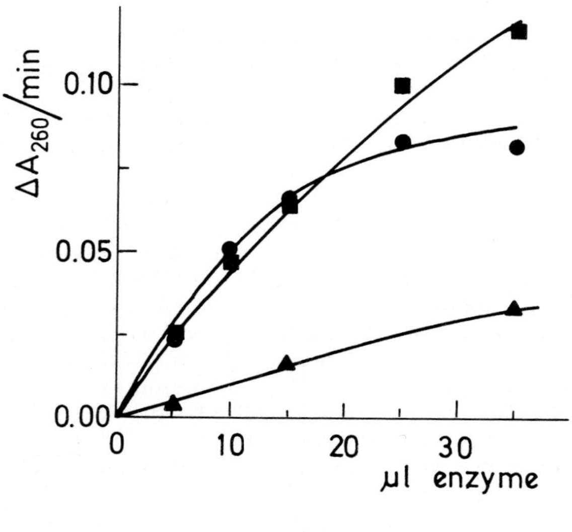

Fig. 3.- Activity of immobilized nuclease versus amount of enzyme. ●, 0 % organic solvent; ■ , 2 % DMSO; ▲ , 5 % DMF.

of organic solvents, was observed in the immobilized enzyme as compared with the native one. As in the case of the soluble enzyme, the insoluble derivative had no activity in 2 % THF.

If the water concentration is reduced in a hydrolytic reaction by including an organic solvent, one would expect that the rate of hydrolysis will diminish because the reaction equilibrium will be affected. In addition, the nature of the organic compound may be responsible for partial or total inactivation of the enzyme. In this respect, the immobilization of the biocatalyst on tosylated agarose serves the useful purpose of stabilizing the protein molecule for catalysis; this stabilization is very important and makes possible the use of the enzyme in the reverse reaction, i. e. the synthesis of oligonucleotides.

ACKNOWLEDGMENTS

We thank F. García Blanco for helpful advice. This work has been supported by the Spanish CAICYT (grant BT85-0043).

REFERENCES

1 S. Fukui and A. Tanaka, Enzymatic reactions in organic solvents, Endeavour, 9 (1981) 10-17.
2 A. Zaks and A. M. Klibanov, Enzyme-catalyzed processes in organic solvents, Proc. Natl. Acad. Sci US 82 (1985) 3192-3196.
3 M. D. Lilly, Two-liquid-phase biocatalytic reactions, J. Chem. Tech. Biotechnol. 32 (1982) 162-169.
4 G. Carrea, Biocatalysis in water-organic solvent two-phase systems, Trends Biotechnol. 2 (1984) 102-106.
5 C.B. Anfinsen, P. Cuatrecasas and H. Taniuchi, Staphylococcal nuclease. Chemical properties and catalysis, in: P. D. Boyer (ed), The Enzymes, 3rd ed., Academic, New York, 1971, vol. 4, pp. 177-204.
6 J.M. Guisán, F. V. Melo and A. Ballesteros, Determination of intrinsic properties of immobilized enzymes, Appl. Biochem. Biotechnol. 6 (1981) 25-36.
7 J.M. Guisán and A. Ballesteros, Hydrolysis of nucleic acids by Sepharose-micrococcal endonuclease, Enz. Mic. Technol. 3 (1981) 313-320.
8 L. E. S. Brink and J. Tramper, Optimization of organic solvent in multiphase biocatalysis, Biotechnol. Bioeng. 27 (1985) 1258-1269.
9 A. Ballesteros, J. M. S. Montero and J. V. Sinisterra, p-toluene-sulfonyl chloride activation of agarose as exemplified by the coupling of lysine and micrococcal endonuclease, J. Mol. Catal. 38 (1986) 227-236.
10 K. Nilsson and K. Mosbach, p-toluenesulfonyl chloride as an activating agent of agarose, Eur. J. Biochem. 122 (1980) 397-402.
11 J.M. Guisán and A. Ballesteros, Preparation of immobilized Sepharose-micrococcal nuclease derivatives, J. Solid-Phase Biochem. 4 (1979) 245-252.
12 R. C. Weast (Ed.), Handbook of Chemistry and Physics, 65th ed., CRC Press, Cleveland, 1984.

C. Laane, J. Tramper and M.D. Lilly (Editors), *Biocatalysis in Organic Media*,
Proceedings of an International Symposium held at Wageningen,
The Netherlands, 7–10 December 1986.

STEROID BIOCONVERSION IN AQUEOUS TWO-PHASE SYSTEMS

R. KAUL and B. MATTIASSON

Department of Biotechnology, Chemical Center, University of Lund, P.O.Box 124, S-221 00 Lund, (Sweden)

SUMMARY

Aqueous two-phase systems have been shown to have potential in bioorganic synthesis. The transformation of hydrocortisone by the cells of *Arthrobacter simplex* was studied in these systems. The bioconversion rate was comparable to that in organic media. The product could be recovered from the reaction medium by adsorption on Amberlite XAD-4. The bacterium was able to grow in the two-phase system, and could be activated after a period of use.

INTRODUCTION

Steroid transformations in media containing organic solvents have demonstrated improved substrate availability to the biocatalyst, and simpler downstream processing as compared to that in aqueous systems (ref 1-2). However, in several cases, especially when viable cells are required for catalyses, operational stability of the biocatalyst may be markedly diminished, thus restricting the application of these solvents. This is apparent by the fact that pseudocrystalline fermentations are still commonly performed (ref. 3).

We have shown that the mass transfer conditions for poorly insoluble compounds in water solutions could be substantially improved and also the operational lifetime of the cells increased by the use of aqueous two-phase systems. Hydrocortisone was transformed to prednisolone by *Arthrobacter simplex.* These biphasic systems are characterized by a difference in hydrophobicity of the phase forming polymers; the phase that is relatively more hydrophobic serves as the steroid reservoir.

METHODS

The induction of the enzyme steroid Δ'-dehydrogenase in *Arthrobacter simplex* ATCC 6946 has been described earlier (ref.4).

Hydrocortisone powder (Serva, FRG), 20 mg was mixed into 20 ml of the aqueous two-phase systems, composed of polyethylene glycol (PEG) (Union Carbide, USA) and dextran (Pharmacia, Sweden) in 0.05 M Tris-HCl, pH 7.0.

The reaction was started by the addition of 22 mg (dryweight) cells, and the reaction mixture stirred at 20°C. Aliquots were removed at different time intervals to determine the amount of substrate and the product, as described previously (ref 4-5).

The phases were allowed to settle after the termination of the reaction, the top phase with the bulk of the product removed and then passed over a column packed with 5 g Amberlite XAD-4 (BDH, England) at a flow rate of 0.5 ml/min. The PEG eluate was returned to the reactor with more substrate for subsequent conversions.

RESULTS AND DISCUSSION

The aqueous solubility of hydrocortisone has been reported to be 0.38 g/L at 37° C (ref. 6). The solubility of the steroid at 20°C in PEG and dextran phases of 25% PEG 8000- 6% Dextran was found to be 1.0 and 0.1 mg/ml respectively (ref. 5). Also the partition coefficient (ratio of concentration in the top phase to that in bottom phase) of hydrocortisone and prednisolone was more favourable towards the PEG rich top phase (see Table 1). The increase in the the concentration of PEG resulted in an increase in the partitioning of the steroids to that phase. Hence, an effective extractive bioconversion would be performed by using high top to bottom phase volume ratios.

TABLE 1

Partitioning of hydrocortisone and prednisolone in PEG-dextran two-phase systems.

PEG %(w/w) :	Dextran %(w/w)	Vol.ratio of top to bottom phase	Partition coefficients of Hydrocortisone	Prednisolone
15	3	12.3:1	2.6	6.4
20	3	15.7:1	2.8	6.6
25	3	19.0:1	2.8	6.8
25	6	8.5:1	3.3	7.1

Aqueous two-phase systems offer extremely favourable mass transfer conditions during bioconversion. In a well-mixed system, the microbial cells are temporarily immobilized within the liquid droplets of the bottom phase, surrounded by the bulk top phase (ref. 7). These droplets can be made extremely small, thus increasing the surface area of contact between the two phases, and requiring the substrate and the product to diffuse only short distances. This is in contrast to the condition achieved in water-organic solvent two-phase system containing an immobilized biocatalyst, where the reaction rate is limited by mass transfer across the various interfaces, especially the liquid-solid interface (ref. 8).

The transformation of hydrocortisone in the aqueous two-phase systems was seen to take place at a rate comparable to that in media employing organic solvents (see Table 2). The reaction rate could be further improved in the presence of emulsions of perfluorochemicals (ref. 5).

TABLE 2
Comparison of hydrocortisone transformation in aqueous two-phase systems with some reported systems containing methanol.

A. simplex in	Hydrocortisone conc. (g/l)	Reaction time (h)	Conversion (%)	Reference
polyacrylamide	0.36	18[a]	100	9
calcium alginate	9.0	24	85	10
photocrosslinkable resin prepolymers	1.0	1	90-100	11
glass slides	0.36	19	89	12
PEG/dextran two-phase systems	1.0	2.5	99.5-100	5

[a]Time required for total hydrocortisone transformation is reduced after activation of the immobilized cells.

The major amount of the product was partitioned to the top phase. Extraction of the product from the reaction system constitutes an important step during downstream processing. In order to make the reuse of PEG possible, a mild process of extraction was followed. Passing the PEG phase with prednisolone over a column of Amberlite XAD-4 led to a complete adsorption of the steroid on to the hydrophobic resin. The eluate of PEG polymer was recycled back to the reactor with more hydrocortisone and the process repeated. Losses of PEG, if any during the adsorption step, could be made up by supplementing more polymer for the bioconversion. The prednisolone was eluted from the sorbent with methanol, and the column used repeatedly for steroid extraction after being washed with Tris-HCl solution, pH 7.0 (ref. 5).

Hydrocortisone levels were increased further up to 5 mg /ml, and conversions higher than 95% were obtained (ref. 5). The bottom phase containing the cells could be used repeatedly for the steroid transformation (ref. 5). The biocompatability of the aqueous two-phase systems was demonstrated by the possibility of culturing the microbial cells in these systems. This is important in situations when the catalytic activity is diminishing.

The studies with Reppal PES 200 (Reppe Glykos AB, Sweden) as the bottom phase polymer in place of dextran also exhibited a similar reaction profile. This was done with the aim of lowering the costs for operation in aqueous two-phase systems, Reppal being about 40 times cheaper than dextran.

ACKNOWLEDGEMENT

The authors gratefully acknowledge the financial support provided by the National Swedish Board for Technical Development.

REFERENCES

1 G. Carrea, Biocatalysis in water-organic solvent two-phase systems, Trends Biotechnol. 2 (4) (1984) 102-106.
2 P.L. Luisi and C. Laane, Solubilization of enzymes in apolar solvents via reverse micelles, Trends Biotechnol. 4 (6) (1986) 153-161.
3 F.B. Kolot, Microbial catalysts for steroid transformations - Part 1, Process Biochem. 17 (6) (l982) 12-18.
4 R. Kaul, P. Adlercreutz and B. Mattiasson, Coimmobilization of substrate and biocatalyst: a method for bioconversion of poorly soluble substances in water milieu, Biotechnol. Bioeng. 28 (l986) 1432-1437.
5 R. Kaul and B. Mattiasson. Extractive bioconversion in aqueous two-phase systems. Production of prednisolone from hydrocortisone using *Arthrobacter simplex* as catalyst, Appl. Microbiol. Biotechnol. 24 (1986) 259-265.
6 J. Kloosterman IV and M.D. Lilly, Effect of supersaturated aqueous hydrocortisone concentrations on the Δ'-dehydrogenase activity of free and immobilized *Arthrobacter simplex,* Enzyme Microb. Technol. 6 (1984) 113-116.
7 B. Mattiasson, Applications of aqueous two-phase systems in biotechnology, Trends Biotechnol. 1(1) (l983) 16-20.
8 M.D. Lilly and J. M. Woodley, Biocatalytic reactions involving water-insoluble organic compounds in: J. Tramper, H.C. van der Plas and P. Linko (Eds.), Biocatalysis in Organic Syntheses, Elsevier, Amsterdam, l985, pp. 179-192.
9 S. Ohlson, P.-O. Larsson and K. Mosbach, Steroid transformation by activated living immobilized *Arthrobacter simplex* cells, Biotechnol. Bioeng. 20(1978) 1267-1284.

10 S. Ohlson, P.-O. Larsson and K. Mosbach, Steroid transformation by living cells immobilized in calcium alginate, Eur. J. Appl. Microbiol. Biotechnol. 7(1979) 103-110.
11 K. Sonomoto, A. Tanaka, T. Omata, T. Yamane and S. Fukui, Application of photocrosslinkable resin prepolymers to entrap microbial cells. Effects of increased cell-entrapping gel hydrophobicity on the hydrocortisone Δ'-dehydrogenation, Eur. J. Appl. Microbiol. Biotechnol. 6(1979) 325-334.
12 M. Mozes and P.G. Rouxhet, Dehydrogenation of cortisol by *Arthrobacter simplex* immobilized as supported monolayer, Enz. Microb. Technol. 6 (1984) 497-502.

C. Laane, J. Tramper and M.D. Lilly (Editors), *Biocatalysis in Organic Media*,
Proceedings of an International Symposium held at Wageningen,
The Netherlands, 7–10 December 1986.

ORGANIC SOLVENTS FOR BIOORGANIC SYNTHESIS

2. Influence of log P and water solubility in solvents on enzymatic activity.

M. RESLOW, P. ADLERCREUTZ and B. MATTIASSON
Department of Biotechnology, Chemical Center, University of Lund, P.O.Box 124, S-221 00 Lund, Sweden.

SUMMARY
The influence of solvents on α-chymotrypsin-catalyzed esterification of N-acetyl-L-phenylalanine with ethanol was investigated . The enzyme was adsorbed on porous glass beads and was used in various water-saturated solvents. Small additions of water increased enzymatic activity. The highest activities were obtained in the hydrophobic solvents, and the activity correlated well with the log P value of the solvent. An even better correlation for solvents in the range of log P 0.5-1.5 was obtained if the log P value was corrected for the water content.

INTRODUCTION

Enzymes can be used successfully in systems having quite low water contents. In such systems hydrolytic enzymes have been used to catalyze esterification and transesterification reactions. The enzymes have been used as solid particles, usually adsorbed on an inert support material. It is often desirable to have a minimun content of water because too much water gives rise to unfavourable thermodynamic equilibrium, aggregation of enzyme-support material, or poorer enzyme stability.

The solvent chosen is important for the enzymatic activity (ref. 1, ref. 2). Different parameters have been used to characterize the solvent. Log P (the logarithm of the partition coefficient of a given component in the octanol-water two phase system) has been useful to predict the activity of whole cell catalysts in different solvents (ref. 3).

In this report the chymotrypsin-catalyzed esterification of N-Ac-Phe with ethanol in different solvents was studied. The reaction rates obtained were correlated with log P values and modified log P values.

MATERIALS AND METHODS

α -chymotrypsin from bovine pancreas (Sigma Chemicals) was dissolved in 4.3 ml 50 mM sodium phosphate buffer pH 7.8 and was mixed with 2.0 g porous glass (CPG-10, mesh size: 200-400, surface area: 7.4 m^2/g; mean pore dia: 2147Å, Sigma Chemicals). The mixture was dried under reduced pressure for 3 h. The water content was determined after drying, by analyzing a part of the preparation gravimetrically. This preparation had an enzyme content of 4.29% (w/w) and a water content of 0.6% (w/w).

The activity of the enzyme was measured by following the esterification of Ac-Phe with ethanol. To 100 mg chymotrypsin-glass in 10 ml stoppered glass bottles, were added 1.9 ml of water-saturated solvent and different amounts of water. The bottles were shaken on a reciprocal shaker (150 rpm) at 20 °C. After 20 min the reaction was started by adding ethanol containing Ac-Phe so that final concentrations of 10 mM Ac-Phe and 1.0 M ethanol were obtained. 20 µl samples were taken at intervals to determine the content of Ac-Phe-OEt by HPLC. (Nucleosil C18-column eluted with H_2O/MeCN/HAc, 55:40:5).

In these systems it is necessary to control the water content and we have controlled the water content by drying the catalyst, and then using water-saturated solvents.

Log P values used in this work were calculated from fragmental hydrophopic constants (ref. 4).

RESULTS AND DISCUSSION.

A rather good correlation was found when our experimentally determined reaction rates were expressed as a function of the log P value of the solvent (Fig. 1a) (ref. 2). However, some discrepancies were found in the log P range 0.5 -1,5. Some ways to further improve the correlation were tried. Solvents with log P values between 0.3 and 2.0 have water solubilities in the range 0.6 - 25 mole%. The water solubilities are not an exact function of log P -values. We have found that if two solvents have the same log P value, the enzyme activity was higher in the solvent with lower water-solubilizing capacity.

We have tried to take into account the water present in the solvents. For mixtures of solvents log P can be calculated using the following semiempirical formula (ref. 6)

$$\text{Log } P_{\text{mixture}} = x_1 \log P_1 + x_2 \log P_2$$

where x_1 and x_2 are the mole fraction of the solvents. In order to calculate corrected log P values for water saturated solvents the following formula was used:

$$\log P_{\text{corr}} = (1-x) \log P_{\text{solv}} + x \log P_{H_2O}$$

where x is the water solubility expressed as mole fraction (ref. 5). A quite

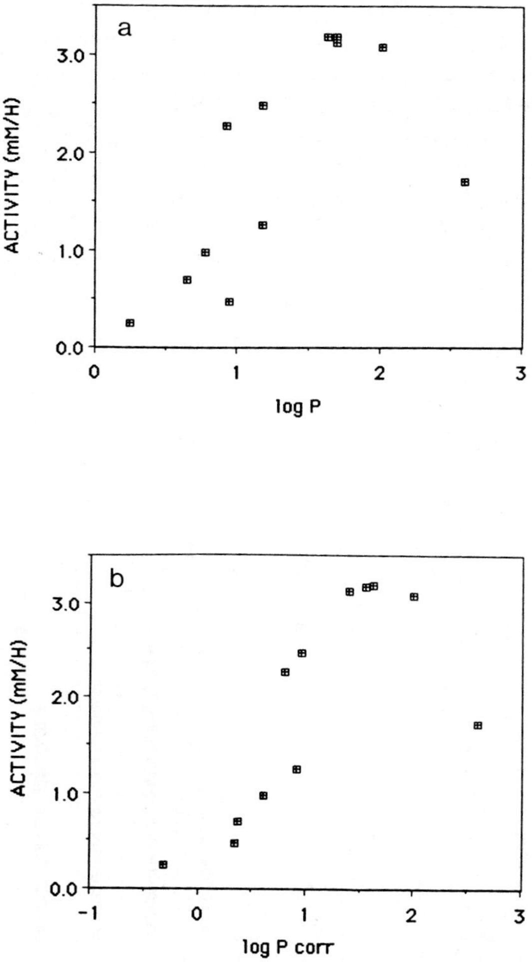

Fig. 1. Activity of chymotrypsin in different solvents at 2.0% (v/v) added water plotted against (a) Log P, (b) Log P_{corr}.

good correlation was found between the experimentally determined reaction rates and the corrected log P values (Fig. 1b). The differences compared to uncorrected log P values are of course larger for solvents having low log P values as these dissolve more water than the more hydrophobic solvents. The correction of log P values presented here seems to be a very simple way to make log P an even more useful parameter in the choice of organic solvents for biocatalysis.

The initial reaction rates for the chymotrypsin catalyzed esterification of Ac-Phe under different conditions are presented in fig. 2. At 0.5% (v/v) water addition highest enzyme activity was obtained in the more hydrophobic solvents (log P_{corr} >1.61). Further water additions in these solvents result in slight increase or decrease in enzyme activity. The enzyme activity in the more hydrophilic solvent increased largely with the amount of water added. This can be explained by the fact that in hydrophobic solvents water partitions more easily to the enzyme than in the more hydrophilic solvents. A certain amount of water is needed to hydrate the enzyme and make it fully active. Consequently less water needs to be added to the more hydrophobic solvents to obtain maximal enzymatic activity.

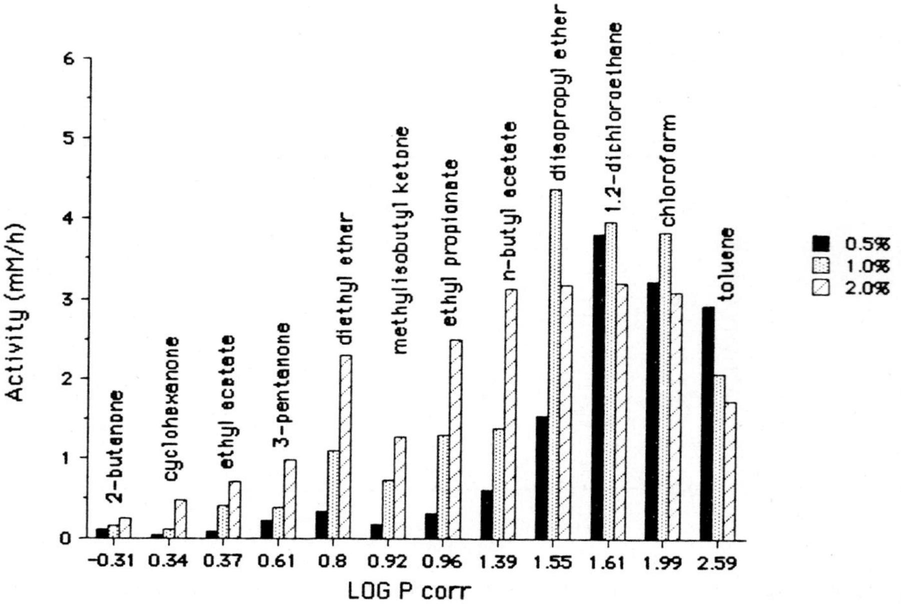

Fig. 2 Initial reaction rates of chymotrypsin in different solvents with various water additions (0.5, 1.0, 2.0% v/v).

REFERENCES

1 L.E.S. Brink and J. Tramper, Optimization of organic solvent in multiphase biocatalysis, Biotechnol. Bioeng. 27 (l985) 1258-1269.
2 M. Reslow, P. Adlercreutz and B. Mattiasson, Organic solvents for bioorganic synthesis. 1. Optimization of parameters for a chymotrypsin catalyzed process. Submitted for publication.
3 C. Laane, S. Boeren and K. Vos, On optimizing organic solvents in multi-liquid-phase biocatalysis. Trends Biotechnol. 3 (1985) 251-252.
4 R.F. Rekker, The hydrophobic fragmental constant. Elsevier Amsterdam, l977.
5 J.M. Sörensen and W.Arlt, Liquid-Liquid Equilibrium Data Collection.

Binary systems. Chemistry Data Series, Vol. V, Part 1, Eds. D. Behrens, R. Eckerman, DECHEMA, Frankfurt, l979.

6 R. Hilhorst, R. Spruijt, C. Laane, and C. Veeger, Rules for the regulation of enzyme activity in reversed micelles as illustrated by the conversion of apolar steroids by 20ß-hydroxysteroid dehydrogenase. Eur. J. Biochem. 144 (1984) 459-466.

C. Laane, J. Tramper and M.D. Lilly (Editors), *Biocatalysis in Organic Media*,
Proceedings of an International Symposium held at Wageningen,
The Netherlands, 7–10 December 1986.

STUDY OF HORSE LIVER ALCOHOL DEHYDROGENASE (HLADH) IN AOT-CYCLOHEXANE REVERSE MICELLES

K. LARSSON, P. ADLERCREUTZ AND B. MATTIASSON
Department of Biotechnology, Chemical Center, University of Lund,
P.O. Box 124, S-221 00 Lund, Sweden.

SUMMARY

Horse liver alcohol dehydrogenase (HLADH, E.C. 1.1.1.1) solubilized in AOT-(dioctyl sulfosuccinate-Na salt) -reverse micelles catalyzed oxidation of ethanol and reduction of cyclohexanone in a coupled substrate co-enzyme regenerating system. The reaction conditions were optimized and the stability of the system towards storage and operation have been studied. HLADH showed good catalytic activity after 2 weeks of operation in reverse micelles. A steroid, eticholan-3ßol-17one was also used as a substrate; the reaction rate in reverse micelles was about 5 times that in buffer.

INTRODUCTION

Reverse micelles have been shown to be a favourable microenvironment for biocatalysts in organic solvents (ref 1). There have been several reports on the solubilization of HLADH in reverse micelles (ref 2-4).

However, there is little data of the operational stability of enzymes in reverse micelles. The aim of this investigation was to study the storage and operational stability of HLADH in microemulsion. We chose to use a coupled-substrate recycling system where ethanol is oxidized to acetaldehyde and cyclohexanone is reduced to cyclohexanol.

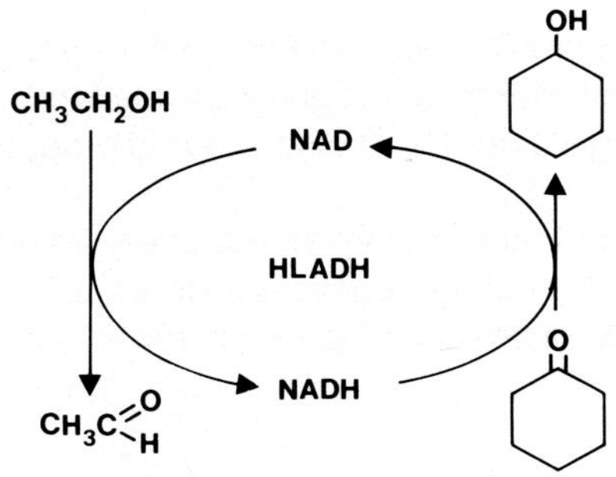

MATERIALS AND METHODS

HLADH (Sigma Chemicals) and NAD (Sigma Chemicals grade 7) or NADH (Sigma Chemicals, grade 3) were solubilized in cyclohexane-AOT (Serva Chemicals) reverse micelles by directly injecting a water-solution of enzyme and cofactor into an AOT-solution in cyclohexane. The organic solvents were of p.a. quality and were dried over molecular sive 3 Å before use. The reaction mixtures were incubated in closed bottles at room temperature without stirring.

The products, cyclohexanol and/or acetaldehyde were determined by gas chromatography, using a Shimadzu gas chromatograph GC-9AM with FID detector. The analyses were carried out on a chromosorb 101 column, of 2.1 m length, 2.6 mm I.D, at 200°C. The HLADH catalyzed oxidation of eticholan-3ßo1-17one was followed by determining the initial reaction rate. The absorbance of the co-enzyme was spectroscopically measured at 340 nm.

RESULTS AND DISCUSSION

Optimization of the reaction parameters

The optimal pH of the reaction in the reverse micelles was found to be around 8. The microemulsions made were clear up to a water content of W_o=16. Optimal catalytic activity was observed at W_o=14. The reaction rate of the co-enzyme regenerating system was constant above a co-enzyme concentration of 10^{-4} M. Activity dependance on concentrations of the substrates has also been studied. Optimal activity was found at an ethanol concentration of 0.5 M and cyclohexanone concentration of 1.0 M. (ref. 5)

Enzyme stability

Fig 1 shows the storage stability in AOT-cyclohexane reverse micelle system. The enzymatic activity decreased rapidly during the first 24 hours and was then stable for at least 2 weeks. The steep fall in HLADH activity after 6 days of incubation was interpreted as being due to bacterial contamination.

The operational stability of HLADH in AOT-cyclohexane reverse micelles is presented in figs 2 and 3. A special dilution technique was used to prevent the reaction from reaching equilibrium, and also to avoid product inhibition.

The enzyme (0.02 U/ml) was added to the system at t=0. Every 24 hour samples were taken out for analysis. Half of the reaction mixture was withdrawn every 48 hour, and the same amount of the fresh microemulsion, containing substrate and co-enzyme was added. The enzyme and product concentrations were thus reduced by a factor of 2 in every dilution step.

Fig 3 shows the specific activity of the enzyme in microemulsion, calculated from the slopes in fig. 2, as a function of time. Here the activity has been compensated for the reduced enzyme concentration.

The specific acticity remained high during the whole measuring period. The decreased product concentration and the very low enzyme concentration towards the end of the experiment was interpreted to cause the apparent increase in specific activity.

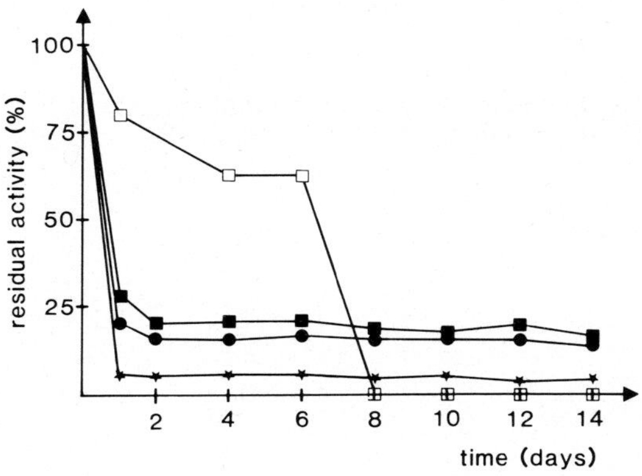

Fig. 1. Storage stability of HLADH
■ microemulsion, W_o=9.4, 60 mM trisbuffer, pH 7
● microemulsion, W_o=9.4, 60 mM trisbuffer, pH 8
★ microemulsion, W_o=9.4, 60 mM trisbuffer, pH 9
□ 60 mM trisbuffer pH 8

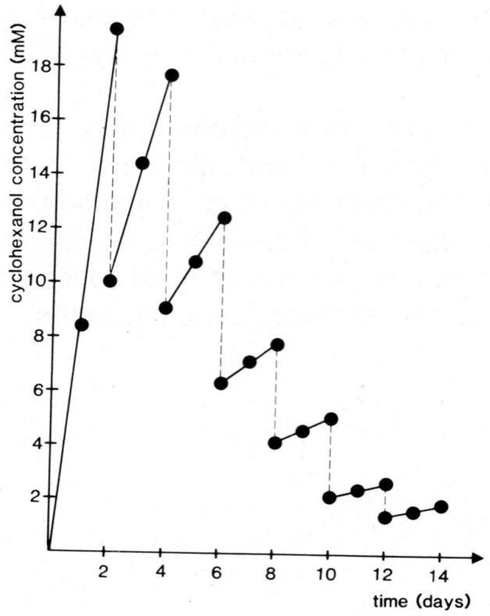

Fig 2. Dilution experiment to investigate the operational stability of HLADH in microemulsion. W_o = 9.4, 60 mM tris buffer pH 9.

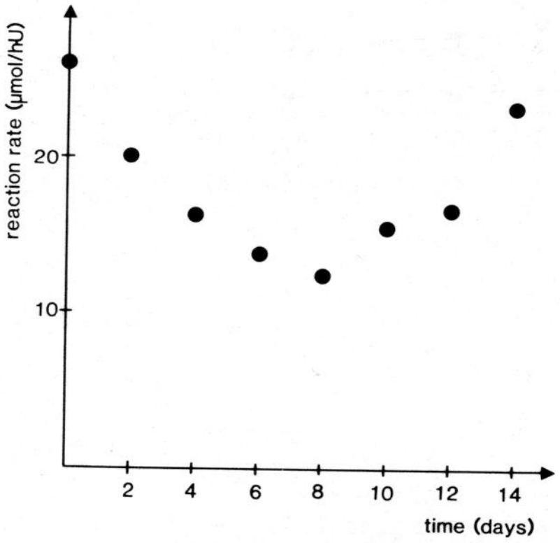

Fig. 3. Specific activity of HLADH operating in microemulsion, calculated from the slopes in Fig. 2.

<u>Steroid conversion</u>

HLADH has isoenzymes which can oxidize 3ß-hydroxyl groups in steroids. In Fig 4 is shown the pH-profile of the initial oxidation rate of eticholan-3ßol-17one in microemulsions and in buffer solutions.

The enzymatic activity was about 5 times greater in microemulsion as compared to the activity in buffer; this increase in catalytic efficiency was probably due to a better solubility of the steroid in the microemulsion.

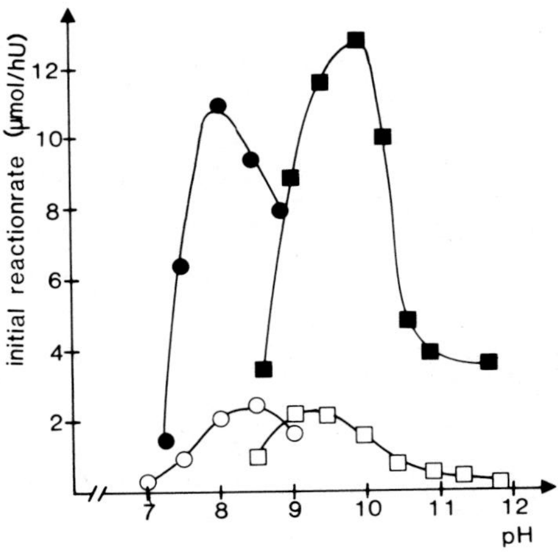

Fig. 4. pH-profile of HLADH for the oxidation of eticholan-3ßol-17one in;
● microemulsion, W_o= 12.1, 60 mM tris buffer
■ microemulsion, W_o= 12.1, 60 mM glycine buffer
○ tris buffer, 60 mM
□ glycine buffer, 60 mM

ACKNOWLEDGEMENTS

This project was supported by the National Swedish Board for Technical Development. Drs. Rajni Kaul and Krister Holmberg are acknowleged for valuable discussion.

REFERENCES

1 P.L. Luisi, L.J. Magid, CRC Critical Reviews in biochemistry 20(4)(l986), 409-474.
2 P. Meier, P.L. Luisi, J. Solid-Phase Biochem. 5(4) (l980), 269-282.
3 I.V. Berezin, and K. Martinek, Ann. N.Y. Acad. Sci. 434 Enzyme Eng. (7) 577-579.
4 K. Martinek, A.U. Levashov, Yu. L. Khmelnitsky, N. L. Klyachko and I.V. Berezin, Science 218(26) (1982), 889-891.
5 K. Larsson, P. Adlercreutz, B. Mattiasson, Activity and Stability of Horse Liver Alcohol Dehydrogenase in AOT-cyclohexane reverse micelles. (Submitted).

C. Laane, J. Tramper and M.D. Lilly (Editors), *Biocatalysis in Organic Media*,
Proceedings of an International Symposium held at Wageningen,
The Netherlands, 7–10 December 1986.

PIG LIVER ESTERASE IN ASYMMETRIC SYNTHESIS. STERIC REQUIREMENTS AND CONTROL OF REACTION CONDITIONS

J. BOUTELJE[1], M. HJALMARSSON[2], P. SZMULIK[2], T. NORIN[2]
AND K. HULT[1]
[1]Department of Biochemistry and Biotechnology, and
[2]Department of Organic Chemistry,
Royal Institute of Technology, S-100 44 Stockholm (Sweden)

SUMMARY

Pig liver esterase catalyses the enantioselective hydrolysis of prochiral and racemic diesters. The stereoselectivity is influenced by the structure of the substrate and by the reaction conditions. Organic cosolvents and the nature of the buffer used affect the enantioselectivity. An imaginary "single" active site model consistent with presented experimental results is discussed.

INTRODUCTION

Many commercial enzymes, in particular readily available hydrolases like chymotrypsin (EC 3.4.21.1) and pig liver esterase (EC 3.1.1.1) have become important tools in asymmetric organic synthesis (ref. 1). Both these enzymes have been the target for several investigations on structure/activity relationships. Unfortunately such studies on pig liver esterase suffer severely from the fact that the precise structure of the enzyme has never been elucidated, due to the difficulties in obtaining homogeneous enzyme preparations. This heterogeneity is not fully understood, but pig liver esterase preparations may contain several isoenzymes, combinations of subunits (refs. 2-3), or various other enzymes with esterolytic properties which are all difficult to separate.

In this paper we will nevertheless fearlessly discuss a hypothetical but applicable model of the pig liver esterase active site. We will also describe how the enantioselectivity of the enzyme can be influenced. Some major advantages, in choosing pig liver esterase for the preparation of optically active substances are the broad substrate specificity, the often high enantioselectivity and the commercial availability of the enzyme. The fact that no cofactor is needed is also a merit when hydrolases are chosen as catalysts for the production of optically pure starting materials for further synthesis (compare e.g. horse liver alcohol dehydrogenase).

Steric requirements

The pig liver esterase catalysed hydrolyses of diesters of 1,2-cyclohexane dicarboxylic acid and related 1,2-diesters of different ring size provide interesting information regarding structural requirements and restrictions for the active site (refs. 4-7).

It was shown that the pig liver esterase catalysed hydrolysis of the dimethyl ester of meso-cis-1,2-cyclohexane dicarboxylic acid (1a) gave the optically pure (1S,2R)-monoester (2a)(Scheme 1). The alcohol moiety is of vital importance, as was shown by the fact that the hydrolysis of the corresponding diethyl ester (1b) gave a racemic product and proceeded at a rate that was two orders of magnitude lower than in the dimethyl ester case. In contrast to this the hydrolyses of the racemic trans-diesters of 1,2-cyclohexane dicarboxylic acids (3a-c) were not very sensitive to structural variations in the ester function of the substrate (ref.4).

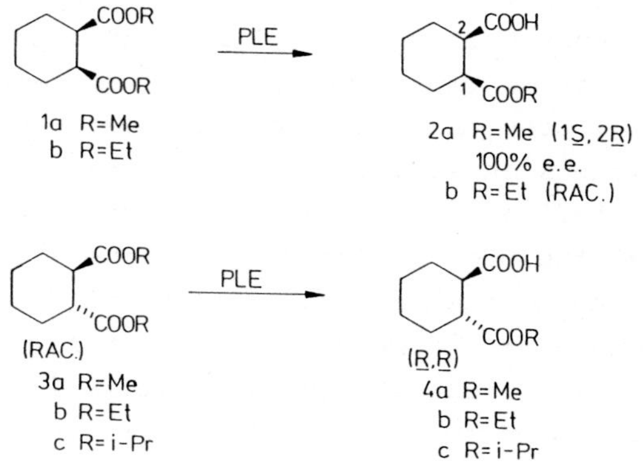

Scheme 1

The demonstration of a ring-size-mediated reversal of the enantioselectivity in the hydrolysis of cyclic meso-1,2-diesters also provides information on the features of the active site of pig liver esterase (refs. 5-7). While the 1,2-cyclohexane dimethyl ester provided (1S,2R)-monoester in high enantiomeric excess (e.e.), the cyclobutane and cyclopropane dimethyl esters gave the (1R,2S)-monoester of equally high optical purity. Accordingly the hydrolysis of the cyclopentane dimethyl ester resulted in an almost racemic monoester.

Another class of substrates where size induced reversal of enantioselectivity occurs, are the dialkylated propanedioic acid diesters (ref. 8). The pro-R ester group was primarily hydrolysed for dimethyl esters disubstituted with a methyl substituent and an ethyl (5a), propyl (5b) or butyl (5c) group in the α-position (Scheme 2). When the larger alkyl substituent was pentyl (5d) or longer chains the enantioselectivity dramatically changed to give hydrolysis of the pro-S ester group. The benzyl substituted analogue (6a) was hydrolysed with a low preference for the pro-R ester group as was expected, since the length of the benzyl group is somewhere in between that of a butyl and a pentyl group (Scheme 3). When the benzyl substituent is extended with e.g. a para-methoxy group (6b,c), the enantioselectivity changes to give hydrolysis of the pro-S ester group as predicted (refs. 4,10,11).

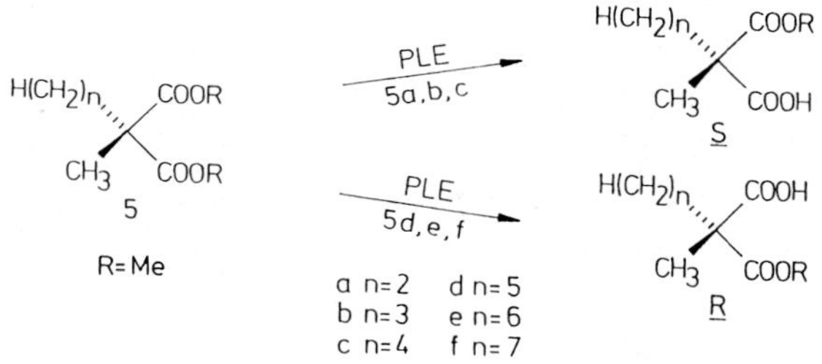

Scheme 2

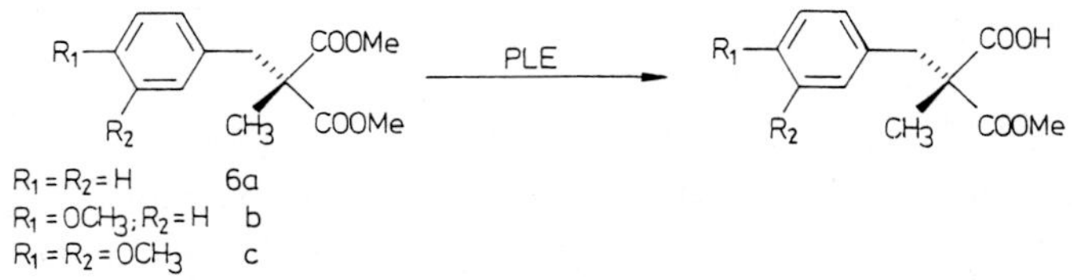

Scheme 3

With all these dialkylated propanedioic acid diesters, the substitution of methoxy for ethoxy in the alcohol part of the diester resulted in very low enantioselectivities (refs. 8-9), as was likewise experienced with the meso-cis-cyclohexane diesters.

Effect of reaction conditions

So far only the steric requirements of the substrates in pig liver esterase hydrolysis have been discussed. The reaction conditions also have an effect on the enantioselectivity of the enzyme. Kinetic measurements on the enzyme catalysed hydrolysis of trans-1,2-cyclohexane-dicarboxylic acid diesters (3) and batch incubations in the pH-stat, were compared to incubations buffered with Tris-HCl (tris(hydroxymethyl)aminomethane). Hydrolysis in the presence of Tris gave products of higher enantiomeric excess than the corresponding unbuffered reactions in the pH-stat (ref. 4).

Another case, where the buffer seems to affect the enantioselectivity, is in the esterase catalysed hydrolyses of pyrrolidine derivatives. Currently the hydrolysis of meso-cis-2,5-dicarbomethoxy-N-benzylpyrrolidine (7) and some structurally related diesters are under investigation (Scheme 4). Our results show that reactions carried out in Tris-buffer give significantly higher enantiomeric excess than the corresponding reactions run without buffer in the pH-stat (refs. 12-13). The reason for this effect is not yet clear. However, it is known that Tris can deacylate acylchymotrypsin (ref. 14) and thus likely also pig liver esterase. Whether the influence on the enantioselectivity is due to a differentiated effect on different isoenzymes, or if Tris affects the tetrahedral intermediate in an enantioselective fashion, or if there is an other complex explanation remains to be shown.

Scheme 4

The enzymatic hydrolyses of esters are normally carried out in aqueous solutions, but often organic solvents have to be added to make the substrates more soluble. Even though the effect of cosolvent addition on the kinetic parameters k_{cat} and K_m has been extensively studied for different enzymes and solvent mixtures, little is known on the influence of organic cosolvents on the stereoselectivity of enzyme catalysed reactions (ref. 11). In connection with studies on the pig liver esterase catalysed hydrolysis of benzyl derivatives of propanedioic acid diesters a notable increase of the enantioselectivities by addition of various organic cosolvents like acetone, dioxane, acetonitrile, dimethylformamide, methanol or dimethylsulphoxide (DMSO) to the medium, was observed (ref. 10 and unpublished work by the authors).

A more detailed investigation on the effect of DMSO on the enantioselectivity of the esterase catalysed hydrolyses of dialkylated propanedioic acid dimethylesters was carried out (ref. 11). The products from hydrolyses of different esters showed large differences in changes of enantiomeric excess with increasing concentration of DMSO. The effect on the enantiomeric ratio, E (ref. 15), was however very similar for all substrates, except for the one with two short alkyl substituents (5a). The substrates all had in common the increased preference for hydrolysis of the pro-S ester group with increasing concentration of DMSO. Another example is the hydrolysis of the 3,4-dimethoxybenzyl derivative (6c) from which an R-monoester of 93% enantiomeric excess at 50% (v/v) DMSO was obtained as compared to 25% e.e. without DMSO (ref. 10).

The striking effect of organic cosolvents on pig liver esterase was also taken advantage of in the hydrolyses of some pyrrolidine derivatives. With 10% (v/v) MeOH or 25% (v/v) DMSO as cosolvent in the buffered reaction medium the hydrolysis of cis-diester 7 gave the optically pure (2R,5S)-monoester 8 (refs. 12-13). Without organic cosolvent the enantiomeric excess was only 17% for the same monoester. The optically pure monoester is a potential starting material for the preparation of pyrrolidine alkaloids or carbapenem antibiotics (ref. 16) and is also an entry to less common R-proline derivatives.

Presently pig liver esterase catalysed hydrolysis of the diester of the corresponding trans-compound is being studied. Kinetic resolution, using the above mentioned means of influencing the enantiomeric excess obtained from this substrate, provides a method for the syntheses of chiral auxiliaries with C_2-symmetry (ref. 17).

Active site model

Commercial pig liver esterase batches are not homogeneous due to the above mentioned separation problems. The elucidation of the three dimensional structures of any of the specific isoenzymes will therefore be a difficult task. A more practical approach is to postulate an empirical model of an imaginary "single" active site, supported by data on the outcome of esterase catalysed reactions. When such a model is to be discussed certain facts that are confusing have to be mentioned. Several authors have pointed out differences between enzyme fractions separated by isoelectric focussing regarding the activity with different substrates or inhibition/activation caused by added organic solvents or inhibitors (refs. 3,18). However, it has recently been reported that the stereoselectivity of such enzyme fractions were fundamentally the same (ref. 19). Variations in the enantiomeric excess of the product obtained with different commercial enzyme batches, which might have differing isoenzymical compositions, have been reported (ref. 20). These variations may be caused by even slightly different reaction conditions when taking into consideration the above mentioned results.

Lately, several groups have reported on the structural requirements of the active site of pig liver esterase (ref. 5) and models for prediction of the stereoselectivity of the enzyme (refs. 8,19,21,22). Taking into account the results referred to in this paper we suggested a practical but not complete model for the active site. According to this model, the active site possesses a hydrophobic binding site with certain steric limitations. The change in enantioselectivity enforced by the length of one of the alkyl groups of the disubstituted propanedioic acid diesters, may be due to an interaction of the short side chains of the substrate with this site. The hydrophobic site is too small for alkyl groups longer than butyl and substrates with longer side chains have therefore to interact with their methyl group in the hydrophobic site, resulting in an inversion of the enantioselectivity (Fig. 1a, b). The low enantioselectivity for the diethyl esters compared to the dimethyl esters can be explained by a greater ability of the former ester function for binding to the hydrophobic site.

The increase of the pro-S selectivity with increasing concentrations of DMSO with dimethyl ester substrates, also supports our model (refs. 10-11). When the hydrophobicity of the solvent is increased, the tendency for the more hydrophobic substrate side chain to bind to the hydrophobic site, becomes weakened and we observe an increase of the (R)-enantiomer.

The model can also be used to explain the enantioselectivity of the hydrolyses of the meso-cis-1,2-cycloalkane dicarboxylic acid diesters. The cycloalkane substrates will bind with as much as possible of the ring situated in the hydrophobic binding site, affording (1S,2R)-monoester when cyclohexane diesters are hydrolysed, while cyclobutane and cyclopropane diesters give (1R,2S)-monoesters (Fig. 1c, d). For the same reason as for the dialkylated propanedioic acid diesters, the dimethyl ester gives much higher enantiomeric excess than the diethyl ester also with the cyclohexane substrate.

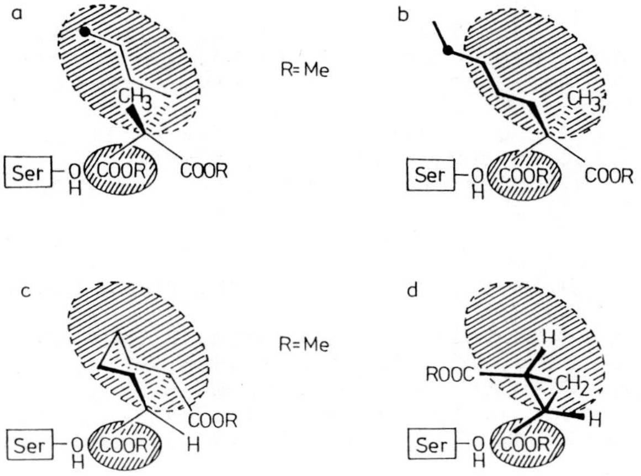

Fig. 1. Model of the pig liver esterase active site indicating the interaction of various dimethyl esters with the hydrophobic site.

The size induced inversion of enantioselectivity for N-substituted ß-aminoglutaric acid dimethyl esters fit into this model as well (ref. 21). The N-acetyl-ß-aminoglutaric acid dimethyl ester can bind with the amide group in the hydrophobic site while larger N-substituents will make the amide too large for binding there. The distance from the carbonyl group of the hydrolysed ester function to the methyl-C of the acetylgroup, is similar to the distance from the corresponding carbonyl group in the butylmethylpropanedioic acid dimethyl ester (5c) to the methyl-C of the butyl group.

A simple model cannot predict the stereochemical outcome in all cases. The possibility remains that the described effects on the enantioselectivity of pig liver esterase are caused by differentiated effects on different isoenzymes. More data on structure/activity relationships and optimization of reaction conditions will make predictions of the enantioselectivity of pig liver esterase more reliable and useful.

REFERENCES

1 J.B.Jones, Tetrahedron, 42 (1986) 3351-3403.
2 E.Heymann and W.Junge, Eur.J.Biochem., 95 (1979) 509-518.
3 D.Farb and W.P.Jencks, Arch.Biochem.Biophys., 203 (1980) 214-226.
4 F.Björkling, J.Boutelje, S.Gatenbeck, K.Hult and T.Norin, Appl.Microbiol.Biotechnol., 21 (1985) 16-19.
5 P.Mohr, N.Waespe-Sarcevic, C.Tamm, K.Gawronska and J.K.Gawronski, Helv.Chim.Acta, 66 (1983) 2501-2511.
6 M.Schneider, N.Engel, P.Hönicke, G.Heinemann and H.Görisch, Angew.Chem.Int.Ed.Engl., 23 (1984) 67-68.
7 G.Sabbioni, M.L.Shea and J.B.Jones, J.Chem.Soc., Chem. Commun., (1984) 236-238.
8 F.Björkling, J.Boutelje, S.Gatenbeck, K.Hult, T.Norin and P.Szmulik, Tetrahedron, 41 (1985) 1347-1352.
9 M.Schneider, N.Engel and H.Boensmann, Angew.Chem.Int.Ed. Engl., 23 (1984) 66.
10 F.Björkling, J.Boutelje, S.Gatenbeck, K.Hult and T.Norin, Tetrahedron Lett., 26 (1985) 4957-4958.
11 F.Björkling, J.Boutelje, S.Gatenbeck, K.Hult, T.Norin, and P.Szmulik, Bioorg.Chem., 14 (1986) 176-181.
12 F.Björkling, J.Boutelje, M.Hjalmarsson, K.Hult and T.Norin, submitted.
13 J.Boutelje, M.Hjalmarsson, K.Hult, M.Lindbäck and T.Norin, in preparation.
14 V.Kasche and R.Zöllner, Hoppe-Seyler's Z.Physiol.Chem., 363 (1982) 531-534.
15 C.S.Chen, Y.Fujimoto, G.Girdaukas, C.J.Sih, J.Am.Chem.Soc., 104 (1982) 7294-7299.
16 M.Kurihara, K.Kamiyama, S.Kobayashi and M.Ohno, Tetrahedron Lett., 26 (1985) 5831-5834.
17 Y.Kawanami, Y.Ito, T.Kitagawa, Y.Taniguchi, T.Katsuki and M.Yamaguchi, Tetrahedron Lett., 25 (1984) 857-860.
18 W.Junge and E.Heymann, Eur.J.Biochem., 95 (1979) 519-525.
19 J.B.Jones, Enzymes in organic synthesis. Some illustrative examples, in: Proceedings, FECS Third International Conference on Chemistry and Biotechnology of Biologically Active Natural Products, Sofia, Bulgaria, September 16-21, 1985, Vol. 1, pp. 18-39.
20 L.K.P.Lam, R.A.H.F.Hui and J.B.Jones, J.Org.Chem., 51 (1986) 2047-2050.
21 M.Ohno, Creation of novel chiral synthons with enzymes: application to enantioselective synthesis of antibiotics, in: R.Porter and S.Clark (Eds.), Enzymes in organic synthesis, Ciba Foundation symposium III, Pitman, London, 1984, pp. 171-183.
22 M.Ohno, S.Kobayashi and K.Adashi, Creation of novel chiral synthons with pig liver esterase: application to natural product synthesis and the substrate recognition, in: Proceedings, NATO Advanced Research Workshop on Enzymes as Catalysts in Organic Synthesis, Reisenburg/Ulm (Donau), F.R.G., June 16-22, 1985, D.Reidel Publish. Co., Dordrecht, 1986, pp. 123-142.

C. Laane, J. Tramper and M.D. Lilly (Editors), *Biocatalysis in Organic Media*, Proceedings of an International Symposium held at Wageningen, The Netherlands, 7–10 December 1986.

A COMPARISON OF THE ENZYME CATALYZED FORMATION OF PEPTIDES AND OLIGOSACCHARIDES IN VARIOUS HYDROORGANIC SOLUTIONS USING THE NONEQUILIBRIUM APPROACH

K. G. I. NILSSON

Swedish Sugar co. Ltd., Research and Development, P.O. Box 6, S-232 00 Arlöv, (Sweden).

SUMMARY

Procedures have been developed for the preparation of a number of biologically active oligosaccharide glycosides, by means of glycosidase catalyzed transglycosidations using various α- or β-glycosides as acceptors. The reactions were carried out with both soluble and immobilized glycosidases in aqueous solvents containing 0 - 45 % N,N-dimethylformamide. The yields of α-galactosidase catalyzed p-nitrophenyl digalactoside and methyl digalactoside formation were found to decrease with the concentration of organic cosolvent. However, the ratio between the formed isomers was largely unaffected. The results are compared with those for the chymotrypsin catalyzed aminolysis of amino acid esters in various hydroorganic solutions.

INTRODUCTION

The unravelling of the various important biological functions of the carbohydrate structures present as glycoconjugates in living organisms (e.g. as bloodgroup determinants, tumor associated antigens, receptors for pathogens) has led to an intense activity for their isolation and synthesis (refs. 1,2). A number of elaborate organic chemical synthetic routes have been devised, but all these are restricted by the many protection and deprotection steps that are necessary to effect regioselective and stereospecific glycoside synthesis (ref. 3). It is well known that glycosidases catalyze not only the hydrolysis, but also the stereospecific formation of glycosidic bonds (ref. 4). The wider application of glycosidases in oligosaccharide synthesis has, however, been hampered by the frequently occuring predominant formation of 1-6 glycosidic linkages, while the often more interesting 1-2, 1-3 or 1-4 linkages have been formed to a lesser extent or not at all (refs. 4-6).

We have recently found that the regioselectivity of glycosidase catalyzed oligosaccharide formation can be changed by using a simple strategy, employing various α- or β-glycosides as glycosyl acceptors (refs. 7,8). A schematic outline of the reactions involved (transglycosidations (ref. 4) i.e. the kinetic approach) is shown below:

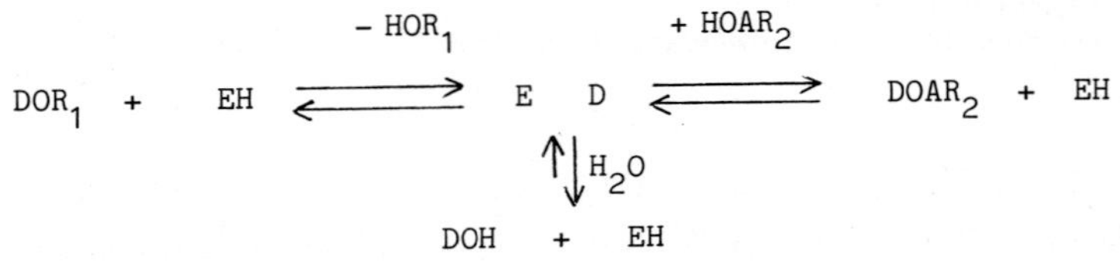

(DOR_1 = glycosyl donor, R = aglycon, HOA = glycosyl moiety of acceptor, EH = glycosidase).

By using this strategy, the predominant formation of glycosidic linkages other than 1-6 can be obtained with glycosidases, which with the corresponding free sugars (i.e. HOA) as acceptors predominantly give 1-6 linkages. (For example, it was possible to manipulate an α-galactosidase to catalyze the predominant formation of either α1-2, α1-3 or α1-6 linked digalactosides (refs. 7,8). Another advantage of the method is that useful glycosides (e.g. methyl, acrylate, nitrophenyl glycosides) suitable for use as inhibitors, enzyme substrates, affinity labels, or for attachment to chromatography media, can be synthesized by chosing an acceptor glycoside (i.e. $HOAR_2$) with the proper aglycon.

Hydrolysis is a competing side-reaction in both glycosidase catalyzed transglycosidation (see scheme above) and in protease catalyzed aminolysis of esters, thus lowering the yield of glycoside and peptide bond formation, respectively (refs. 4,9). In a previous study it was shown that the yield of peptide catalyzed by chymotrypsin using the kinetic approach increases considerably when the water activity is lowered by addition of organic cosolvent (ref. 10). Thus, it could be expected that the use of organic cosolvent also should favor transglycosidation. The results obtained with various acceptor glycosides are discussed.

MATERIAL AND METHODS

α-Galactosidase (EC 3.2.1.22; coffee bean), β-galactosidase (EC 3.2.1.23; E. coli, grade VIII) and all the monosaccharide glycosides were from Sigma (St. Louis, Mo, USA). α-Mannosidase (EC 3.2.1.24; jack bean) was from Boehringer. All reagents used were of analytical grade and used as supplied. All sugars were of the D-configuration. The reactions were performed in mixed solutions of buffer and DMF (N,N-dimethylformamide, 0-45%, V/V). The reactions were followed by TLC, HPLC and spectrophotometric measurement (400 nm) of liberated nitrophenol. The products were isolated by LC on silica (Merck, Kieselgel 60, 230-400 mesh) and Sephadex G10 (Pharmacia) and their purity (usually >97%) and structure determined with HPLC (Waters equipment Spherisorb, 5 µm, NH_2-silica; Nucleosil, 5 µm, C_{18}-silica) and NMR (^{13}C, ^{1}H; Varian XL 200), (ref. 11). The correlation with literature data where available, was satisfactory. Complete assignment of protons and their coupling constants were achieved on peracetylated products (ref. 11). Methylation analyses of the isolated products were carried out when NMR literature data were not available in order to confirm the NMR determinations.

In a typical experiment, 0.07 M p-nitrophenyl α-mannopyranoside (Man(α)-OC_6H_4-NO_2-p), was prepared in a mixture of 60 ml 0.05 M sodium phosphate, pH 6.5, and 20 ml of DMF. α-Mannosidase (5 U) was added and the reaction was continued at room temperature. The reaction was stopped after 72 h by heating. (Note that in this type of synthesis the donor and acceptor were the same compound). The product was isolated by LC on silica (eluent CH_2Cl_2:MeOH:H_2O, 6:4:0.5 V/V/V), peracetylated with pyridin and acetic anhydride, chromatographed on a silica column and deacetylated in methanol with a catalytic amount of sodium methoxide. The product was dried and analyzed as described above.

In a typical experiment with α-galactosidase, 0,1 M p-nitrophenyl α-D-galactopyranoside and 0.45 M methyl α-D-galactopyranoside (Gal(α)-OMe) were prepared in 100 ml 0.03 M sodium phosphate, pH 6.5. α-Galactosidase (100 U) was added and the reaction was continued at room temperature. The reaction was stopped after 68 h by heating. The product (Gal(α1-3)Gal(α)-OMe) was isolated and characterized by the same steps as described for the p-nitrophenyl dimannoside. The reactions referred to in table 1 were performed in 25-30 % DMF.

The yield of products as a function of time and DMF-concentration was followed by HPLC (C_{18}-silica for the nitrophenyl glycosides and NH_2-silica for the methyl glycosides, see above). The yield was calculated from the initial glycosyl donor concentration.

RESULTS AND DISCUSSION

The effect on the regioselectivity of α-galactosidase, β-galactosidase and α-mannosidase catalyzed glycoside bond formation when various α- or β-glycosides are used as acceptors is summarized in Table 1.

Many of the disaccharide sequences synthesized occur widely in various glycolipids (e.g. Gal(β1-3)Gal, and Gal(α1-3)Gal, which also is a fragment of blood group determinant B), and in N-linked oligosaccharides (Man(α1-2)Man). The glycosides can be used as enzyme substrates, inhibitors, or following modification of the nitrophenyl group as affinity labels, or for coupling to chromatography media and proteins.

TABLE 1

Synthesis of disaccharide glycosides with α-galactosidase, β-galactosidase and α-mannosidase.

Enzyme	Glycosyl donor	Glycosyl acceptor	Main glycosides formed[a]	Yield %
α-Galactosidase				
	Gal(α)-$OPhNO_2$-p	Gal(α)-OMe	Gal(α1-3)Gal(α)-OMe	27
	Gal(α)-$OPhNO_2$-p	Gal(β)-OMe	Gal(α1-6)Gal(β)-OMe	18
	Gal(α)-$OPhNO_2$-p	Gal(α)-$OPhNO_2$-p	Gal(α1-3)Gal(α)-$OPhNO_2$-p	14
	Gal(α)-$OPhNO_2$-o	Gal(α)-$OPhNO_2$-o	Gal(α1-2)Gal(α)-$OPhNO_2$-o	6
β-Galactosidase				
	Gal(β)-$OPhNO_2$-o	Gal(α)-OMe	Gal(β1-6)Gal(α)-OMe	14
	Gal(β)-$OPhNO_2$-o	Gal(β)-OMe	Gal(β1-3)Gal(β)-OMe	22
α-Mannosidase				
	Man(α)-$OPhNO_2$-p	Man(α)-OMe	Man(α1-2)Man(α)-OMe	18
	Man(α)-$OPhNO_2$-p	Man(α)-OMe	Man(α1-2)Man(α)-$OPhNO_2$-p	8

a) Abbreviated nomenclature of glycosides accordning to IUB-IUPAC recommendations (J. Biol. Chem. (1982) 257, 3347-3351).

As can be seen from the table, both the structure of the aglycon and the anomeric configuration of the glycosidic linkage between the glycosyl moiety and the aglycon have a pronounced influence on the regioselectivity. Thus, in the case of α-galactosidase, the methyl β-galactoside predominantly leads to 1-6 linked digalactoside, while the methyl α-galactoside almost exclusively gives 1-3 linked digalactoside. The reverse result is found with β-galactosidase. Furthermore, with α-galactosidase, 1-2 linkage and a small amount of an unidentified linkage are obtained with the o-nitrophenyl α-galactoside as acceptor, while with the corresponding p-nitrophenyl glycoside 1-3 linkage is predominantly obtained and only a small amount of 1-2 linkage.

The above effects can be rationalized if one considers the reported hydrophobicity of the glycosyl acceptor sites of glycosidases (ref. 12). The aglycons of the above acceptor glycosides are expected to interact with these acceptor sites and cause a perturbed binding of the glycosyl moiety. The perturbation should be different if the acceptor is bound in α- or β-configuration. In line with this explanation is also the fact that the bulky, more hydrophobic aryl groups gave larger effects than the corresponding methyl glycosides.

The above experiments were carried out in hydroorganic solutions (25-30% DMF). These were used since previous studies on protease catalyzed aminolysis of ester substrates (i.e. the kinetic (non equilibrium) approach) showed that the yield of peptide increased considerably with increased organic cosolvent concentration (ref. 10). This is expected since water is compet-

TABLE 2.

Effects of DMF and butanediol on chymotrypsin-agarose catalyzed peptide syntheses[a]

Solvent	%	Acceptor	Concentration M	Yield %
DMF	10	Gly-NH_2	0.1	14
DMF	30	Gly-NH_2	0.1	19
DMF	50	Gly-NH_2	0.1	37
DMF	50	Gly-NH_2	0.2	69
Butanediol	50	Gly-NH_2	0.2	40
Butanediol	80	Gly-NH_2	0.2	65
Butanediol	90	Gly-NH_2	0.2	86
Butanediol	90	D-Ala-NH_2	0.1	33

a) The concentration of Ac-Phe-OMe was 0.1 M. DMF=N,N-dimethylformamide. Data obtained with permission from reference 10.

ing with the amino acid acceptor for the acyl-enzyme intermediate. A good yield of a D-amino acid containing dipeptide (Ac-L-Phe-D-Ala-OMe) was obtained when a high concentration of polyol (90% butanediol) was used as cosolvent (see table 2). However, solvents such as acetonitrile, dioxan, and DMF had a favorable effect on peptide yield at a lower concentration than the 1,4-butanediol (ref. 10). Dipolar aprotic solvents have been advocated to have a greater tendency to bind to proteins and may thus lower the water activity around the protease more effectively than the polyol (ref. 13).

However, the yield of α-galactosidase catalyzed transglycosidation was found to decrease with the concentration of DMF (see table 3). In fact the yield of Gal(α1-3)Gal(α)-OPhNO$_2$-p increases by a factor of three when the DMF-concentration is decreased from 45 to 0%. A possible explanation for this may be that the acceptor binds to the enzymes' acceptor binding site with a much lower affinity the higher the concentration of organic cosolvent is. The conformation of the enzyme does not seem to be distorted by the cosolvent, because the

TABLE 3

Effect of DMF on -galactosidase catalyzed glycoside bond formation[a]

Acceptor	Concentration of cosolvent %	Main glycoside formed	Yield %
Gal(α)-OMe	0	Gal(α1-3)Gal(α)-OMe	32
Gal(α)-OMe	10	Gal(α1-3)Gal(α)-OMe	27
Gal(α)-OMe	20	Gal(α1-3)Gal(α)-OMe	24
Gal(α)-OMe	30	Gal(α1-3)Gal(α)-OMe	21
Gal(α)-OPhNO$_2$-p	0	Gal(α1-3)Gal(α)-OPhNO$_2$-p	32
Gal(α)-OPhNO$_2$-p	15	Gal(α1-3)Gal(α)-OPhNO$_2$-p	18
Gal(α)-OPhNO$_2$-p	30	Gal(α1-3)Gal(α)-OPhNO$_2$-p	14
Gal(α)-OPhNO$_2$-p	45	Gal(α1-3)Gal(α)-OPhNO$_2$-p	10

a) Gal(α)-OPhNO$_2$-p (0.1 and 0.15 M, respectively) was used as a glycosyl donor.

regioselectivity of the enzyme, i.e. the ratio between the formed α1-2, α1-3 and α1-6 linked p-nitrophenyl digalactosides, remains largely constant (ca 5:30:1) with the different DMF concentrations used.

Another glycosidase, α-amylase, has been reported to bind its substrate with a lower affinity in aqueous cosolvents than in water (ref. 14). This was attributed to a weakened hydrophobic interaction between CH and CH_2 groups of the substrate and amino acid side chains of the enzyme binding sites upon addition of the cosolvent. It may thus be that a hydrophobic interaction between the acceptor and the acceptor binding site is weakened when DMF is used in the transglycosidation reactions with α-galactosidase, thus leading to a decreased binding of the acceptor to the glycosyl enzyme intermediate with a lower yield of product glycosides as a consequence. This is supported by that a more pronounced effect on the glycoside yield was obtained with the more hydrophobic glycoside (i.e. the nitrophenyl digalactoside, see table 3). The explanation is also in line with the above suggested rationalization of the changed regioselectivity of glycosidase catalyzed oligosaccharide formation employing various α- or β-glycosides as acceptors.

An advantage of miscible organic solvent-water systems compared with aqueous solutions without cosolvent is that the solubility of glycosides with hydrophobic aglycons normally are enhanced in the former, thus allowing the use of higher concentrations of such glycosides as acceptors. Alternatively, higher reaction temperatures can be used with aqueous solutions and this, in addition, allows the use of high concentrations of cheap, but at room temperature less reactive, glycosyl donors such as methyl glycosides and oligosaccharides. This and principles for the in situ preparation of the acceptor glycosides are under investigation.

REFERENCES

1 Biology of Carbohydrates, vol.2(1984), V. Ginsburg and P. W. Robbins (eds.) Wiley, New York.
2 S. Hakomori, Ann. Rev. Biochem., 50(1981) 733-764.
3 H. Paulsen, Chem. Soc. Rev., 13(1984) 15-45.
4 K. Wallenfels and R. Weil, The Enzymes, 7(1972) 617-663.
5 L. Hedbys, P.- O. Larsson, K. Mosbach, and S. Svensson, Biochem. Biophys. Res. Commun., 123(1984) 8-15.
6 E. Johansson, L. Hedbys, P.- O. Larsson, K. Mosbach, A. Gunnarsson, and S. Svensson, Biotechnol. Lett., 8(1986) 421-424.
7 K. G. I. Nilsson, presented at the 8th International Carbohydrate Symposium, Ithaca, N.Y., USA, August 10-15, 1986.
8 K. G. I. Nilsson, submitted for publication.
9 J. S. Fruton, Adv. Enzymol., 53(1982) 239-306.
10 K. G. I. Nilsson and K. Mosbach, Biotechnol. Bioeng. 26(1984) 1146-1154.
11 J. Dahmén, T. Frejd, T. Lave, F. Lindh, G. Magnusson, G. Noori, and K. Pålsson, Carbohydr. Res., 113(1983) 219-224.
12 R. E. Huber and M. T. Gaunt, Arch. Biochem. Biophys., 220(1983) 263-271.
13 K. Gekko and S. N. Timasheff, Biochemistry, 20(1981) 4667-4675.
14 H. Nagamoto, T. Yasuda, and H. Inoue, Biotechnol. Bioeng.,28(1986) 1172-77.

C. Laane, J. Tramper and M.D. Lilly (Editors), *Biocatalysis in Organic Media*,
Proceedings of an International Symposium held at Wageningen,
The Netherlands, 7–10 December 1986.

OPTICAL RESOLUTION OF PHENYLALANINE BY ENZYMIC TRANSESTERIFICATION

E. Flaschel and A. Renken

Institut de génie chimique, Ecole Polytechnique Fédérale de Lausanne, Chimie-Ecublens, CH-1015 Lausanne (Switzerland)

SUMMARY

The acyl transfer of phenylalanine to methanol by means of chymotrypsin starting from phenylalanine propyl ester has been studied. This transesterification turns out to be useful for producing optically pure phenylalanine esters as well as for the optical resolution of phenylalanine. The L-phenylalanine methyl ester may be obtained with yields exceeding 80 %. The reaction medium required for transesterification represents a special case with respect to the use of organic solvents in enzymic catalysis, since methanol can be regarded as a water miscible organic solvent, but behaves, in addition, as a co-substrate.

INTRODUCTION

Since the classical experiments of M.L. Bender et al. (ref. 1) it is well established that chymotrypsin forms an acyl intermediate during hydrolysis. He was able to show that other nucleophiles like amino- and hydroxyl groups can compete with water for accepting the acyl radical. The aminolytic route is used successfully for peptide synthesis (e.g. ref. 2). The alcoholysis, however, also called transesterification, seems to have found only little attention for practical applications in amino acid or peptide chemistry. For synthetic purposes, the transferring activity most often leads to better results than the reversal of hydrolysis, because higher rates as well as yields exceeding chemical equilibrium may be achieved (e.g. ref. 3, 4).

An optimization of the enzymic productivity for the preparation of the L-phenylalanine methyl ester from the racemate of the phenylalanine propyl ester was undertaken to study the influence of the main operating variables on the transesterification reaction catalyzed by chymotrypsin.

EXPERIMENTAL

Soluble chymotrypsin was purchased from SIGMA (Type II, C 4129). Phenylalanine propyl ester was obtained in the form of its hydrogen propyl sulfate salt. All other reagents were of analytical grade.

The reaction was carried out in discontinuous mode in a jacketed stirred tank which was kept at constant temperature. The desired pH was established with 5 M NaOH solution prior to adding the soluble chymotrypsin dissolved in 1 ml of water. During reaction the pH was automatically kept constant by adding 1 M NaOH solution.

Aliquots were taken in intervals of 10 min and analysed by HPLC with a column of 25 cm in length and 4.6 mm in diameter filled with Spherisorb 5 C8. The eluant contained 10 g KH_2PO_4 and 1 g tetrabutylammonium hydrogen sulfate in 1.2 l of water. This solution was adjusted to pH 2.1 with concentrated phosphoric acid prior to adding 0.8 l of methanol. A UV detector was used at a wavelength of 256 nm.

RESULTS AND DISCUSSION

Since optimization was not possible without specifying a desired degree of conversion, the optimum attainable yield of the methyl ester was taken as the reference quantity. The transformation always has to go through a maximum yield of methyl ester since both esters are hydrolysed at the same time the transesterification takes place. The yield/time profiles of the methyl ester and phenylalanine were modelled according to:

$$X = t\,(a_1 + a_2t + a_3t^2)^{-1} \quad (1)$$

The time necessary to obtain the maximum methyl ester yield was:

$$t_{opt} = (a_1/a_3)^{1/2} \quad (2)$$

from which the optimum yield (X_{opt}) is obtained by inserting Eq. (2) in Eq. (1). The productivity of methyl ester formation is given by:

$$L_{opt} = S_{Lo}X_{opt}/Et_{opt} \quad (3)$$

The main operating variables varied were the temperature (T_c), the pH, the volumetric portion of methanol (v_{Me}), and the initial substrate concentration (S_{Lo}). The optimization was started with 6 intuitive settings of these variables. Then a simplex method according to Nelder-Mead (ref. 5) was applied. With 13 measurements the strategy was changed to an interpolation procedure with a linear model with additive polynomials of third order for each variable defined around the center in the space of variables. The maxima predicted by this simple model were investigated to end up with 36 data sets.

Some of these data are given in Tab. 1. It does not intend to show how the optimization took place, but how the variables influenced the optimum yield, the optimum methyl ester concentration (P_{opt}) and the productivity. The data in Table 1 are given in the order of increasing productivity.

The optimization led to pH values of 6.5 to 7 and methanol concentrations of less than 10 %. Both the pH and the methanol portion were the main variables to optimize the productivity. The influence of the pH can be seen in Fig. 1, where all the data are gathered. It has to be emphasized that the data were obtained by a certain strategy and are, in consequence, not statistically distributed with respect to the other operating variables. Lowering the methanol concentration in general led to lower methyl ester yields as shown in Fig. 2. The optimization strategy favoured operating conditions, which led to moderate maximum yields, since attaining yields in excess of 80 % was normally combined with low rates of reaction. The temperature accelerated both the rate of hydrolysis and of methanolysis. The initial substrate concentration had to be kept low, certainly due to the pronounced inhibition by the D-ester and the counter-ion propyl sulfate (ref. 6).

When the combined hydrolysis and alcoholysis is considered as a general method for optical resolution, the sum of the yields of the ester and the acid formed has to be considered. Two examples are shown in Fig. 3 and Fig. 4 for operating conditions corresponding to entry No. 2 and 15, respectively, of Tab. 1. Besides the experimental data, the estimated time course according to the simple model of Eq. (1) is given. The rate of depletion of the L-propyl ester follows the productivity profile of the methyl ester. Rather high rates and high degrees of conversion may be obtained by using the combined effect of hydrolysis and alcoholysis.

TABLE 1

Examples of transesterification experiments

No.	T_c °C	pH -	v_{Me} vol %	S_{Lo} mM	X_{opt} %	P_{opt} mM	L_{opt} mmol/g min
1	15.0	6.00	25.0	57.6	85.7	49.3	3.86
2	18.8	5.82	22.5	72.0	86.5	62.2	3.87
3	25.0	5.75	16.0	100.7	80.3	80.9	4.98
4	28.5	5.62	20.0	72.0	80.6	58.0	5.21
5	25.2	5.96	19.6	72.5	80.8	58.6	7.54
6	28.0	7.50	24.0	92.1	78.5	72.3	8.80
7	28.0	7.00	24.0	115.1	78.9	90.8	10.61
8	28.0	6.50	24.0	92.1	80.7	74.3	10.74
9	28.0	6.50	18.0	86.3	79.5	68.7	12.39
10	28.0	7.00	24.0	92.1	80.5	74.1	12.89
11	30.0	6.75	16.0	80.6	73.5	59.2	12.90
12	29.9	7.00	12.1	72.0	68.8	49.5	13.66
13	32.1	6.96	7.0	128.4	53.6	68.8	14.49
14	32.1	6.96	7.0	74.0	60.0	44.4	14.82
15	35.2	6.68	9.0	70.2	63.3	44.4	17.05

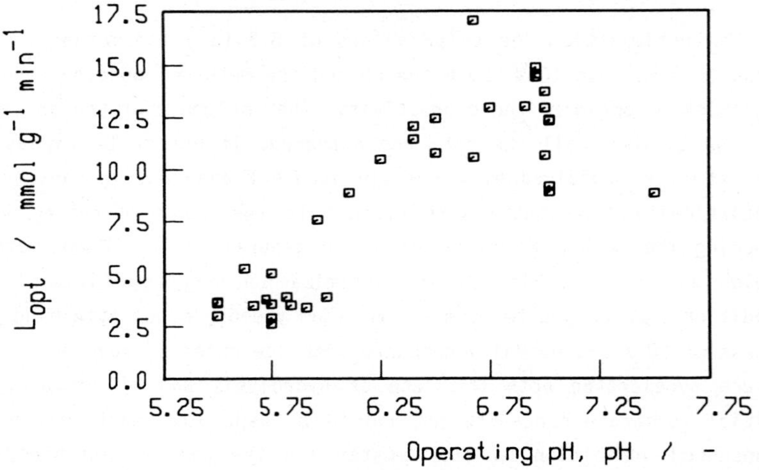

Fig. 1. Influence of the operating pH on maximum transesterification performance

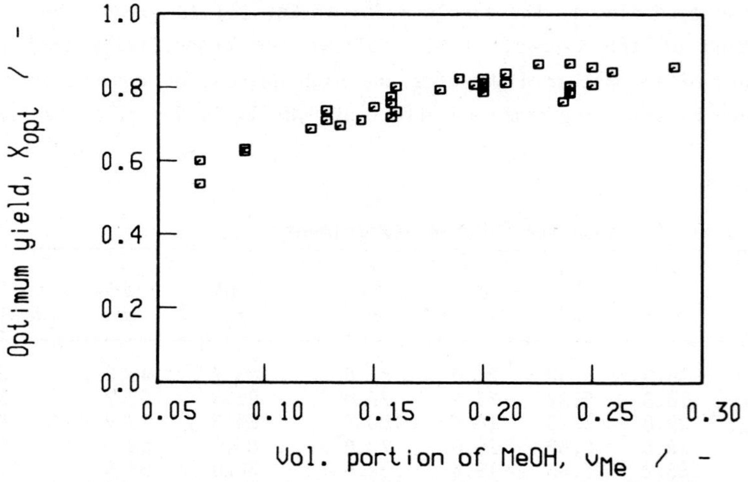

Fig. 2. Influence of the methanol concentration on the maximum methyl ester yield

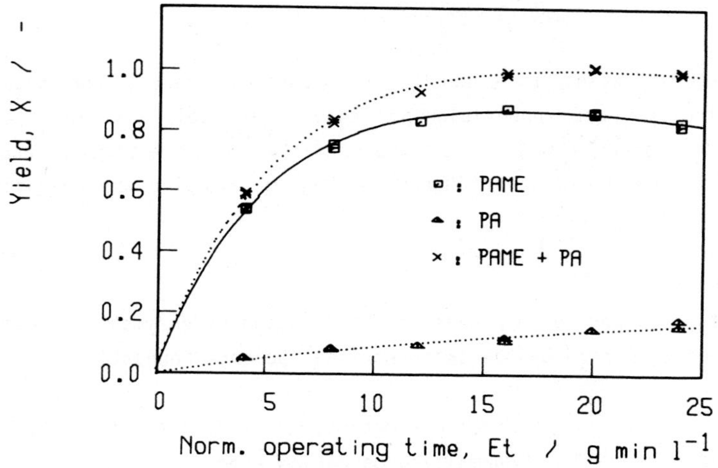

Fig. 3. Example of a transesterification reaction giving a high methyl ester yield (See Table 1, No. 2)

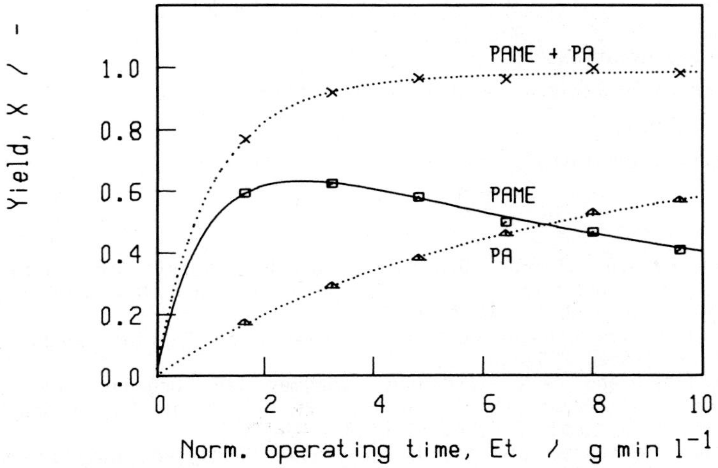

Fig. 4. Example of a transesterification reaction giving a high methyl ester productivity (See Table 1, No. 15)

CONCLUSION

Transesterification is shown to be a useful process for producing enantiomerically pure phenylalanine esters as well as for optical resolution of phenylalanine. The productivity of chymotrypsin and the maximum ester yield can be influenced mainly by changing the operating pH and the alcohol concentration.

ACKNOWLEDGEMENT

This study is part of a project on utilization of enzymes in two-liquid phase systems supported by the Swiss National Science Foundation.

SYMBOLS

a_i	-	Model parameters, i = 1-3, PAME; i = 1-2, PA
Et	g min l^{-1}	Normalized operating time
L	mmol g^{-1} min^{-1}	Productivity of methyl ester
P	mmol l^{-1}	Product (methyl ester) concentration
S_{Lo}	mmol l^{-1}	Initial substrate (propyl ester) concentration referred to the L-component of the racemate
t	min	Operating time
T_c	$^\circ$ C	Operating temperature
v_{Me}	-, vol %	Volumetric portion of methanol
X	-, %	Yield, (= P/S_{Lo})

Indices and abbreviations

opt	Referring to maximum yield of methyl ester
PA	Phenylalanine
PAME	Phenylalanine methyl ester

REFERENCES

1 M.L. Bender, G.E. Clement, G.R. Gunter and F.J. Kézdy, The Kinetics of α-Chymotrypsin Reactions in the Presence of Added Nucleophiles, J. Amer. Chem. Soc., 86 (1964) 3697-3703
2 J.S. Fruton, Proteinase-Catalyzed Synthesis of Peptide Bonds, Adv. Enzymol. 53 (1982) 239-306
3 K. Martinek and A.N. Semenov, Enzymes in Organic Synthesis: Physicochemical Means of Increasing the Yield of End Product in Biocatalysis, J. Appl. Biochem. 3 (1981) 93-126
4 V. Kasche, Mechanism and Yields in Enzyme Catalysed Equilibrium and Kinetically Controlled Synthesis of β-Lactam Antibiotics, Peptides and other Condensation Products, Enz. Microb. Technol. (1986(8)) 4-16
5 J.A. Nelder and R. Mead, A Simplex Method for Function Minimization, Computer J., 7 (1965) 308-313
6 E. Flaschel and A. Renken, unpublished results

C. Laane, J. Tramper and M.D. Lilly (Editors), *Biocatalysis in Organic Media*, Proceedings of an International Symposium held at Wageningen, The Netherlands, 7–10 December 1986.

THE EFFECTS OF SOLVENTS ON THE KINETICS OF FREE AND IMMOBILIZED LIPASE

T.UÇAR[1], H.I.EKİZ[2], S.S.ÇELEBİ[3] and A.ÇAĞLAR[2]

[1]Lassa Research and Development Department, İzmit (Turkey)

[2]Chemical Engineering Department, Fırat University, Elazığ (Turkey)

[3]Chemical Engineering Department, Hacettepe University, Ankara (Turkey)

SUMMARY

The effects of solvents on the kinetics of enzymatic hydrolysis of lipids by free and immobilized lipase of *Candida cylindracea* have been investigated.

In the hydrolysis of tributyrin, non-polar solvents do not affect the pH and temperature optima, however, increasing amounts of these solvents decrease the activitys of free and immobilized lipase. n-Heptane was found to be the optimum non-polar solvent. Increasing amounts of polar solvents also decrease lipase activity but do not interfere with the pH optimum. Acetone has been found to have the least inhibitory effect for free lipase up to 30 % (v/v).

Triacetin, which is water soluble up to 6 % (v/v) has been used for the study of interfacial effects on lipase activity In the soluble region, the variation of the hydrolysis rate by triacetin concentration shows a sigmoid behaviour, due to formation of micelles. Various non-polar solvents have been tested to create interfacial area in the soluble region. Non-polar solvents show an activating effect, enhance free lipase activity up to 250-360 % for n-alkanes 5 to 10 carbons.

INTRODUCTION

The hydrolysis of lipids is carried out in aqueous media with the substrate in emulsified form (ref. 1). Insoluble liquid lipids, emulsified in aqueous solutions are easily hydrolyzed by lipases. On the other hand, solid lipids and some simple liquid lipids of soluble concentrations in water are hydrolyzed only at low rates (ref. 2-3). When solid lipids and simple liquid lipids are dissolved in suitable organic solvents, the rate of hydrolysis increases noticeably. However, similar solvents cause decreases in the rates of hydrolysis of water insoluble liquid lipids such as tributyrin. The loss of activity and the change in the behaviour of the enzyme in the presence of organic solvents deserve a detailed investigation. Therefore, the effects of various organic solvents on the hydrolysis of tributyrin and triacetin (especially at the soluble concentrations) by a candidal lipase were investigated. Of solvents investigated, n-heptane was found to be the most suitable solvent (ref.3) The effects of n-heptane on the rate of hydrolysis of tributyrin have been investigated in detail using free and immobilized lipase.

The hydrolysis of triacetin has been investigated in both soluble and insoluble regions. In the soluble region, the rate of hydrolysis varies with triacetin concentration, as a sigmoid curve. The response of this type was shown to be the result of formation of triacetin micelles in aqueous media, which starts to form between 3.50 % and 3.75 % (v/v) of triacetin. The effects on the initial hydrolysis rates of the various members of the homologous series of n-hydrocarbons were also investigated. All of the members of this series, having 5 to 10 carbon atoms in the hydrocarbon chain, show an activating effect in the soluble region.

MATERIAL AND METHODS

Experimental Set-up :

A batch reactor with controlled pH, temperature and stirring rate has been used. In monitoring the reaction rate, a pH-stat system, a dose integrator and a recorder were used.

Preparation of Lipase Solution :

The lipase of Candida cylindracea used in all the experiments is in powder form with an activity of 50.6μmoles acid formed/min/mg protein. Lipase dissolved in a 0.01M phosphate buffer has been used in this study.

Immobilization Technique :

Controlled-pore glass used in all the experiments had a mesh size of 80-115, pore diameter 729 Å and a surface area of 33.9 m^2 /g. It was silanized by refluxing 3 % (v/v) solution of γ-aminopropyltriethoxysilane in toluene for 24 hours (ref. 4). The enzyme was immobilized by cross-linking with glutaraldehyde.

Substrate Solutions :

The reaction media have consisted of 0.25-8. 00 % (v/v) tributyrin or 0.10-10.00 % (v/v) triacetin. In aqueous phase, the concentration of NaCI is 0.137M, of $CaCI_2 . 4H_2O$ is 3.75 x 10^{-3}M and of sodium butyrate (or of sodium acetate in the hydrolysis of triacetin) is 2.5 x 10^{-3}M.

Experimental Conditions :

The total volume of the reaction mixtures was 350 ml and the stirring rate for all was 600 rpm. Unless otherwise noted, the pH value was 6 and the temperature was 35 °C. Lipase concentrations in the reaction media were expressed as protein: 5 mg/l of free lipase for pure tributyrin, 10 mg/l of free lipase both pure triacetin and triglycerides dissolved in n-heptane and 50 mg of immobilized lipase.

RESULTS AND DISCUSSIONS

Effect of pH :

In the hydrolysis of tributyrin by free and immobilized lipase, the pH profiles are similar, the maxima being around a pH value of 6. The maximum operational stability in both cases lies between pH values of 5 and 6. (ref. 3).

Effect of Temperature :

The activity of free lipase in the hydrolysis of tributyrin shows a maximum between 35 and 40° C, with rapid loss of activity is around 55 °C. In both cases the operational stability decreases with increasing temperatures. On the other hand, considerable increase in thermal stability is realized with immobilization (ref. 3).

Solvent Effects on The Hydrolysis of Tributyrin :

In order to determine the effects of solvents on the lipase activity, where tributyrin as the model substrate, various polar and non-polar solvents were used in the range of 0 to 50 % (v/v). All of the solvents caused a decrease in activity of free lipase in varying magnitudes proportional to increasing amounts of the solvents. Of those tried, n–heptane was found to be most suitable non-polar solvent, causing the minimum loss in activity. Acetone, a polar solvent, has been determined to have the least inhibitory effect up to 30 % (v/v)In with normal alcohols, the activity loss of lipase increases with chain length, methanol causing the minimum decrease in the reaction rate.

The presence of n-heptane and acetone does not alter the pH and temperature profiles of free lipase. The maximum operational stability lies in the pH interval of 5 to 5.5 but as the temperature increases in the range of 30-50 °C there is a proportional decrease in enzyme stability. When n-Heptane is absent, the activities of free and immobilized lipase show maxima at tributyrin concentration of 2-3 % (v/v) and then decrease with the increasing concentrations of tributyrin. In the presence of n-heptane , no such drop in activities is observed in the range of substrate concentrations studied (0.5-8.0 %, v/v). Even if the concentration of tributyrin is increased up to 16 % (v/v) the activity of immobilized lipase does not vary but small decrease is observed in free lipase activity (Fig.1), (ref. 5).

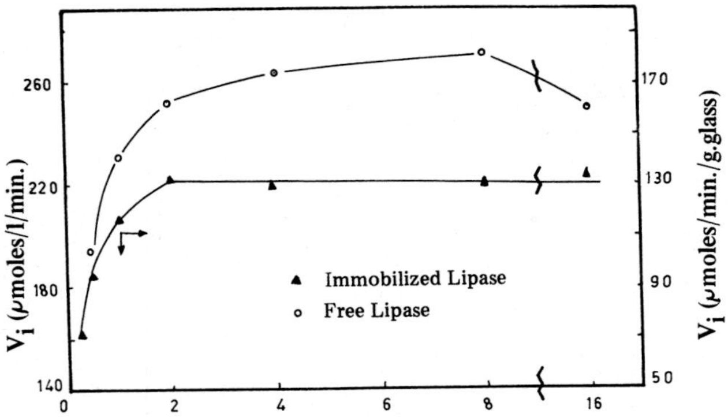

Fig. 1: The effects of tributyrin concentration on the initial rate in the presence of n-heptan (5 %, v/v).

The values of apparent Michaelis constant (K_{Ma}) are given for free and immobilized lipase in Table I. In the presence and the absence of heptane, very similar values of K_{Ma} obtained for both forms of lipase may indicate that the immobilization does not alter the affinity of the enzyme to the substrate. On the other side, K_{Ma} values increase in the presence of n-heptane for both cases (ref. 5).

Table I. K_{Ma} values for free and immobilized lipase in the reaction media with/without n-heptane.

Heptan (% by volume)	0	5
K_{Ma} (Tributyrin, % v/v) For Free Lipase	0.075	0.2196
K_{Ma} (Tributyrin, (% v/v) For Immobilized Lipase	0.072	0.2671

Hydrolysis of Triacetin in The Absence of Organic Solvents :

Triacetin, a simple lipid, is soluble in water up to 6 % (v/v) but at the higher concentrations on oil-water emulsion occurs. In the water soluble region, the variation of the hydrolysis by substrate concentration conforms to the Michaelis-Menten kinetics until a critical concentration of triacetin. After this critical concentration, the variation of the reaction rate shows a sigmoid behaviour. This kind of behaviour was shown to be the result of the formation of the triacetin micelles in aqueous media, which start to form between 3.50 and 3.75 % (v/v) of triacetin concentration (critical micelle concentration, CMC). The increase in the reaction rate proportional to the micelle formation, was discussed with respect to allosteric behaviour and catalytic properties of the candidal lipase. This sigmoid behaviour continues up to the solubility limit, and beyond this limit the hydrolysis rate reaches to a plateau with increasing triacetin concentration (Fig. 2).

At the lower concentrations than CMC, the reaction rate is controlled by the substrate at the monomer state. The rate of hydrolysis increases due to the formation of micelles and micelle-lipase complex with increasing triacetin concentrations starting from CMC, when the solubility limit is also passed the observed reaction rate is a summation of the activites of the adsorbed lipase on the oil/water interface, the lipase-micelle complex and lipase substrate complex at the monomer state.

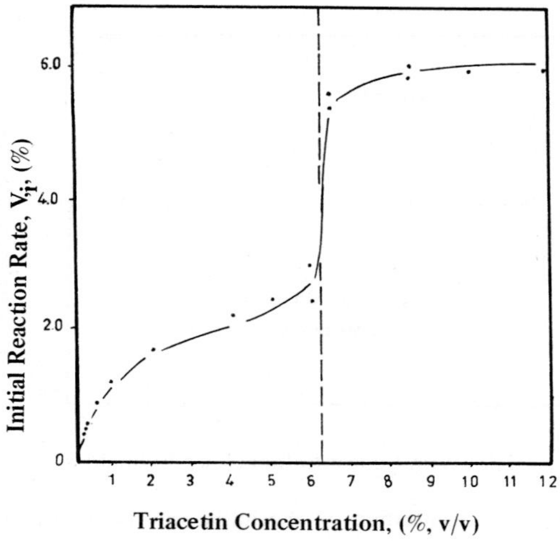

Fig. 2: The variation of the inital rate of hydrolysis by triacetin concentration.

The Effects of Organic Solvents on The Hydrolysis of Triacetin :

In the hydrolysis of triacetin, the effects of the organic solvents were investigated especially in the soluble region and it has been observed that the non-polar solvents cause an increase in the reaction rate noticeably. However, all of the polar solvents show the inhibitory effect on the lipase activity as seen in Table - II.

Table - II: Solvent effects on the hydrolysis of triacetin

Solvent	Solvent Conc. (%,v/v)	Triacetin Conc. (%,v/v)	Physical State of Medium	Variation ofHydrol. Rate,(%)*
n-Pentane	1.0	2.0	Emulsion	+ 276
n-Hexane	1.0	2.0	Emulsion	+ 260
n-Heptane	1.0	2.0	Emulsion	+ 300
n-Nonane	1.0	2.0	Emulsion	+ 322
n-Decane	1.0	2.0	Emulsion	+ 357
Benzene	1.0	2.0	Emulsion	+ 52
Chloroform	1.0	2.0	Emulsion	+ 7
Methanol	5.0	1.0	Homogen	-12.3
Methanol	5.0	3.0	Homogen	-34.3
Methanol	5.0	8.0	Homogen	-45.9
Ethanol	5.0	3.0	Homogen	-40,3
n-Propanol	5.0	3.0	Homogen	-42.0
n-Butanol	5.0	3.0	Homogen	-61.3
Diethylether	5.0	2.0	Homogen	-69.0

*The variation percentage is calculated according to the corresponding reference state contains no solvent.

According to the data given in Table-II, n-alkanes show an activating effect on the lipase in the soluble region. For a pH value of 6 and at 35°C, in the presence of 2 % (v/v) triacetin and 10 mg/l lipase (as protein), addition of 1 % (v/v) of C_5-C_{10} n-hydrocarbons provides 250-360 % increase in the initial rate of hydrolysis of triacetin. n-Heptane of tested n-alkanes was chosen as a model non-polar solvent and its effect was studied in detail. On the other hand, all of the polar solvents cause the activity loss in the range of 10-70 %. Moreover increasing the number of carbon atoms in alcohols, the rate of the hyrolysis of triacetin decreases like as tributyrin hydrolysis.

The Effects of n-Heptane on the Hydrolysis of Triacetin in the Soluble Region:

The effects of triacetin concentration on the reaction rate have been studied at the various-constant concentrations of n-heptane (Fig. 3).

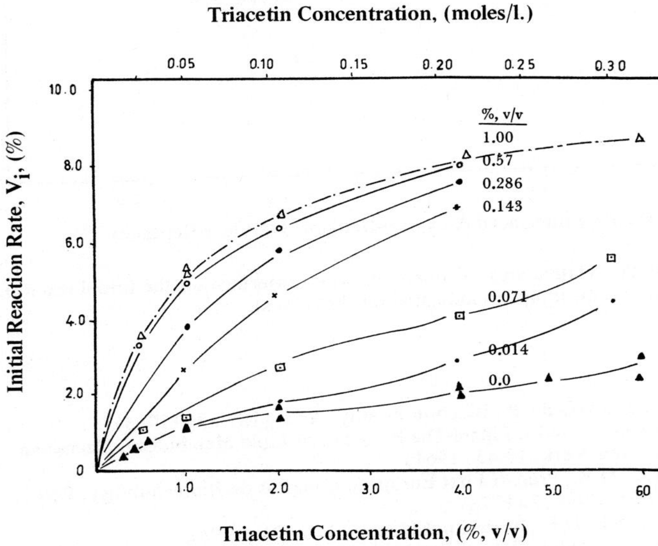

Fig.3: The variation of reaction rate with triacetin concentration for various n-heptane concentrations (as indicated on the curves-%, v/v)

The increase in interfacial area of emulsion stemming from n-heptane concentration increase causes the loss of the sigmoid behaviour trend on the hydrolysis rate. This case indicates that the reaction occured at the interface of emulsion turns to be dominant over the others in the overall reaction rate.

The reaction rate observed at low n-heptane concentrations consists of mainly the reactions of substrate present in monomer and micelle forms and the reaction of the substrate at the n-heptane/water interface with a small contribution. On the other hand, at constant concentrations of triacetin any enhancement in the rate of hydrolylsis can not be accomplished by increasing of the concentration of n-heptane more than 1 % (v/v) (Fig.4). This case due to the fact that the relative area of n-heptane/water interface does not increase at greater concentrations of n-heptane than 1 % (v/v).

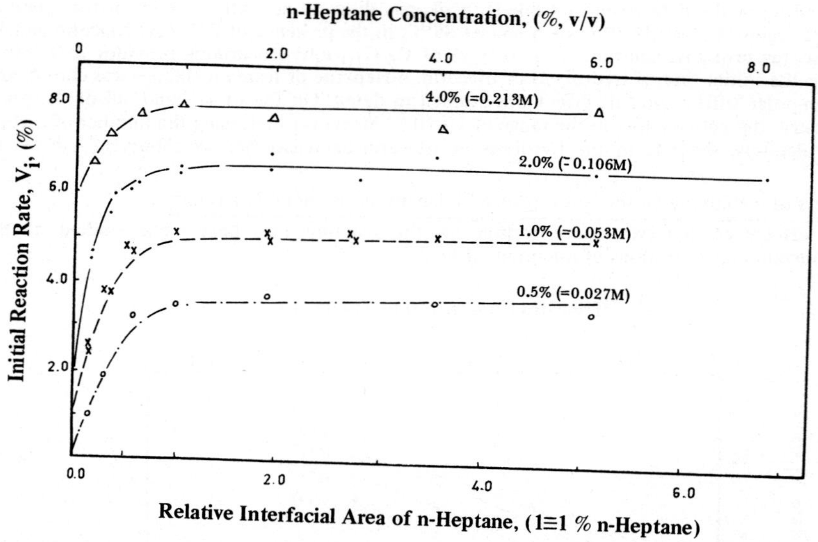

Fig. 4: The effect of the relative area of n-heptane/water interface on the initial reaction rate for the soluble triacetin concentrations (as indicated on the curves).

REFERENCES

1– Benzonana, G., and Desnuelle, P., Biochim.Biophys.Acta., 121-136 (1965).
2– Wills, E.D., Desnuelle, P., (ed.), Lipase-The Enzymes of Lipid Metabolism, Macmillan (Pergamon Press), New York, 13-43 (1961)
3– Çelebi, S.S., Çağlar, M.A., Proc.of First European Congress on Biotechnology, Part 2, Dechema, Frankfurt, 126-127 (1978)
4– Liebermann, R.B., Ollis, D.F., Biotechnol.Bioeng., 17-1401 (1975).
5– Çelebi, S.S., Uçar, T., Çağlar, M.A., Advances in Biotechnology Vol.I, (ed.M.M.Young) Pergamon Press, Toronto, 691-697 (1981).

C. Laane, J. Tramper and M.D. Lilly (Editors), *Biocatalysis in Organic Media*,
Proceedings of an International Symposium held at Wageningen,
The Netherlands, 7–10 December 1986.

STEREOSELECTIVE SYNTHESIS OF S-(-)-B-BLOCKERS VIA MICROBIALLY PRODUCED EPOXIDE INTERMEDIATES

S L JOHNSTONE[1], G T PHILLIPS[1], B W ROBERTSON[1], P D WATTS[1],
M A BERTOLA[2], H S KOGER[2] and A F MARX[2].

[1]SHELL RESEARCH LIMITED, SITTINGBOURNE RESEARCH CENTRE SITTINGBOURNE
KENT ME9 8AG

[2]GIST-BROCADES, RESEARCH AND DEVELOPMENT P O BOX 1 2600 MA DELFT
THE NETHERLANDS

SUMMARY

This paper describes the use of micro-organisms as stereoselective epoxidation catalysts for the conversion of arylallyl ethers into (+)-arylglycidyl ethers.

These intermediates are converted into S-(-)-3-substituted-1-alkylamino-2-propanols which are the physiologically active components of the β-adrenergic receptor blocking drugs. The method has been used to synthesise S-(-)-Metoprolol and S-(-)-Atenolol with enantiomeric purities (S-R/S+R) of 95.4-100% and 97% respectively.

INTRODUCTION

β-Adrenergic Receptor Blocking Agents are a group of compounds used widely in chemical medicine for the treatment of many conditions including hypertension and angina pectoris. The majority have a common structural feature of a 3-aryloxy- or 3-heteroaryloxy-1-alkylamino-2-propanol. Although these drugs have historically been made as racemates there is an increasing tendency to market the single enantiomers e.g. Timolol, Penbutolol and Moprolol. This follows from the observation that the biological activity resides in the laevorotatory isomer. In this context (-)-Propranolol is 60-100 times as active as (+)-Propranolol (ref. 1) and (-)-Metoprolol is 270-380 times as active as (+)-Metoprolol (ref. 2). It is also known in some cases that the laevorotatory enantiomer of 3-substituted-1-alkylamino-2-propanols corresponds with the S configuration e.g. Propranolol (ref. 3), Timolol (ref. 4) and Oxprenolol (ref. 5).

Currently documented routes to single enantiomers of β-blockers generally involve chemical resolutions or rather lengthy chemical syntheses from chiral intermediates using e.g. R-isopropylideneglycerol from 2R,5R-mannitol, R-glycidol from R-isopropylideneglycerol, R-epichlorohydrin from R-glycidol or 2,3-dichloropropanyl acetate, R-glyceraldehyde or S-3-alkyl-5-hydroxymethyl-1,3-oxazolidin-2-ones.

We now report an alternative route to single enantiomer β-blockers which uses stereoselective microbial epoxidation of arylallyl ethers to give S-arylglycidyl ethers which are reacted with amines to give S-(-)-β-blockers as depicted in Scheme 1. The rationale for the scheme is based on the realisation that during the epoxidation of linear olefins the oxygen is introduced stereoselectively on the si side of the double bond giving epoxides with optical purities ranging from 70-100% (refs. 6-9). The stereochemistry of the transformation is consistent with that required to produce S-arylglycidyl ethers in the event that organisms could be found that epoxidise the complex arylallyl ethers related to β-blockers.

RESULTS

Several organisms were grown on the requisite medium and screened for their ability to epoxidise arylallyl ethers. The resulting arylglycidyl ethers were isolated and their optical purities measured using proton magnetic resonance spectroscopy, prior to their chemical conversion into the β-blocker. The enantiomeric purities of the β-blockers were determined after separating either the diastereoisomeric derivatives or enantiomeric derivatives by HPLC.

Micro-organisms screened for epoxidation activity were selected from bacteria belonging to the genera Rhodococcus, Mycobacterium, Nocardia and Pseudomonas. Species of Pseudomonas gave the best activities but there was variation between the individual members and Pseudomonas oleovorans was the most active organism (Table I). The activity was further enhanced when the transformation was carried out in the presence of a cosubstrate such as glucose.

The results illustrated in Table II show that several organisms catalyse the synthesis of S-(+)-4-(2-methoxyethyl)phenylglycidyl ether (IIb) with high enantiomeric purity. The model S-(+)-phenylglycidyl ether (IIa) (prepared using Mycobacterium WMI in a two-phase system with iso-octane) was of lower optical purity. (Table II). The measurement of optical purity is carried out using proton magnetic resonance spectroscopy and is based on the resolution of the two non-equivalent geminal protons when the enantiomeric pair is subjected to the influence of Europium shift reagent.

The enantiomeric purities of the β-blockers, S-(-)-Metoprolol (IIIc) and S-(-)-Atenolol (IIIc) and the analogue (-)-3-phenoxy-1-isopropylamino-2-propanol (IIIa) have been determined using several methods. Both racemic and enantiomerically enriched Metoprolol (IIIb) have been converted into the

S-leucyl diastereomeric amide derivatives and separated by HPLC using previously described conditions (ref. 10) which identified the S isomer as eluting first. In this way S-(−)-Metoprolol (IIIb) prepared using P. oleovorans has been shown to have an optical purity of 98.4% and (−)-3-phenoxy-1-isopropylamino-2-propanol (IIIa) prepared using Mycobacterium WMI NCIB11626 shown to have an optical purity of 80%. Similarly the diastereoisomeric esters formed between a different preparation of Metoprolol (IIIb) and (R,R)-O,O-dibenzoyl tartaric anhydride (ref. 11) have been separated and showed an enantiomeric purity at 99.4%. This method was also used to ascertain the purity of the S-(−)-Atenolol (IIIc) (Table II). Finally the microbially produced S-(−)-Metoprolol (IIIb) was converted into the 2-oxazolidone (ref. 12) and shown to be enantiomerically pure following chromatography on a chiral HPLC column in which N-formyl-L-isoleucine is covalently bound to aminopropylated silica (S5 CHIRAL, UMIST Batch).

CONCLUSIONS

The feasibility is demonstrated of producing physiologically active laevorotatory enantiomers of two commercially important β-blockers, by the integrative approach of linking a stereospecific microbial epoxidation step with a chemical transformation.

SCHEME 1

Synthetic route to S-(−)-3-aryloxy-1-alkylamino-2-propanols incorporating stereospecific microbial epoxidation

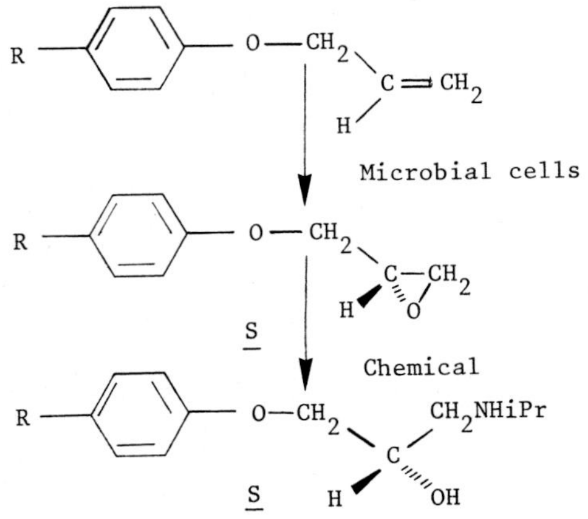

(Ia, Ib, Ic)

(IIa, IIb, IIc)

(IIIa, IIIb, IIIc)

Ia = IIa = IIIa : R = H- : IIIa = ANALOGUE

Ib = IIb = IIIb : R = $CH_3OCH_2CH_2$- : IIIb = METOPROLOL

Ic = IIc = IIIc : R = NH_2COCH_2- : IIIc = ATENOLOL

TABLE I
Catalytic activity of Pseudomonas strains for the epoxidation of 4-(2-methoxyethyl)phenylallyl ether (Ib)

Organisms		Epoxide (g/l) [time h]
P. aeruginosa[1]	NCIB 12036	0.84 [24]
P. aeruginosa[1]	NCIB 8704	0.45 [24]
P. putida[1]	NCIB 9571	1.62 [24]
P. oleovorans[1]	ATCC 29347	1.44 [5]
P. oleovorans[2]	ATCC 29347	7.3 [6]

[1]Cells were grown for 24 h on minimal medium containing 0.75% lactate and 0.05% diethoxymethane at 30°C, concentrated tenfold and incubated with 1% 4-(2-methoxyethyl)phenylallyl ether (Ib) at 37°C.
[2]Cells grown as above but with 0.75% glycerol and incubated with 2.5% 4-(2-methoxyethyl)phenylallyl ether (Ib) in the presence of 0.5% glucose at 37°C.

TABLE II
Micro-organisms catalysing epoxidation reactions leading to the synthesis of S-(-)-Metoprolol (IIIb) and S-(-)-Atenolol (IIIc)

Organism	$[\alpha]^{25}_{D}$	Optical purity (PMR) %	$[\alpha]^{D}_{25}$	Optical purity (HPLC) %
	S-(+)-4-(2-methoxyethyl)-phenylglycidyl ether (IIb)		Metoprolol (IIIb)	
R. equi NCIB 12035	+8.11°	100	-5.46°	95.4
N. corallina ATCC 31338	-	100	-	-
P. aeruginosa NCIB 12036	+8.21°	100	-5.49°	98.0
P. oleovorans ATCC 29347	+8.02°	100	-5.27°	98.4
P. putida NCIB 9571	-	100	-5.0°	98.0
P. aeruginosa NCIB 8704	-	100	-4.93°	98.8
	S-(+)-4-(acetamidomethyl)-phenylglycidyl ether (IIc)		Atenolol (IIIc)	
P. oleovorans ATCC 29347	-	-	-3.9°	97.0
	S-(+)-phenylglycidyl ether (IIa)		3-phenoxy-1-isopropyl-amino-2-propanol (IIIa)	
M.WMI NCIB11624	+11.38°	74	-6.51°	80.0

REFERENCES

1 R. Howe and R.G. Shanks, Optical Isomers of Propranolol, Nature, 210 (1966) 1336-1338.

2 N. Toda, S. Hayashi, Y. Hatano, H. Okunishi and M. Miyazaki, Selectivity and Steric Effects of Metoprolol Isomers on Isolated Rabbit Atria, Arteries and Tracheal Muscles, J. Pharm. Exp. Ther., 207 (1978) 311-319.

3 M. Dukes and L.H. Smith, β-Adrenergic Blocking Agents 9, Absolute Configuration of Propranolol and of a Number of Related Aryloxypropanolamines and Arylethanolamines, J. Med. Chem., 14 (1971) 326-328.

4 L.M. Weinstock, D.M. Mulvey and R. Tull, Synthesis of the β-Adrenergic Blocking Agent Timolol from Optically Active Precursors, J. Org. Chem., 41 (1976) 3121-3124.

5 W.L. Nelson and T.R. Burke, Absolute Configuration of Glycerol Derivatives 5, Oxprenolol Enantiomers, J. Org. Chem., 43 (1978) 3641-3645.

6 S.W. May and R.D. Schwartz, Stereoselective Epoxidation of Octadiene Catalysed by an Enzyme System of Pseudomonas oleovorans, J. Amer. Chem. Soc., 96: 12 (1974) 4031-4032.

7 H. Ohta and H. Tetsukawa, Microbial Epoxidation of Long-Chain Terminal Olefins, J.C.S. Chem. Comm., (1978) 849-850.

8 M.J. de Smet, B. Witholt and H. Wynberg, Practical Approach to High-Yield Enzymatic Stereospecific Organic Synthesis in Multiphase Systems, J. Org. Chem., 46 (1981) 3128-3131.

9 A.Q.H. Habets-Crutzen, S.J.N. Carlier, J.A.M. de Bont, D. Wistuba, V. Schurig, S. Hartmans and J. Tramper, Stereospecific formation of 1,2-epoxypropane, 1,2-epoxybutane and 1-chloro-2,3-epoxypropane by Alkane Utilizing Bacteria, Enzyme Microb. Technol., 7 (1985) 17-21.

10 J. Hermansson and C.J. von Bahr, Determination of (R)- and (S)-Alprenolol and (R)- and (S)-Metoprolol as their Diastereomeric derivatives in Human Plasma by Reversed-Phase Liquid Chromatography, J. Chromatogr., 227 (1982) 113-127.

11 W. Lindner, Ch. Leitner and G. Uray, Liquid Chromatographic Separation of Enantiomeric Alkanolamines via Diastereomeric Tartaric Acid Monoesters, J. Chromatogr., 316 (1984) 605-616.

12 J. Hermansson, Resolution of Racemic Aminoalcohols (β-blockers), Amines and Acids as Enantiomeric Derivatives using a Chiral α_1-acid Glycoprotein Column, J. Chromatogr., 325 (1985) 379-384.

C. Laane, J. Tramper and M.D. Lilly (Editors), *Biocatalysis in Organic Media*, Proceedings of an International Symposium held at Wageningen, The Netherlands, 7–10 December 1986.

THE Δ^1–DEHYDROGENATION OF HYDROCORTISONE BY *ARTHROBACTER SIMPLEX* IN ORGANIC–AQUEOUS TWO–LIQUID PHASE ENVIRONMENTS

M. D. HOCKNULL and M. D. LILLY

Department of Chemical and Biochemical Engineering, University College London, Torrington Place, London, WC1E 7JE, U.K.

SUMMARY

In shaken flask experiments, Δ^1–dehydrogenation activity of free *Arthrobacter simplex* cells is related to the log P of the organic solvent in a similar way to that proposed by Laane *et al*. (ref. 1). In stirred tank reactors, however, all organic solvents affect the stability of the cell biocatalyst. This decrease in stability of the cell in high log P solvents is related to the increased agitation rate in the reactors.

Results from further experiments in which the electron transport chain was replaced with the artificial electron acceptor phenazine methosulphate (PMS) indicate that organic solvents exert their primary affects on the electron transport chain. With some solvents, however, there is a secondary loss of enzyme activity which cannot be overcome by PMS.

INTRODUCTION

Laane *et al* (ref. 1) have proposed that the retention of activity of biocatalysts in organic solvent/aqueous two–liquid phase environments is related to the logarithm of the partition coefficient (log P) of the organic solvents. They concluded that retention of activity was low in solvents of log P $\leq$ 2, high only in solvents of log P $\geq$ 4 and that solvents of log P 2–4 may be used, but harmful affects can be expected. So far, there have been no reports in the literature of how solvents cause inactivation of biocatalysts. We are using the Δ^1–dehydrogenation system of *A. simplex* to study biocatalyst–organic phase interactions. Δ^1–dehydrogenation is a cofactor dependent reaction, and regeneration of the cofactor (FAD^+) is necessary for continued activity. Medentsev *et al* (ref. 2) have shown that this regeneration proceeds via the electron transport chain. It is possible however to regenerate cofactor artificially using PMS. Thus a loss of activity due to a loss of Δ^1–dehydrogenase activity itself or due to an affect on the electron transport chain may be distinguished.

METHODS

Reactions were done in small stirred tank reactors (70 ml working volume), stirred at 750 rpm and maintained at 30 °C in a heated water bath. 35 mg of hydrocortisone was dissolved in 35 ml of organic solvent and then combined with 30 ml of 50 mM Tris HCl buffer, pH 7.8. The two liquids were left to equilibrate for 10 mins under operational conditions. 70 mg (wet weight) of frozen cells were thawed in 5 ml Tris HCl buffer pH 7.8. The reaction was started by the addition of the cell suspension to the reactor. The final organic/aqueous phase ratio was therefore 1. For experiments involving PMS, this was dissolved in the aqueous phase at a concentration of 0.4 mM. Samples were withdrawn periodically for steroid analysis, which was done by HPLC. Shaken flask experiments were done under similar conditions in 250 ml baffled flasks, agitation was 300 rpm. Log P values were obtained from Leo *et al* (ref. 3).

RESULTS

In shaken flasks, plots of steroid conversion against Log P show a similar S−shaped curve to that proposed by Laane *et al.* (ref. 1). Activity was low in solvents of Log P ∠ 2, intermediate in solvents of Log P 2−4 and high in solvents of Log P ≥ 4.

In bioreactors, the curve is displaced such that no activity is seen in solvents of Log P ∠ 2 and high activity was seen only in one solvent, di−n−octylphthalate (Log P = 9.6).

Plots of initial activity in the bioreactors against Log P however show that the curve is in its expected position with high activity retention seen in solvents of Log P ≥ 4.

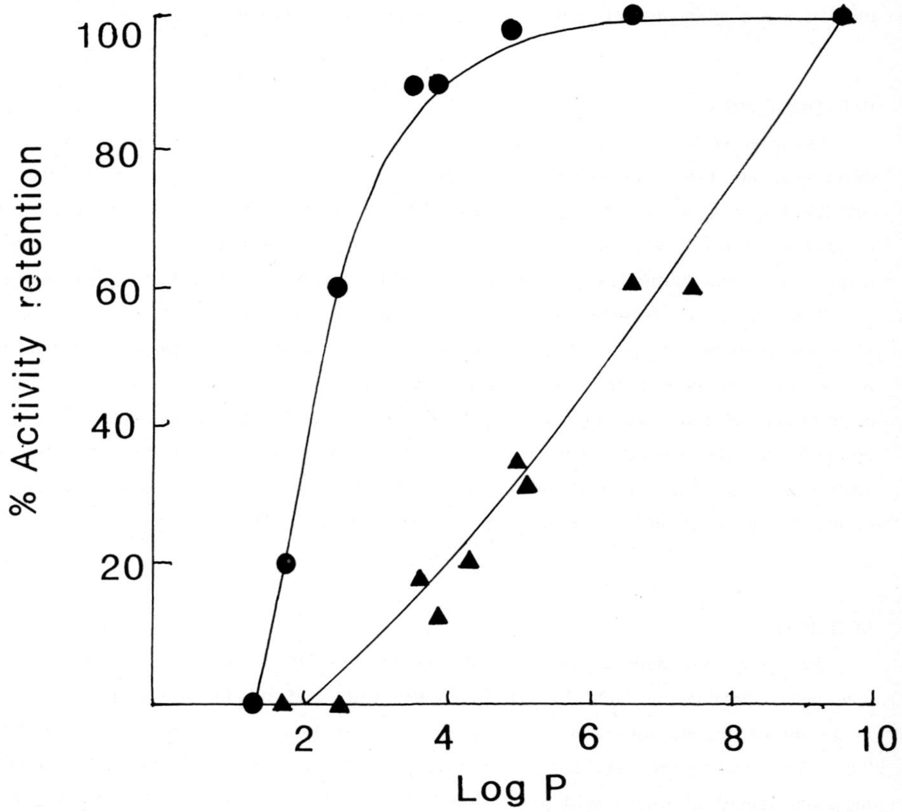

Fig. 1. Retention of Δ'−dehydrogenation activity of *A. simplex* in two−liquid phase environments versus Log P (●) shaken flasks, (▲) reactors. Measurements made after 2h.

The reason for the displacement of the activity retention curve in reactors is shown in Fig. 2. In the reactor, although initial activity is high, cell inactivation is rapid. In the shaken flask a lower activity is seen, presumably due to mass transfer limitations, but the activity is more stable.

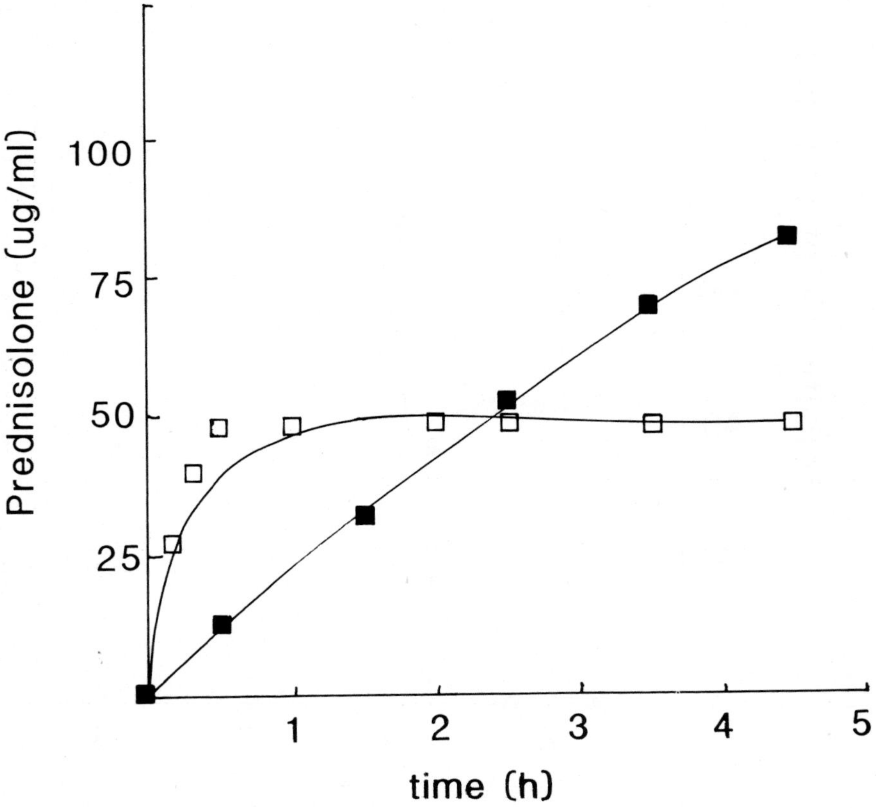

Fig. 2. Δ'−dehydrogenation reaction in a reactor and shaken flask with di−iso−pentyl ether (Log P = 3.9) as the organic phase. Initial activity in the reactor was high (140 ug/mg/h) but the reaction stops after 0.3h. In the shaken flask activity is lower (20 ug/mg/h) but more stable.

In bioreactor experiments where PMS was dissolved in the aqueous phase, Δ'-dehydrogenation was possible in all solvents tested (Log P = 1.3−9.6), except for hexanol (Log P = 1.8). PMS replaces the need for a fully functioning electron transport chain. The high initial activities observed in all organic solvents in the presence of PMS show that all solvents exert their primary affects on the electron transport chain. In the presence of PMS, initial activity is independent of solvent Log P for solvents of Log P $\geq$ 2.5. There is, however, a further loss of activity seen in low Log P solvents (less than 3.6) which is not overcome by the addition of PMS. This loss of activity must result from a loss of enzyme activity.

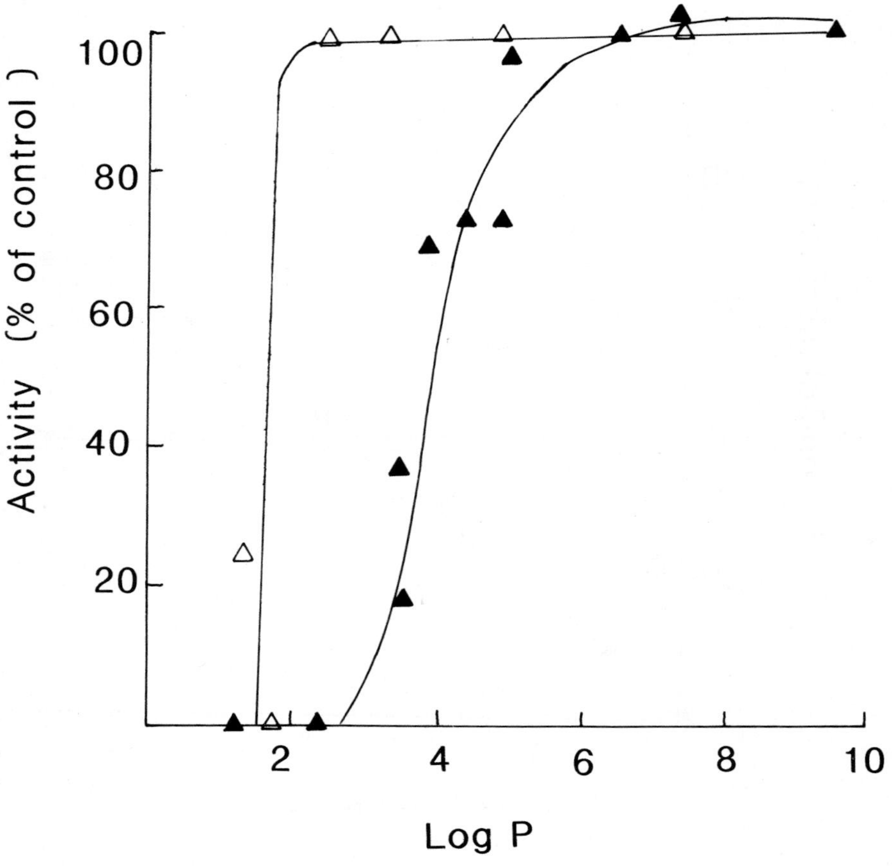

Fig. 3. Initial Δ'-dehydrogenation activity in reactors in the presence of PMS (△) and without PMS (▲). Calculations made over the first 5 min. of the reaction.

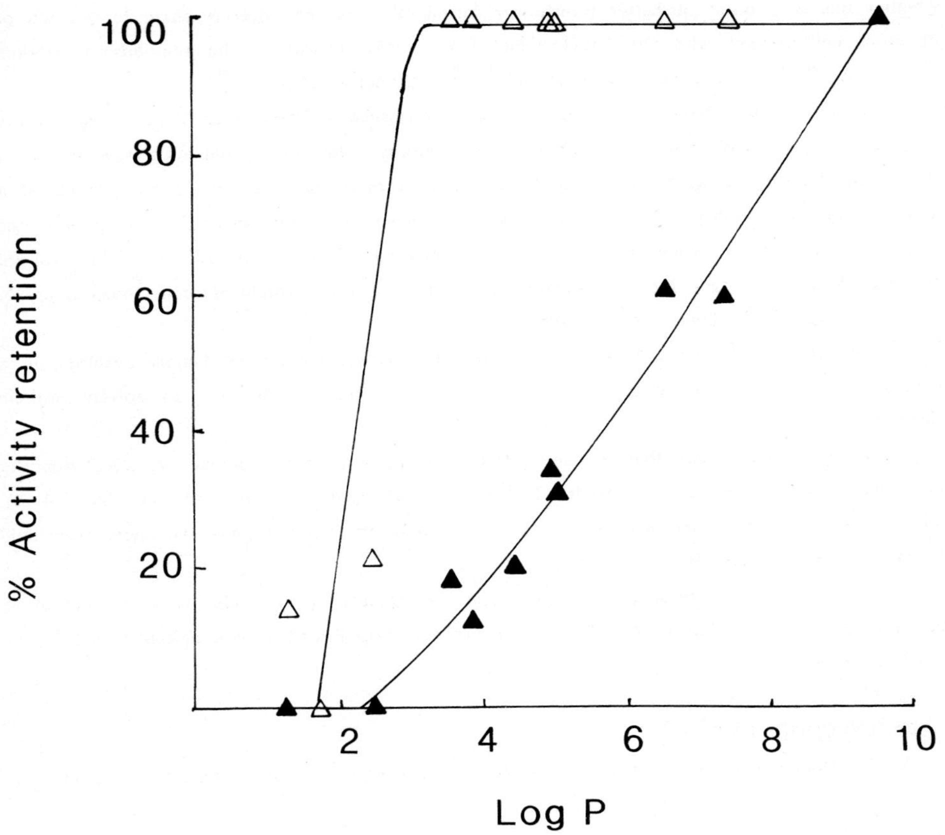

Fig. 4. Δ'–dehydrogenation activities in the presence of PMS (△) and without PMS (▲). Measurements taken at 2h.

DISCUSSION

Bioreactor studies have shown that organic solvents of Log P $\geq$ 4 also affect the activity of the cell biocatalyst, though at a slower rate than is seen with lower Log P solvents.

Since the operating conditions of the shaken flasks and the bioreactor experiments were similar in all respects except for agitation rate, we suggest that the displacement in the activity *v* Log P curve seen in the bioreactor is a function of the increased agitation rate. This displacement represents a decrease in the stability of the cell at higher agitation rates, though whether this is a result of better transfer of solvent through the aqueous phase to the cell or of more cell contact with the liquid–liquid interface is unclear. The relationship between solvent Log P and cell activity retention implies the former mechanism.

Two affects of organic solvents have been distinguished. There is an initial affect on the electron transport chain and a second later affect causing a loss of Δ'–dehydrogenase activity.

The electron transport chain can be impaired in three ways, either (i) as a result of a loss of the permeability barrier caused by low solvent concentrations partitioning into the membrane, (ii) as a result of more severe membrane damage caused by higher solvent concentrations partitioning into the membrane, interfering with protein–protein interactions, or (iii) as a result of solvent protein interactions.

Loss of enzyme activity may be due to either severe membrane damage resulting in a change of the bound enzymes micro–environment or it may be due to solvent protein interactions.

It has been found that alcohols affect the rate of sugar uptake in *Saccharomyces cerevisiae* without affecting the affinity of the transport protein for the substrate (refs. 4–5). This is consistent with the suggestion that partitioning of alcohols into the membrane lipid bilayer causes loss of activity.

In the light of our results, we propose that the stability of free cells in two–liquid phase environments is a function of both the organic solvent Log P and of the agitation rate in the reaction vessel.

ACKNOWLEDGEMENT

M. D. Hocknull wishes to thank the SERC for allocating a research studentship to him.

REFERENCES

1 C. Laane, S. Boeren and K. Vos, On optimizing organic solvents in multi–liquid–phase biocatalysis, Trends in Biotechnology, 3 (10) (1985), 257–252.

2 A. G. Medentsev, A. Y. Arinbasarova, K. A. Koshcheyenko, V. K. Akimenko and G. K. Skryabin, Regulation of 3 ketosteroid–1–en–dehydrogenase activity of *Arthrobacter globiformis* cells by a respiratory chain, J.Steroid Biochem., 23 (3) (1985), 365–368.

3 A. Leo, C. Hansch and D. Elkins, Partition coefficients and their uses, Chem.Rev. 71 (1971), 525.

4 C. Leao and N. van Uden, Effects of ethanol and other alkanols on the glucose transport system of *Saccharomyes cerevisiae*, Biotechnol.Bioeng. 24 (1982), 2601–2604.

5 M. C. Laurerio–Dias and J. M. Peinado, Effect of ethanol and other alkanols on the maltose transport system of *Saccharomyces cerevisiae*, Biotechnology Letters 4 (1982), 721–724.

C. Laane, J. Tramper and M.D. Lilly (Editors), *Biocatalysis in Organic Media*,
Proceedings of an International Symposium held at Wageningen,
The Netherlands, 7–10 December 1986.

DENATURATION AND INHIBITION STUDIES IN A TWO–LIQUID PHASE BIOCATALYTIC REACTION: THE HYDROLYSIS OF MENTHYL ACETATE BY PIG LIVER ESTERASE

A. C. WILLIAMS, J. M. WOODLEY, P. A. ELLIS and M. D. LILLY

Department of Chemical and Biochemical Engineering, University College London, Torrington Place, London, WC1E 7JE, England

SUMMARY

A study of the effect of stirrer speed on activity and stability of the enzyme, pig liver esterase, in a two–liquid phase environment is presented. Several mechanisms by which activity could be lost are examined and discussed. Finally a comparison is made between the use of the free enzyme and cellular enzyme as biocatalysts for the hydrolysis of menthyl acetate.

INTRODUCTION

One of the most important considerations when designing a biocatalytic reactor is the activity and stability of the catalyst (ref. 1). Factors affecting the activity and stability can, to an extent, be predicted and at University College London we are developing a system that will enable such predictions to be made for all types of biocatalyst which would ultimately enable us to choose the most suitable conditions for optimum biocatalysis in an industrial process.

As part of this programme we sought to compare the activity and stability of an enzyme, pig liver esterase, which catalyses the hydrolysis of menthyl acetate with that of a bacterial cell *B. subtilis* which also catalyses the hydrolysis of menthyl acetate (albeit stereospecifically). This reaction occurs by necessity in a two–liquid phase environment because the substrate is an organic liquid, sparingly soluble in water. An outline of the system is shown in Fig. 1.

The conversion using *B. subtilis* has been studied previously (refs. 2 and 3) and it was shown that an increase in stirrer speed resulted in increased specific activity at speeds less than 13.3 sec^{-1}. At higher stirrer speeds there was no further activity increase. No loss of stability was observed over the time period of the experiment.

Employing the pig liver esterase to catalyse the reaction we found at lower stirrer speeds similar behaviour to the cell, with respect to activity. However, on increasing stirrer speed we found that activity decreased markedly. We also noticed a loss of stability over the whole range of stirrer speeds.

In this paper we seek to elucidate the mechanisms for loss of activity of the pig liver esterase. These are discussed and a comparison is made between the enzyme and *B. subtilis* as biocatalyst.

MATERIALS AND METHODS

Racemic menthyl acetate was obtained from Haarmann and Reimer. Pig liver esterase was obtained from Boehringer and used without further purification.

A standard assay was performed in a 100 ml baffled reactor equipped with a rotating turbine impeller (Citenco) and immersed in a water bath at 30 °C. The phase ratio was 0.3 i.e. 22.5 mls of menthyl acetate were added to 52.5 mls of 0.1M phosphate buffer at pH 7.0 containing enzyme (0.1 mg/ml aqueous phase). Samples were taken by withdrawing 0.4 mls of the emulsion and diluting it into 0.8 mls of ice−cold isopropanol (Sigma). This both terminated the reaction and created a single liquid phase. Product and substrate were separated by HPLC (Perkin Elmer) isocratically on a C18 reversed phase column and detected by refractive index and measured relative to standards.

A standard pretreatment was performed on the enzyme solution by stirring for 24 hours under the desired conditions. If a second liquid phase was present the dispersion was allowed to settle and the clear aqueous phase removed using a separating funnel. This was assayed for enzyme activity using the standard procedure.

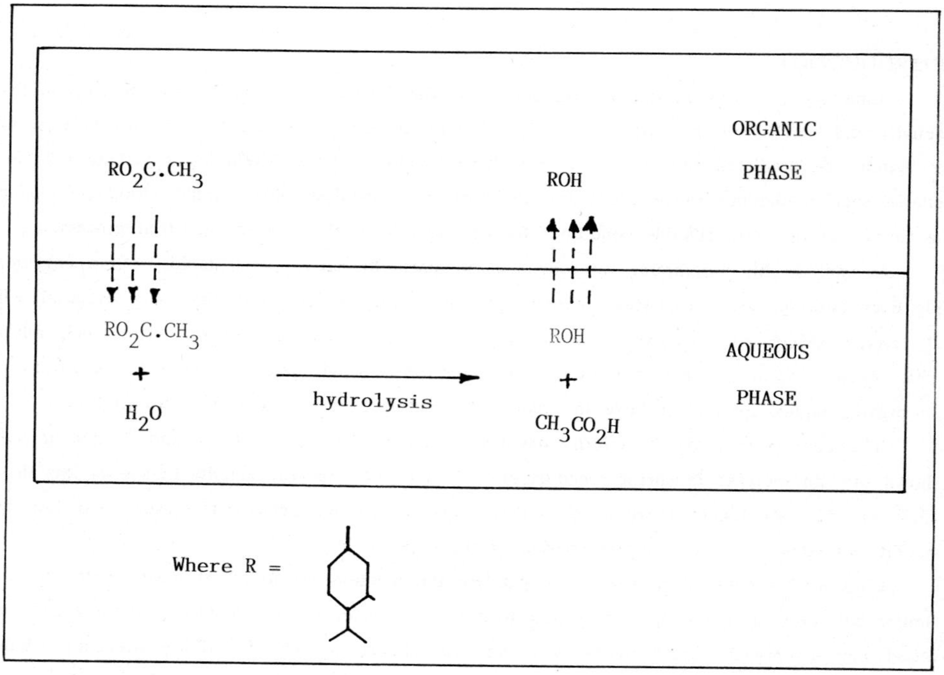

Fig. 1. Distribution of reactants and products in the reaction system.

RESULTS

Employing the esterase to hydrolyse the menthyl acetate revealed that an increase in reactor stirrer speed while increasing the activity at low stirrer speeds, reduced activity at high stirrer speeds, Fig. 2. We have attempted to elucidate the mechanism causing this loss of activity. Five mechanisms have been proposed; gas–liquid interfacial denaturation, product inhibition, pH change from optimum, liquid–liquid interfacial denaturation and substrate inhibition.

(i) Gas–liquid interfacial denaturation. In a highly agitated system it is inevitable that air will become entrained in the liquid phase(s) to form a gas–liquid interface. Enzymes are generally known to denature at interfaces (refs. 4 and 5), so this was considered a possible

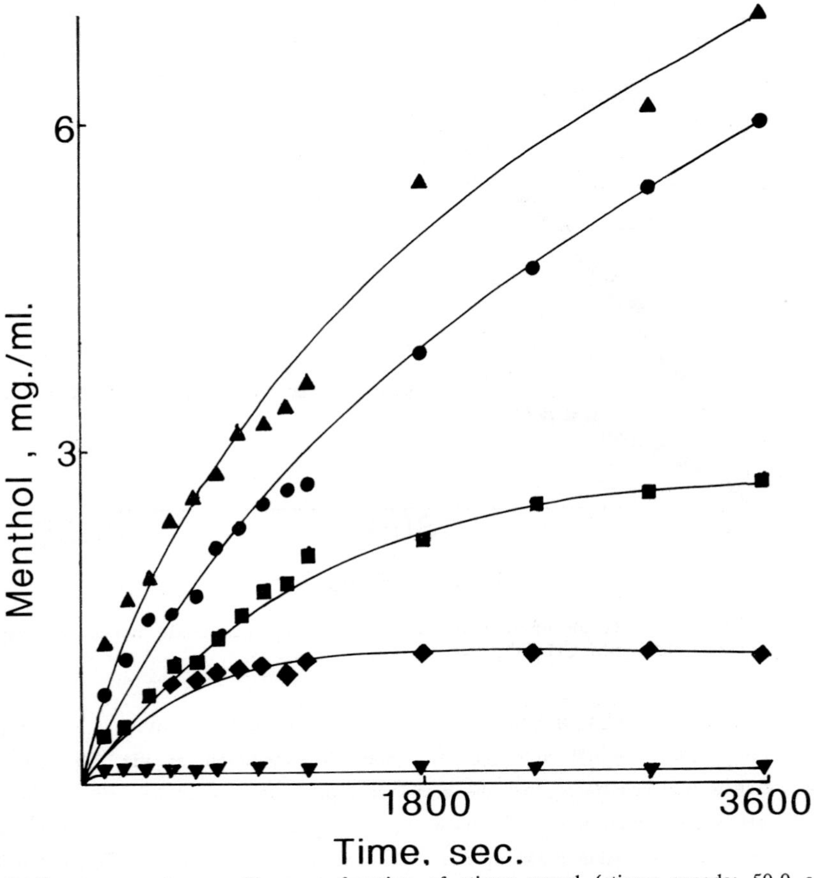

Fig. 2. Enzyme reaction profile as a function of stirrer speed (stirrer speeds: 50.0 sec^{-1} ◆, 41.7 sec^{-1} ■, 25.0 sec^{-1} ▲, 16.7 sec^{-1} ●, 8.3 sec^{-1} ▼).

cause for the loss of activity at the higher stirrer speeds. When the pig liver esterase was treated for 24 h by stirring at a speed of 12.5 sec^{-1} it retained its activity during the assay.

In comparison, when it was stirred at 41.7 sec^{-1} over the same time period, it lost some, but not all, of its activity.

(ii) Product inhibition. The hydrolysis of menthyl acetate produces two products, menthol and acetic acid. Using pretreatments again menthol was shown to have no deleterious effect on enzyme activity during an assay when added at reaction concentration levels, however acetic acid was shown to cause some loss in activity, as shown in Fig. 3.

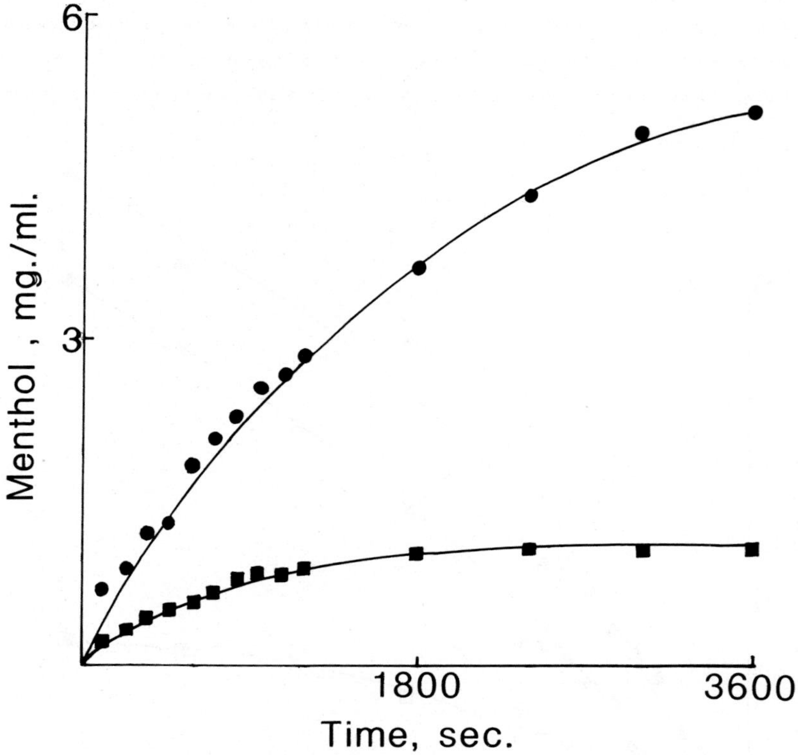

Fig. 3. Effect of CH_3CO_2H on enzyme reaction profile (aqueous phase control ●, aqueous phase pretreated with 45.0 mM CH_3CO_2H ■).

(iii) A pH change from optimum. Pig liver esterase operates within a comparatively narrow pH range (ref. 5) and although the system was buffered at pH 7.0 with 0.1M phosphate buffer it was considered that the acetic acid production could be sufficient to cause a change in pH which would lower the enzyme activity. However, no change in the reaction profile was observed when using 0.5M in place of 0.1M phospate buffer at a stirrer speed of 41.7 sec^{-1} and it was concluded that a pH change was not therefore causing a loss of activity at that stirrer speed.

(iv) Liquid–liquid interfacial denaturation. As enzymes are generally denatured at an interface it was considered that the presence of the liquid–liquid interface was likely to be a major contribution to the loss of activity. Experiments which tried to examine this effect alone however were prone to error since a change in the size of the liquid–liquid interfacial area also resulted in a change in the dissolved substrate concentration within the aqueous phase (on account of a change in mass transfer). Nevertheless, an attempt to 'plug' interface by adding protein (bovine serum albumen) prior to the enzyme did result in a reduced loss of activity, and this was more marked at the higher stirrer speeds where the interfacial area was larger, Fig. 4.

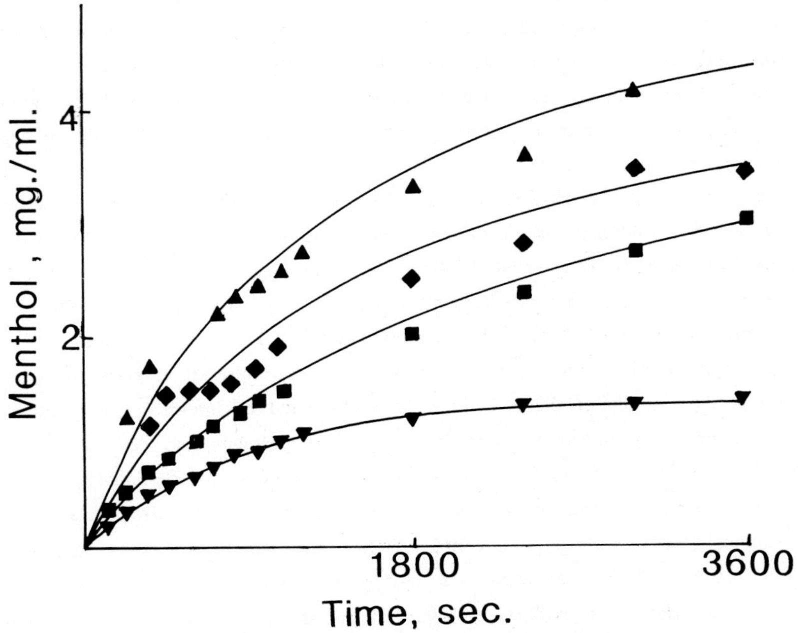

Fig. 4. Effect of bovine serum albumen (BSA) addition on the enzyme reaction profile (stirrer speeds: 41.7 sec^{-1} with BSA ▲, 41.7 sec^{-1} without BSA ▼, 16.7 sec^{-1} with BSA ◆, 16.7 sec^{-1} without BSA ■).

(v) Substrate inhibition. The organic phase solvent which partitions into the aqueous phase in this system is the substrate, menthyl acetate. As a result pretreatment experiments to examine the effects of substrate within the aqueous phase have to be carried out in the presence of the liquid–liquid interface since otherwise all the substrate is converted to product and the enzyme solution is pretreated with product rather than substrate.

A method of solving the problem would be to study the aqueous phase kinetics i.e. the kinetics of the dissolved substrate where the effect of inhibitors can thus be elucidated. Such studies are now in progress.

DISCUSSION

Our studies of the hydrolysis of menthyl acetate by pig liver esterase in a two–liquid phase environment have led us to conclude that there are several factors contributing to loss of enzyme activity: gas–liquid interfacial denaturation, liquid–liquid interfacial denaturation and acetic acid inhibition.

It is interesting to note that for the enzyme pig liver esterase both activity and stability are being lost in the two–liquid phase environment. This is in contrast to *B. subtilis* where stability did not appear to be affected over the 1 h time period.

Bar has argued (ref. 7) that with cells solvent effects can be thought of as resulting from two phenomena; the first acting at the molecular level (a 'molecular' effect) and the second acting as a result of the presence of the second liquid phase (a 'phase' effect). The mechanisms suggested for how the two types of effect act on a cell could also be applied to an enzyme. The 'phase' effect could be expressed in the enzyme by a disruption of the protein bonding and/or a distortion of shape while the 'molecular' effect could be expressed by inhibition, for example.

If we apply these definitions to the effects we observed for the pig liver esterase we could say we have evidence for an example of the 'phase' effect i.e. liquid–liquid interfacial denaturation and one of a 'molecular' effect, i.e. inhibition by acetic acid. Thus we could conjecture that the 'molecular' effect might cause a loss of activity on account of its kinetic effect, whereas the 'phase' effect might be more crucial to longer term stability.

In this system, where the organic phase solvent is the substrate (menthyl acetate) we have observed that the stability of the *B. subtilis* is not affected by interface. In contrast pig liver esterase is affected by interface which has led us to postulate the cell membrane may be protecting the cellular enzyme from 'phase' effects. However, two–liquid phase biocatalytic reactions often require organic phase solvents which are not the substrate (ref. 8).

Although the mechanisms affecting activity and stability of cells and enzymes are necessarily specific, Bar's classification is useful in presenting guidelines for describing the general characteristics of biocatalysts which need to be understood in order to exploit them fully.

REFERENCES

1 M. D. Lilly, Two–liquid–phase biocatalytic reactions, J.Chem.Tech.Biotechnol., 32 (1982), 162–169.
2 B. W. Smith, T. J. Narendranathan and M. D. Lilly, unpublished results.
3 I. K. Brookes, M. D. Lilly and J. W. Drozd, Stereospecific hydrolysis of d,l–menthyl acetate by *B. subtilis*: mass transfer–reaction interactions in a liquid–liquid system, Enzyme Microb.Technol., 8 (1986), 53–57.
4 C. R. Thomas, A. W. Nienow and P. Dunnill, Action of shear on enzymes: studies with alcohol dehydrogenase, Biotechnol.Bioeng., 21 (1979), 2263–2278.
5 C. R. Thomas and P. Dunnill, Action of shear on enzymes: studies with catalase and urease, Biotechnol.Bioeng., 21 (1979), 2279–2302.
6 D. L. Barker and W. P. Jencks, Pig liver esterase; some kinetic properties, Biochemistry, 8 (10) (1969), 3890–3897.
7 R. Bar, Phase toxicity in multiphase biocatalysis, TIB Tech., 4 (7) (1986), 167.
8 L. E. S. Brink and J. Tramper, Optimization of organic solvent in multiphase biocatalysis, Biotech.Bioeng., 23 (1985), 1258–1269.

C. Laane, J. Tramper and M.D. Lilly (Editors), *Biocatalysis in Organic Media*,
Proceedings of an International Symposium held at Wageningen,
The Netherlands, 7–10 December 1986.

EFFECT OF WATER MISCIBLE ORGANIC SOLVENTS ON ACTIVITY AND THERMOSTABILITY OF THERMOLYSIN

P.B. RODGERS[1], I. DURRANT[2] and R.J. BEYNON[2]

1. Tate and Lyle Group R & D, P.O. Box 68, Reading, RG6 2BX (U.K.)
2. Department of Biochemistry, University of Liverpool, P.O. Box 147, Liverpool, L69 3BX (U.K.)

SUMMARY

A range of water-miscible organic solvents have been tested for their effects on hydrolytic and peptide synthetic activities, and thermostability, of thermolysin. All solvents inhibited the hydrolytic activity; the magnitude of the effect appeared to be related to solvent dielectric constant. At a concentration of 2M all solvents except methanol inhibited peptide synthesis but there was no correlation of effect with solvent dielectric constant. Most solvents stabilised thermolysin against thermal inactivation.

INTRODUCTION

Much interest has been shown in the use of the metalloprotease thermolysin as a catalyst in the synthesis of a precursor of the dipeptide sweetener aspartame. The aspartame precursor, N-(benzyloxycarbonyl)-L-aspartyl-L-phenylalanine methyl ester (Z-AspPheOMe), can be synthesised by enzymic condensation of the constituent amino acid derivatives:

$$\text{Z-L-Asp} + \text{L-PheOMe} \rightarrow \text{Z-AspPheOMe} + H_2O \quad (1)$$

When the reaction is performed with an excess of L-PheOMe the condensation product is obtained as a salt of the dipeptide with the excess amine substrate (ref.1). This salt is highly insoluble in water, so when the reaction takes place in an aqueous system the reaction equilibrium is displaced towards synthesis by precipitation of the product.

Although product yields are good in the aqueous system the formation of a precipitate causes problems when trying to operate a large scale continuous process. An alternative is to use immobilised thermolysin and a water-immiscible organic solvent in which the product salt is soluble (ref.2). However the enzyme has poor stability under these operating conditions (ref. 2).

We have investigated the use of water-miscible organic solvents for Z-AspPheOMe synthesis because many of these solvents stabilise proteins against thermal inactivation and so may offer process advantages. In the initial phase of the study, effects of a range of solvents on thermolysin activity and thermostability were investigated; these results are reported here. Other aspects of this work have already been published (refs. 3-5).

METHODS

Hydrolytic activity

The hydrolytic activity of thermolysin was assayed using azocasein as substrate. Azocasein (20mg/ml) was incubated in a volume of 1.1ml at 37°C in 0.1M N-2-hydroxyethylpiperazine-N^1-2-ethanesulphonic acid (Hepes), pH 7.4. The reaction was initiated by the addition of thermolysin in a volume of 10 µl. At various times samples (250 µl) were withdrawn and added to 1ml of 5% v/v trichloroacetic acid. After centrifugation at 10000g for 3 minutes the acid-soluble dye-containing peptides were determined by measurement of the absorbance at 340nm. The rate of hydrolysis was determined by linear regression of the curve; in all cases the correlation coefficient was greater than 0.97.

Synthetic activity

The synthesis of Z-AspPheOMe was performed at 37°C in 0.05 M 4-morpholine ethane sulphonic acid (Mes), pH6.2. The reaction mixture (final volume 1.0ml) contained 0.1M Z-L-Asp, 0.2M L-PheOMe and 1 mg of thermolysin. The reaction was monitored by serial sampling of the reaction mixture. Samples (10 µl) were removed and were diluted 100-fold. A 50 ul portion of the diluted mixture was analysed by reverse-phase HPLC on C-18 crosslinked silica. The column was developed isocratically at 2 ml/min in a solvent comprising 60% (v/v) methanol and 0.15% (v/v) perchloric acid. The eluted materials were monitored at 254 nm.

Stability studies

Thermolysin (0.1mg) was incubated at 85°C in 0.05M Hepes buffer, pH 7.4, containing 25mM Ca^{2+} and 2M organic solvent where applicable. Samples (10 µl) were removed at appropriate intervals and assayed immediately using 3-(2-furyl) acryloyl-glycyl-leucinamide (FAGLA) as substrate. The substrate was dissolved in 0.1M Hepes buffer, pH 7.4, at an initial concentration of 0.45mM and the rate of hydrolysis at 25°C was monitored continuously at 340nm. The first-order rate constant for inactivation was derived by nonlinear curve fitting of the (time, activity) data to a parameter accuracy of better than 0.1%.

RESULTS

The effects of increasing concentrations of methanol, ethanol and propan-1-ol on the hydrolytic activity of thermolysin are depicted in Fig 1. Similar curves were obtained for six other water-miscible solvents. The results are summarised in Table 1 by comparing the amount of hydrolytic activity remaining in the presence of 2M cosolvent. An inverse correlation is apparent between inhibitory effect and cosolvent dielectric constant (but note the exception of methanol).

A similar comparison was performed for the synthetic activity of thermolysin (Table 1). By comparison with hydrolytic activity, synthetic activity of thermolysin was less inhibited by organic cosolvents, and in fact 2M methanol stimulated Z-AspPheOMe synthesis. There was no simple correlation between solvent dielectric constant and effect on synthetic activity.

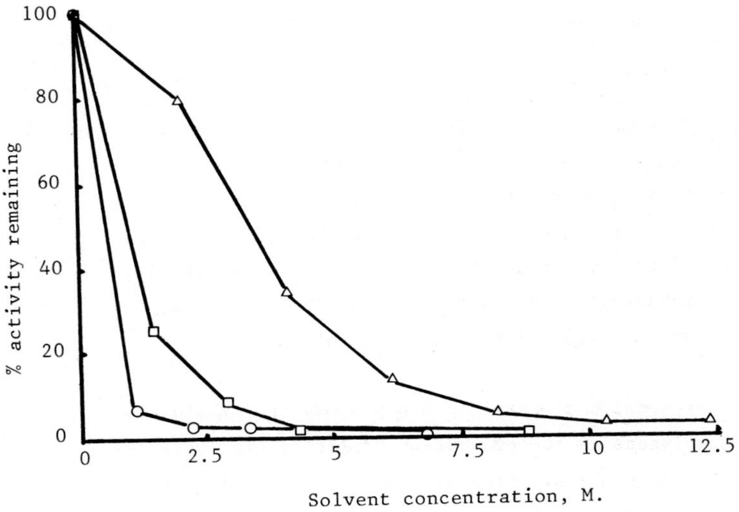

Fig. 1. Diagram showing the effects of methanol (△ — △), ethanol (□ — □) and propan-1-ol (○ — ○) on thermolysin hydrolytic activity.

TABLE 1

Comparison of the effects of organic cosolvents on activity and thermal stability of thermolysin

Solvent	Dielectric constant (at 25°C)	% activity remaining[a] Hydrolysis	Synthesis	Thermal inactivation rate constant ($10^{-4}s^{-1}$)[b]
Water	78	100	100	4.2
Glycerol	43	52.5	67.4	1.0
Ethanediol	38	52	73.7	1.7
Propane 1, 3-diol	35	30	29.5	2.5
Methanol	33	66	128.4	3.4
Propane 1, 2-diol	33	24	40	2.2
butane 1, 4-diol	30	14	25.3	6.3
Ethanol	24	18	77.9	4.9
Propan-1-ol	20	3.5	10.5	20.0

a % activity in 2M cosolvent compared to that observed in the absence of cosolvent.

b In the presence of 25mM Ca^{2+}, and 2M cosolvent where applicable.

The effect of organic cosolvents on thermostability of thermolysin was studied in the presence of Ca^{2+} ions which protect the enzyme against thermal inactivation (Table 1). Solvents of low dielectric constant, butane 1,4-diol, ethanol and propan-1-ol, destabilised thermolysin but all other solvents tested enhanced the thermal stability of the enzyme. Glycerol was the most effective stabilising agent.

DISCUSSION

In general organic cosolvents are inhibitory towards thermolysin but the extent of inhibition varies with the reaction catalysed. This observation suggests that solvent interactions with substrates and/or products should be taken into account when seeking to explain solvent effects on enzyme activity. It should be noted that initial activity rates only have been measured; the effects of organic solvents on reaction equilibria have not been investigated in this study.

In designing a production process for Z-AspPheOMe the inhibitory effect of organic cosolvents is to some extent offset by the increased thermostability observed in the presence of most of these solvents. Glycerol is thought to stabilise proteins by causing preferential hydration of the protein (ref. 6); it is possible that other stabilising solvents also have this effect. The results outlined in this study suggest that preferential hydration of proteins may cause inhibition of enzyme activity. It is not possible to advance a mechanism for this effect. However, if the rational design of operating conditions for the optimal use of biocatalysts in organic media is to become a reality, greater effort must be directed towards understanding the inter-relationships between solvent effects, enzyme activity and stability.

REFERENCES

1 Y. Isowa, M. Ohmori, T. Ichikawa and K. Mori, The thermolysin-catalysed condensation reactions of N-substituted aspartic and glutamic acids with phenylalanine alkyl esters, Tet. Lett., 28 (1978) 2611-2612.

2 K. Nakanishi, T. Kamikubo and R. Matsuno, Continuous synthesis of N-(Benzyloxycarbonyl)-L-aspartyl-L-phenylalanine methyl ester with immobilised thermolysin in an organic solvent, Bio/Tech. 3 (1985) 459-464.

3 I. Durrant, R.J. Beynon and P.B. Rodgers, The effect of inhibitors on thermolysin - catalysed peptide bond synthesis, Biochem, Soc. Trans. 14 (1986) 143.

4 I. Durrant, R.J. Beynon and P.B. Rodgers, The effect of metal ion substitutions on thermolysin - catalysed peptide bond synthesis, Biochem. Soc. Trans. 14 (1986) 957-958.

5 I. Durrant, R.J. Beynon and P.B. Rodgers, The effect of glycerol on thermolysin-catalysed peptide bond synthesis, Arch. Biochem. Biophys. 250 (1986) 280-285.

6 K. Gekko and S.N. Timasheff, Mechanism of protein stabilisation by glycerol; preferential hydration in glycerol-water mixtures, Biochem. 20 (1981) 4667-4676.

C. Laane, J. Tramper and M.D. Lilly (Editors), *Biocatalysis in Organic Media*,
Proceedings of an International Symposium held at Wageningen,
The Netherlands, 7–10 December 1986.

STEREOSPECIFIC REDUCTIONS OF BICYCLOHEPTENONES CATALYSED BY 3α,20β - HYDROXYSTEROID DEHYDROGENASE IN ONE, TWO AND THREE PHASE SYSTEMS.

J.LEAVER, T.C.C.GARTENMANN*, S.M.ROBERTS**, and M.K.TURNER.

Glaxo Group Research Ltd., Greenford, Middx., UB6 OHE, UK
*Current address-University of Zurich, Chemisch-Organisches Institut, Winterthurerstr. 190, CH-8057 Zurich. **Current address-University of Exeter, Dept. of Chemistry, Exeter, Devon, EX4 4QD, UK.

SUMMARY

A series of racemic bicyclo[3.2.0]hept-2-en-6-ones substituted at C7 were reduced with 3α,20β-hydroxysteroid dehydrogenase(HSDH). $NAD^+$was recycled with yeast alcohol dehydrogenase(YADH) and ethanol. Where the ketone was substituted with various combinations of chloride and methyl functions, the reductions were both regioselective (only the endoalcohol being formed) and enantioselective(enantiomeric excess >90%). HSDH was immobilised on Eupergit beads with and without YADH, the specificity of the reaction being unchanged. A second phase of octan-1-ol or n-hexane will act as a carrier for the ketones which are sparingly soluble in water. HSDH is also active in positively charged reverse micelles. A simple three-phase enzyme reactor was constructed. Bicycloheptenones were dissolved in octanol and dispersed in water containing ethanol and NAD^+. The emulsion was pumped through a column of coimmobilised HSDH and YADH. A total input of 1950mg of dichlorobicycloheptenone yielded 185mg of endoalcohol.

INTRODUCTION.

Bicyclic ketones are intermediates in the synthesis of a number of natural products including boonein, hirsutic acid, prostaglandins and the orphiobolin skeleton. A source of enantiomerically pure synthon is required in these syntheses but no general method is available to separate the enantiomers present in the racemic ketone. Baker's yeast will selectively reduce one enantiomer of bicyclo[3.2.0]hept-2-ene-6-one (1) but the products are toxic to the organism and difficult to recover from such a crude reducing system. An enzyme-mediated resolution based on 3α,20β-hydroxysteroid dehydrogenase(HSDH, EC1.1.1.53) from <u>Streptomyces hydrogenans</u>, was therefore developed.

MATERIALS AND METHODS.

Baker's yeast alcohol dehydrogenase(EC1.1.1.1, YADH, 300-400units/mg) was obtained from Sigma Chemical Co Ltd., Poole, Dorset. HSDH(10units/mg) was obtained from both Sigma and Boehringer Mannheim Ltd., Lewes, East Sussex.

Bicycloheptenone was obtained from Dr R.F.Newton, Glaxo Group Research Ltd., Ware, dichlorobicycloheptenone from Dr H.Fazakerley, Glaxo Group Research Ltd., Greenford, and dimethylbicycloheptenone from Dr J.A.Winders, Dept. of Chemistry, University of Salford. Other substrates were prepared by the cycloaddition of an halogenoketene and a cyclic alkene. Where necessary the substrates were redistilled by H.G.Davies, Glaxo Group Research Ltd., Greenford.

Coimmobilisation of HSDH and YADH.

Dialysed HSDH(90units) and YADH(5000units) were diluted to 4ml with 0.1M-potassium phosphate buffer(pH7.5) and added to 1g of Eupergit beads(Rohm Pharma GMBH, Weiterstadt, W.Germany). The flask was sealed, allowed to stand for 72 hours at 4°C and the beads were then washed with buffer, resupended in MOPS buffer(200mM, pH7.2) containing DTT(10mM) and p-hydroxybenzoic acid ethyl ester(HAEE, 500ppm) and stored at 4°C.

Three-phase reductions with immobilised enzymes.

A reaction mixture containing MOPS(1.85mmoles, pH7.2), DTT(92.5µmoles), NAD^+(50mg), ethanol(16mmoles), HAEE(500ppm) water(9.25ml), octanol(5ml) and (II) or (III) (500µl) was stirred vigorously at room temperature using a magnetic stirrer and pumped through a 1ml column(5cm x 0.5cm i.d) of Eupergit-immobilised enzymes at 300ml/min. At the end of the reaction period, the column and tubing were washed with 2ml of a well-stirred mixture of octanol/MOPS (1 to 1v/v) and the washings combined with the reaction mixture.

Analytical techniques.

(i)HPLC of reaction mixtures. Samples(2-10µl) of the reaction mixtures were chromatographed on an ODS-2 column(170mmx4.6mm id) either with 10%-acetonitrile in water (dihydrobicycloheptenone) or with 40%-acetonitrile(all other substrates). The flow rate was 2ml/min and detection was at 200nm.

(ii)Gas chromatography of reaction mixtures. The concentration of ketone and alcohol in extracts(see Table 2), were determined by gas chromatography over a column of 10%PEG 20M on 100-120 mesh Chro.W.HP at 200°C and a helium flow-rate of 30ml/min.

(iii)Calculation of the enantiomeric excess (ee). After separating residual ketone in the extract from alcohol by column chromatography over silica in light petroleum, the ee was determined by gas chromatography over a chiral column of Chirasil-Valine. The helium flow-rate was 1.7ml/min and the column temperature increased by 3°C/min from 110-200°C.

The absolute configuration of some of the alcohols was determined by conversion to compounds of known configuration(2,3).

RESULTS.

Single-phase reductions with soluble enzyme in water.

HSDH readily reduces bicycloheptenones in the presence of NADH, although the affinity of the enzyme for these ketones varies widely, and is much lower than for cortisone (Table 1). Bicycloheptenone concentrations greater than about 8mg/ml inhibit the reaction, and if HSDH is incubated with bicycloheptenone it forms an inhibited complex which addition of NADH does not break. Incubation with NADH in the absence of ketone causes the loss of 75% of the activity. This does not happen when cortisone is the substrate and the steroid will prevent the enzyme forming the inhibited complex when it is added together with a bicycloheptenone in the absence of NADH.

Surprisingly, the alcohol formed from the most active substrate(III) is not oxidised by HSDH, nor does it inhibit the enzyme.

TABLE 1

Kinetics of the reduction of cortisone and bicycloheptenones catalysed by HSDH in water.

Substrate	App.Km(mM)	Vmax(μmol/min/mg)
Cortisone	0.17	27
7,7-Dihydrobicycloheptenone(I)	80	18
7- Endochlorobicycloheptenone(II)	7	36
7,7-Dichlorobicycloheptenone(III)	1.8	71
7,7-Dimethylbicycloheptenone(IV)	50	36
7,7-Exochloroendomethylheptenone(V)	10	36
7,7-Endochloroexomethylheptenone(VI)	4	71

Substrates were reduced in a mixture of ethanol(0.2ml) and water(0.8ml) which contained MOPS(0.16mmole adjusted to pH7.2 with HCl), NADH(0.9μg) and substrate at various concentrations. The reduction was initiated, at 22°C, by adding HSDH and followed by monitoring the decrease in the absorbance at 340nm.

In larger-scale reactions with 1 to 2g of substrate, the NADH was recycled through the oxidation of ethanol catalysed by YADH. Horse-liver alcohol dehydrogenase is not a satisfactory catalyst because it can also reduce (I) directly. The most economical way of using the HSDH was to add it in small amounts at intervals, often together with NADH and YADH. This was continued until further additions resulted in only a small increase in the level of alcohol.

The specificity of the system was good. Only the endoalcohol was formed, the remaining substrate being recovered as unchanged ketone. Essentially, only one enantiomer of the 7-substituted bicycloheptenones(II-VI) was reduced (ee>90%, Table 2). For (II),(III) and (IV), the product is known to be the 6-S enantiomer(Fig 1), but for (V) and (VI), the absolute stereochemistry has not been determined.

TABLE 2

The stereochemistry of HSDH-catalysed bicycloheptenone reductions.

Substrate	$[\alpha]^{23}_{D}$	Enantiomeric Excess
(I)	+4.7	8
(II)	-190	>90
(III)	-155	>95
(IV)	-145	>95
(V)	-377	not determined
(VI)	-151	>98

Substrates were reduced in a mixture of ethanol(35ml) and water(315ml), containing MOPS(60mmoles, pH7.2), DTT(3mmoles) and ketone(1g). The mixture also included, as a reducing agent, NAD (72μmoles) and YADH(1600units). After stirring for 10min at 22°C to reduce the NAD^+, HSDH(9units) was added. Fresh portions of HSDH, YADH and NAD^+ were added at intervals as the reaction slowed. When reduction as a result of addition of fresh enzymes reached a low level, 70g of NaCl were added and the mixture was extracted four times with 100ml of dichloromethane.

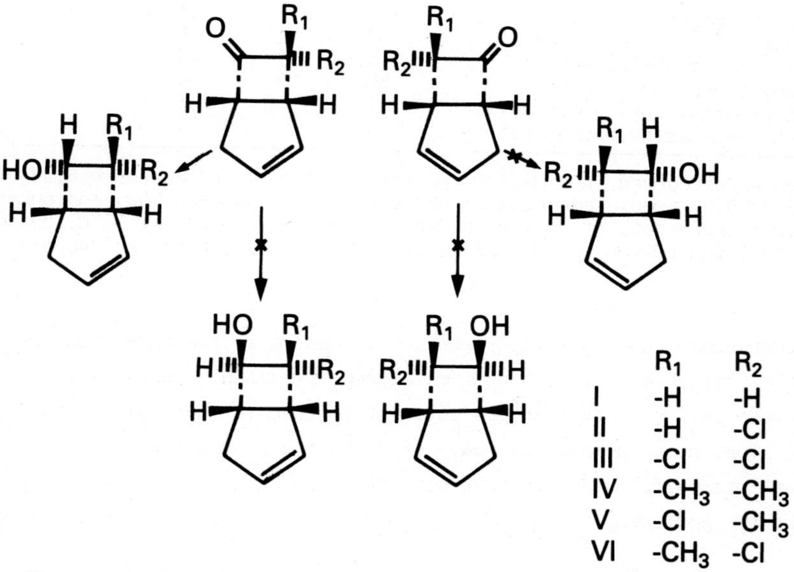

Fig.1. The structure of racemic bicyclo[3.2.0]hept-2-en-6-ones and the possible and probable products of the enzymatic reduction.

The efficiency of the system was also good. After 24 hours the conversion was typically about 70% of the theoretical with 0.36g of alcohol being synthesised from 1g of racemic (III) in a reaction catalysed by 3.6mg HSDH, 12mg YADH and 80mg NAD^+. Under optimum conditions 1 mole of enzyme reduced over 600,000 moles of ketone.

Two-phase reductions with soluble enzyme in water and organic solvents.

HSDH does not reduce the bicycloheptenones as efficiently in the presence of a second phase as it does in a single aqueous phase(Table 3). With dichloromethane, which efficiently extracts the heptenones from water, the reduction is particularly poor.

TABLE 3

Reduction of dichlorobicycloheptenone(III) in two-phase systems.

Solvent	Vol.solvent (ml)	Vol.water (ml)	Reduction (% of theoretical)
None(control)	0	9.9	68
Octanol	2	7.9	34
Hexane	2	7.9	25
Chloroform	2	7.9	12
Ethyl acetate	2	7.9	12
Dichloromethane	2	7.9	1

Sealed flasks containing MOPS(1.4mmoles, pH7.2), DTT(70µmoles), NAD^+ (3.5mg), ethanol(0.14ml) dichlorobicycloheptenone(III)(70µl), YADH(3000units) and HSDH(0.4units) were stirred vigorously overnignt at 22° C. The reaction products were extracted as described above.

Two-phase reductions with immobilised enzymes.

Immobilisation of the enzymes on Eupergit did not affect the specificity of the reaction (e.e. of product from the reduction of II and III was 82% and 95% respectively).

Two phase reductions with reverse micelles.

HSDH is active in reverse micelles formed with the cationic detergent cetyltrimethylammonium bromide (CTAB), but the reaction is slower than it is in water alone (Table 4). In contrast, the enzyme is totally inactive in micelles formed from the anionic detergent dioctyl sulphosuccinate (Aerosol OT).

TABLE 4

Kinetics of reductions in reverse micellar and aqueous systems.

Substrate	System	App.Km(mM)	Vmax(μmoles/min/mg)
(I)	Aqueous	80	18
	Micellar	263	3
(III)	Aqueous	1.8	71
	Micellar	66	25

The reverse micelles were formed with CTAB, hexan-1-ol and octane . The aqueous phase which was incorporated at a concentration of 15μl/ml, contained MOPS(50mM, pH7.2), DTT(2.5mM), NADH(0.21mM) and HSDH(9μg). Various amounts of ketones (I) and (III) were added to the organic phase.

Three-phase reductions with immobilised enzymes.

Some preliminary work has been performed to investigate the feasibility of using columns of immobilised enzymes to catalyse ketone reductions.

If an aqueous dispersion of (III) is pumped through a column of coimmobilised enzymes, droplets of the ketone soon coalesce and do not enter the column. Even high flow rates and vigourous mixing do not prevent this. However the substrate is effectively transported into the bed of the column if a suitable carrier solvent such as octanol (see Table 3) is included.

A simple three-phase system of this kind is far from satisfactory. In extended runs octanol and bicycloheptene nucleus both evaporate. After reductions lasting about 24 hours only about 50% of the octanol and about 20% of the bicycloheptenone nucleus were recovered. No attempt was made to reduce these losses.

Despite these shortcomings, (III) was reduced by the immobilised HSDH, and the NAD^+ produced was recycled by the coimmobilised YADH. Over three 24 hour runs, an input of 1950mg of ketone yielded 185mg of the endo- alcohol. Even assuming that the recovery of the alcohol was quantitative and making no allowance for the evaporative losses, this is equivalent to a 19% conversion of heptenone to chiral alcohol.

After washing the column, a fresh reaction mixture containing (II) was also reduced.

DISCUSSION.

Specificity of the reaction.

Although at first sight it was rather surprising that HSDH, a steroid dehydrogenase, should catalyse the reduction of molecules as small as the bicycloheptenones, the degree of crowding around their keto function is reminiscent of that around the C20 ketone of cortisone. In the bicycloheptenone series, significant changes are tolerated in the substitution of C7, and are necessary before any selectivity between the enantiomers occurs. This C7 plays an important role in the orientation of the substrate in the active site of the enzyme and probably explains why the reduction of the unsubstituted (I) is not selective (Table 2). The racemic alcohol probably represents the thermodynamic product of a reaction in which the kinetic separation of the various possible reductions is small. Electron withdrawing groups increased, and electron donating groups decreased, the reactivity of the keto-function in the bicycloheptenones (Table 1). The effect on the equilibrium is large enough for two chlorine atoms at C7 (III) to make the reaction practically irreversible.

Chirality of products.

In reaction mixtures containing octanol, where the reduction is satisfactory, it is not easy to separate the products, and we have assumed, we believe reasonably, that the degree of chirality in the product is similar to that attained in single-phase reductions in water (Table 2). The chirality of the reaction is not affected by immobilisation. In the single-phase system, three of the substrates (II), (III) and (IV), are known to be reduced to the 6-S enantiomers (2,3), but the absolute stereochemistry needs to be determined for a range of other substrates since it is possible that different substitutions at C7 could reverse the absolute stereochemistry of the alcohol. Any such effects will need to be understood before it is possible to build a sufficiently good model of the active site of the enzyme to make it a predictable synthetic catalyst.

Two- and three-phase reductions.

The inefficient reduction in some two-phase systems (Table 3) probably reflects the unfavourable partition of the substrate between the aqueous and the organic phases. However, we believe that the three-phase system with coimmobilised enzymes will prove to be particularly useful, although YADH is not an ideal choice for recycling NAD^+. The column carrying a mixture of YADH and HSDH slowly lost activity, and was inactive after five runs over a nine day period at 22 °C. However, ketone was still reduced when a fresh reaction mixture containing soluble YADH was passed through the column. This shows the YADH to be less stable than the HSDH, but it should not prove too difficult to find a more stable system for recycling the NAD^+.

ACKNOWLEDGEMENTS

We thank W.M.Blows and K.P.Ayres, Glaxo Group Research Ltd, Greenford, for G.C. analyses.

REFERENCES.

1 M.J.Dawson, G.C.Lawrence, G.Lilley, M.Todd, D.Noble, S.M.Green, T.W.Wallace, R.F.Newton, M.C.Carter, P.Hallet, J.Paton, D.P.Reynolds, S.Young, Reduction of bicyclo[3.2.0]hept-2-en-6-one with dehydrogenase enzymes in whole cell preparations of some fungi and yeasts, J.Chem.Soc.Perkin Trans. 1 (1983) 2119-2125.

2 S.Butt, H.G.Davies, M.J.Dawson, G.C.Lawrence, J.Leaver, S.M.Roberts, M.K.Turner, B.J.Wakefield, W.F.Wall, J.A.Winders, Reduction of bicyclo [3.2.0]hept-2-en-6-one and 7,7-dimethylbicyclo[3.2.0]hept-2en-6-one using dehydrogenase enzymes and the fungus Mortierella ramanniana, Tet. Letts.26(41) (1985)5077-5080.

3 H.G.Davies, T.C.C.Gartenmann, J.Leaver, S.M.Roberts, M.K.Turner, Reduction of 7-chlorobicyclo[3.2.0]hept-2-en-6-ones catalysed by 3α,20β-hydroxysteroid dehydrogenase, Tet.Letts.,27(9) (1986) 1093-1094.

4 R.Hilhorst, R.Spruijt, C.Laane, C.Veeger, Rules for the regulation of enzyme activity in reversed micelles as illustated by the coversion of apolar steroids by 20β-hydroxysteroid dehydrogenase, Eur.J.Biochem., 144 (1984) 459-466.

C. Laane, J. Tramper and M.D. Lilly (Editors), *Biocatalysis in Organic Media*,
Proceedings of an International Symposium held at Wageningen,
The Netherlands, 7–10 December 1986.

L-PHENYLALANINE PRODUCTION PROCESS UTILIZING ENZYMATIC RESOLUTION IN THE PRESENCE OF AN ORGANIC SOLVENT

SAMUN K. DAHOD and MARK W. EMPIE*

Eastern Research Center, Stauffer Chemical Company, Dobbs Ferry, New York 10522 (USA)

SUMMARY

An enzymatic resolution process for production of L-phenylalanine is described. The resolution process uses unacetylated D,L-phenylalanine ethyl ester as the starting material and chymotrypsin as the stereospecific catalyst. Various difficulties inherent in the ester resolution were overcome by the use of an organic solvent, such as toluene, in conjunction with the hydrolytic activity of the enzyme. The solvent allowed us to circumvent three major problems with such ester resolutions: (1) The instability of the esters with respect to nonspecific hydrolysis in the presence of water (2) The synthetic activity of chymotrypsin enzyme that forms an insoluble polymer from phenylalanine esters and (3) The inhibitory effect of D-phenylalanine ester towards chymotrypsin activity. A two-phase reactor system was designed in which the organic phase was used as the reservoir for the ester substrate. The enzymatic reaction took place in the aqueous phase under controlled conditions such that all of the above stated problems were overcome.

INTRODUCTION

The stereoselective hydrolysis of amino acid esters was known as early as 1906(ref. 1). The use of pancreatic enzymes for this purpose was later elucidated for several amino acid esters including phenylalanine (refs.2-4). The pH optimum and the substrate specificity of α-chymotrypsin make it the most desirable enzyme for the resolution of aromatic amino acids (ref. 5). Although an extensive literature exists on resolution of L-phenylalanine and other aromatic amino acid esters with chymotrypsin, no commercial process has been developed based on this enzyme. There are several reasons for this lack of interest.

1. Economical chemical processes for the synthesis of DL-amino acids are lacking.
2. The amino acid esters are unstable in aqueous solutions. As a result, background hydrolysis and concurrent loss in purity are difficult to avoid.

*present address: Dow Chemical Company, Wayland, MA 01778 (USA)

3. In concentrated solutions of esters, α-chymotrypsin catalyzes the formation of peptide bonds between amino acid molecules producing an insoluble polymer.
4. The nonreactive D-amino acid ester inhibits α-chymotrypsin activity. As a result, the enzyme activity in concentrated solutions of DL-ester is low.
5. A practical resolution process requires recycling of the unreacted D-ester after its racemization. Racemization of D-esters is difficult to carry out in aqueous solution without concurrent hydrolysis of the ester bond.

At Stauffer Chemical Company, we have designed a practical L-phenylalanine production process that overcomes these inherent difficulties in ester resolution processes. The overall process includes an economical synthesis route for DL-phenylalanine, a two-phase enzymatic reaction system and a solvent based racemization procedure for D-phenylalanine ester. The part of this process involving two-phase enzymatic reaction system designed to overcome the problems 2, 3 and 4 mentioned above is described.

MATERIALS AND METHODS

The enzyme α-chymotrypsin, L-phenylalanine reference, and DL-phenylalanine and its esters were purchased from Sigma Chemical Company, St. Louis, MO, USA. The immobilization enzyme matrix was a Pharmacia Inc. (Piscataway, NJ, USA) product called Sepharose an ion exchange chromatography medium. Toluene and other chemicals were reagent grade.

The enzyme was immobilized on Sepharose using a literature procedure (ref.6). The enzyme activity was measured titrimetrically using automatic pH-stat instrument equipped with a recorder (Metrohm AG CH-9100). L-phenylalanine methyl ester was used as the substrate for this assay. The activity was determined by manual slope determination from the titrant versus time plots. The concentrations of phenylalanine and its esters in reaction phases were determined using a reverse phase column on HPLC equipment equipped with a UV detector. The optical purity of L-phenylalanine product was determined by the optical rotation measurement of a 2% solution of the amino acid in water. A Perkin-Elmer model 241 polarimeter was used. The reference rotations were obtained using 2% solutions of air-dried commercial (Sigma Chemicals) L-phenylalanine samples.

RESULTS AND DISCUSSION

As mentioned in the introduction part the chymotrypsin catalyzed resolution of amino acids has not been developed for commercial production due to some inherent difficulties with such processes. For large-scale production it is necessary that the enzymatic reactions be carried out on concentrated ester solutions. In concentrated solutions the background hydrolysis of the ester although small is substantial. The high ester concentration also leads to polymer formation due to the reverse peptidase activity of chymotrypsin. The polymer appears as a white residue in the reaction mixture. A third problem is the inhibition of the enzymatic activity by the D-ester in solution. Figure 1 shows the inhibitory effect of the D-phe-OEt on the hydrolysis of L-phe-OEt. A 40% drop in activity is observed at 0.1M D-ester concentration. This activity loss is 78% when the D-ester concentration is 0.4M. Thus the chymotrypsin catalyzed resolution is difficult in concentrated aqueous solution of the ester. To overcome this problem a reaction scheme can be designed where the ester is stored in a solvent reservoir and slowly transferred to the aqueous phase for reaction. If the aqueous phase concentration can be controlled at relatively low levels, all three of the above limitations can be overcome.

We screened several solvents for use as the ester reservoirs. Toluene was selected for its immiscibility with water, for the observed low partition coefficient of the DL-phe-OEt between water and toluene and for good stability of chymotrypsin in the presence of toluene. The partitioning of DL-phe-OEt between water and solvent was tested by equilibrating a 1M solution of the ester with an equal volume of the solvent. The equilibrium compositions of the two phases were then determined. The equilibrium concentration of the ester in the aqueous phase when contacted with toluene under different aqueous phase pHs is shown in Figure 2. The titration curve for titration of the amine group on the ester is also given. It can be noted that at pH 6.5 where chymotrypsin displays high activity only 15% of the total ester is present in the aqueous phase while 85% remains in the toluene phase. Thus this phase system provides a means of avoiding background hydrolysis, polymer formation and enzyme inhibition.

The two-phase reaction was implemented as given in Figure 3. The ester-hydrochloride (DL-phe-OEt.HCl) was dissolved in water

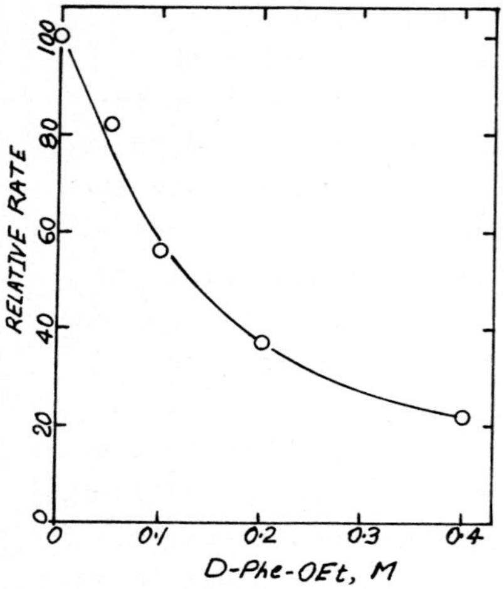

Figure 1. Inhibition of α-chymotrypsin activity by D-phenylalanine ethyl ester. Activity measured at pH 6.0 using 0.1M L-ester as substrate.

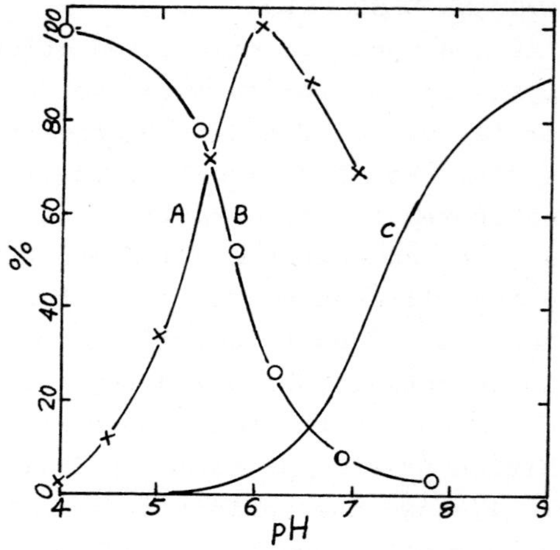

Figure 2. A. pH-rate profile for α-chymotrypsin catalyzed hydrolysis of 0.1M L-phe-OEt in the aqueous phase. B. Ester concentration in aqueous phase when 1M ester solution is equilibrated with equal volume of toluene. C. Titration curve for $-NH_2$ group of L-phe-OEt.

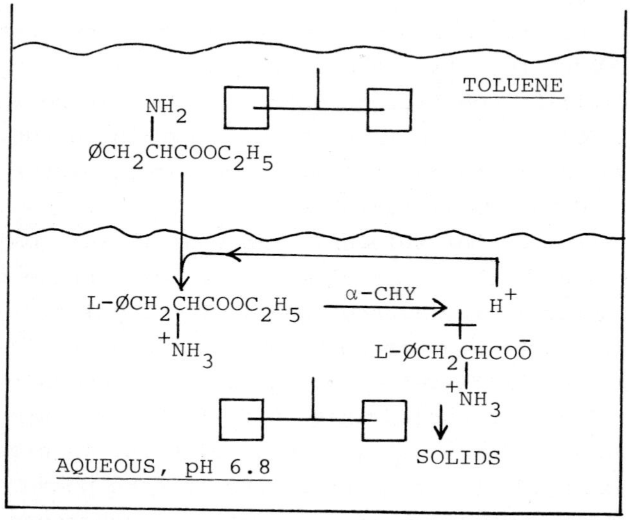

Figure 3. Schematics of two-phase reaction system for α-chymotrypsin catalyzed resolution of phenylalanine esters.

(typically at 1M concentration). Then equal volume of toluene was added and pH of the aqueous phase brought up gradually to 6.5-6.8 using sodium hydroxide while stirring two phases together. Once the aqueous phase pH was 6.8 the enzymatic reaction was started either by adding soluble enzyme to the aqueous phase or by recirculating the aqueous phase through an immobilized chymotrypsin column. During the reaction the phases were kept separated but well stirred to allow mass transfer between the phases. As the hydrolysis reaction progressed the aqueous phase was depleted of the L-ester. The depletion of the L-ester allowed transport of more L-ester from toluene due to a shift in the ester partition equilibrium. This way the reaction could be sustained under low substrate conditions where the enzyme is most active and the polymer formation is neglegible.

There are two other advantages of the two-phase reaction system of Figure 3. First, it does not require pH control. The acid produced due to the hydrolysis reaction is neutralized internally by the migration of the free base ester from toluene to the aqueous phase. Secondly, since L-phe solubility in water at 25°C is only 0.2M, it precipitates in the aqueous phase as the reaction proceeds. In the case of the soluble enzyme reaction the solids can be filtered off and the mother liquor recycled to the next reaction. In the case of the immobilized enzyme reaction an in-line filter is required to remove the product before the aqueous phase is cycled through the immobilized enzyme reactor. Some precipitation does occur in the enzyme column and the circulation lines. As a result, in large scale systems a temperature differential of about 10°C may be required between the immobilized enzyme column and the two-phase reservoir.

The data from a typical enzymatic resolution are given in Figure 4. After phase equilibrium the concentration of the ester in the toluene phase was adjusted to 1M. The reaction was carried out using an external enzyme column. As expected, the toluene phase ester concentration dropped until it reached 0.44M and leveled off at that value. The ester concentration went below the theoretical 0.5M due to the increased solubility of the ester in the aqueous phase as L-phe concentration in the aqueous phase increases. The solids precipitated from the reaction were 95+% L-phe with 5% contaminating D-phe.

CONCLUSION

α-chymotrypsin enzyme can be used for resolving L-phenylalanine isomers if a two-phase reaction system is used. The organic phase is used as the reservoir for the L-phenylalanine where the ester is protected from nonspecific hydrolysis. The enzymatic reaction takes place in the aqueous phase under low ester concentration so that chymotrypsin catalyzed polymer formation is avoided and the enzymatic activity is not inhibited by a large concentration of the D-ester. In addition, the two-phase system provides for an internal pH control mechanism in the aqueous phase reaction zone. Due to its low solubility the product L-phenylalanine precipitates in the aqueous phase and can be readily recovered. The Stauffer L-phenylalanine production process utilizes this resolution concept along with an economical chemical synthesis of DL-phenylalanine and a racemization technique that allows the D-ester to be racemized to DL-ester in toluene solution without hydrolysis to the amino acid.

REFERENCES

1 O. Warburg, J. Physiol. Chem. 48, 205 (1906).
2 K. A. J. Wretlind, J. Biol. Chem. 186, 221 (1950).
3 K. A. J. Wretlind, Acta Chem. Scand. 6 (1952).
4 P. N. Rabinovich, N. L. Pridorogin and M. A. Guberniev. Dokl, Akad, Nauk SSSR. 85, 117 (1952).
5 V. Svedas and I. V. Galaev, Russian Chemical Reviews, (translated from Uspekhi Khimii. 52, 2039 (1983); 52, 1184 (1983).
6 P. Guatnecassas, Anal. Biochem. 64, 149 (1974).

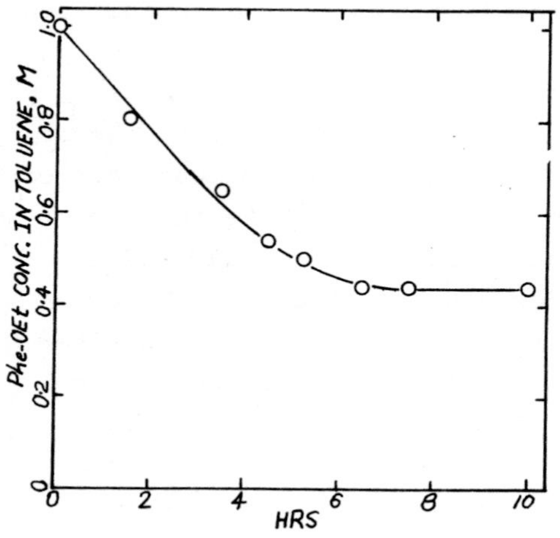

Figure 4. Concentration of ester in toluene phase as function of time in a two-phase reaction run. The reaction was carried out by circulating the aqueous phase through an immobilized α-chymotrypsin column.

Author Index

Adlercreutz, P. 107, 349, 355
Aires Barros, M.R. 185
Alcantara, A. 337
Alvaro, G. 331
Armstrong, D.W. 227

Ballesteros, A. 337
Baltussen, J.W.A. 285
Bar, R. 147
Bartsch, O. 241
Bello, M. 97
Bertola, M.A. 387
Beynon, R.J. 405
Bijsterbosch, B.H. 285
Biton, J. 247
Blanco, R.M. 331
Boeren, S. 65, 303
Boutelje, J. 361
Brazier, A.J. 3
Brink, L.E.S. 133

Cabaret, D. 253
Cabral, J.M.S. 185, 323
Caglar, A. 381
Carrea, G. 157
Celebi, S.S. 381
Cremonesi, P. 267
Critchley, P. 173

D'Angiuro, L. 267
Dahod, S.K. 419
De Boer, E. 317
De Bont, J.A.M. 295
De Rosa, M. 273
Dekker, M. 285
Doddema, H.J. 87
Duine, J.A. 279
Durrant, I. 405

Ekiz, H.I. 381
Ellis, P.A. 399
Empie, M.W. 419

Fernandez I Busquets, X. 43
Flaschel, E. 375
Fonseca, M.M.R. 323
Franssen, M.C.R. 289
Freedman, R.B. 119
Fukui, S. 21

Galliani, S. 267
Gambacorta, A. 273
Gancet, C. 261
Gartenmann, T.C.C. 411
Ghisalba, O. 43
Giesel-Bühler, H. 241
Gillies, B. 227

Grande, H.J. 87
Guignard, C. 261
Guisan, J.M. 331

Halling, P.J. 125
Heras, M.A. 337
Hilhorst, R. 65, 285 303
Hjalmarsson, M. 361
Hocknull, M.D. 3, 393
Huijberts, G.N.M. 295
Hult, K. 361

Iida, T. 21

Johnstone, S.L. 387
Jongejan, J.A. 279

Kaul, R. 107, 343
Klein, J. 51
Klibanov, A.M. 115
Kneifel, H. 241
Koger, H.S. 387

Laane, C. 65, 285, 289, 303
Larreta-Garde, V. 247
Larsson, K. 355
Leaver, J. 411
Legoy, M.D. 97
Liaw, E.-T. 233
Lilly, M.D. 3, 393, 399

Marinas, J.M. 337
Marx, A.F. 387
Mattiasson, B. 107, 343, 349, 355
Montero, J.M.S. 337

Nilsson, K.G.I. 369
Norin, T. 361

Oldfield, C. 119
Oliveira, A.C. 185
Oyama, K. 209

Philippi, M.Chr. 279
Phillips, G.T. 387
Plat, H. 317
Pulvin, S. 97

Raia, C.A. 273
Ramos Tombo, G.M. 43
Rees, G.D. 119
Rella, R. 273
Renken, A. 375
Reslow, M. 349
Rezessy-Szabó, J.M. 295
Ribeiro, M.H. 323

Roberts, S.M. 411
Robertson, B.W. 387
Robinson, B.H. 119
Rodgers, P.B. 405
Rossi, M. 273

Sahm, H. 241
Schaer, H.-P. 43
Schmid, R. 241
Shaw, J.-F. 233
Sinisterra, J.V. 337
Snijder-Lambers, A.M. 87
Steinert, H.-J. 51
Szmulik, P. 361

Tanaka, A. 21
Thomas, D. 97, 247
Tramper, J. 133, 311
Trincone, A. 273
Turner, M.K. 411

Ucar, T. 381

Van 't Riet, K. 285
Van der Plas, H.C. 289
Van Lelyveld, P.H. 87
Veeger, C. 65
Verlaan, P. 311
Vincken, J.P. 289
Vorlop, K.D. 51

Wakselman, M. 253
Watts, P.D. 387
Weijnen, J.G.J. 289
Wever, R. 317
Williams, A.C. 3, 399
Wolters, I. 311
Wong, C.-H. 197
Woodley, J.M. 3, 399

Xu, Z.F. 247

Yamazaki, H. 227